Perspectives on the History of Chemistry

Series Editor

Seth C. Rasmussen, Department of Chemistry and Biochemistry, North Dakota State University, Fargo, USA

Commonly described as the "central science", chemistry and the chemical arts have an extremely long history that is deeply intertwined with a wide variety of other historical subjects. Perspectives on the History of Chemistry is a book series that presents historical subjects covering all aspects of chemistry, alchemy, and chemical technology.

Potential topics might include:

- An updated account or review of an important historical topic of broad interest
- Biographies of prominent scientists, alchemists, or chemical practitioners
- Translations and/or analysis of foundational works in the development of chemical thought

The series aims to provide volumes that advance the historical knowledge of chemistry and its practice, while also remaining accessible to both scientists and formal historians of science. Volumes should thus be of broad interest to the greater chemical community, while still retaining a high level of historical scholarship. All titles should be presented with the aim of reaching a wide audience consisting of scientists, chemists, chemist-historians, and science historians.

All titles in the book series will be peer reviewed. Titles will be published as both printed books and as eBooks. Both solicited and unsolicited manuscripts are considered for publication in this series.

Arthur Greenberg · David E. Lewis
Editors

The Birth of the 3rd Dimension in Chemistry

Springer

Editors
Arthur Greenberg
Department of Chemistry
University of New Hampshire
Durham, NH, USA

David E. Lewis
Department of Chemistry and Biochemistry
University of Wisconsin-Eau Claire
Eau Claire, WI, USA

ISSN 2662-4591 ISSN 2662-4605 (electronic)
Perspectives on the History of Chemistry
ISBN 978-3-031-97742-8 ISBN 978-3-031-97743-5 (eBook)
https://doi.org/10.1007/978-3-031-97743-5

This Springer imprint is published by the registered company Springer Nature Switzerland AG
The registered company address is: Gewerbestrasse 11, 6330 Cham, Switzerland

Preface

During the second half of 1874, Jacobus Henricus van 't Hoff (age 22) and Joseph Achille Le Bel (age 27), working independently in the laboratory of Charles Adolph Wurtz in the École de Médicine in Paris, published the first specific—and explicit—proposals for incorporating a third (spatial) dimension in the structures of organic compounds. There were, of course, discoveries that set the stage for this breakthrough. These included Jean Baptiste Biot's 1815 work on the optical activities of enantiomorphic quartz crystals as well as liquids including oil of turpentine and aqueous sugar solutions. In his doctoral thesis of 1847 and publications of 1848, Louis Pasteur published his study of the asymmetric, hemihedral "paratartate" salts. His refined crystallographer's eye noticed that the salts were mixtures of non-superimposable, mirror-image crystals. When he physically separated the enantiomorphic crystals and dissolved them, he discovered that they exhibited equal and opposite rotations of plane (circularly) polarized light. This led him to the realization that the dissymmetry was somehow a character of the molecules themselves. At that time, there were no rules of valence nor tangible insights into molecular structure. Yet, Pasteur reasoned that such molecules could be helical spirals or asymmetrical tetrahedra, i.e., non-superimposable mirror images.

In 1835, Friedrich Wöhler wrote to his friend Jöns Jacob Berzelius: "Organic chemistry appears to me like a primeval forest in the tropics, full of the most remarkable things." Pathways through the dense forest begin to emerge in 1852 when Edward Frankland conceived of the concept of "valence" and during 1857 and 1858 as August Kekulé and Archibald Scott Couper came to the independent conclusion that the valence of carbon is four. Three years later, Alexandr Mikhailovich Butlerov postulated that the chemical properties of each compound were defined by a unique chemical structure and organic chemists started to represent compounds graphically using two-dimensional figures, based upon experimental data, usually based on the concept of a *chemical* atom, which was defined by the chemical properties of an element. Regardless, Frankland's 1866 textbook *Lecture Notes for Chemical Students: Embracing Mineral and Organic Chemistry* (London, 1866) included alkane structures recognizable to any modern first-year chemistry student. Structural representations suggested new inquiries: for example, how did the ethyl radical traverse a pathway from OH to Cl in the conversion of ethyl alcohol to ethyl chloride? In 1865, Kékule solved one major problem by accounting for the structure of benzene by assessing its substitution chemistry. But therein a new problem: two structures required to represent a single molecule.

Although the term "isomer" had been introduced by Berzelius in 1830, it could only be very vague until structural organic chemistry emerged in the 1860s. But seemingly anomalous observations of optical activity and structural chemistry soon become obvious. Perhaps the most prominent example of this period was lactic acid. Synthetic lactic acid as well as lactic acid obtained from milk fermentation were optically inactive, while lactic acid from meat juice ("paralactic acid") was dextrorotatory. In 1869, Johannes Wislicenus, accepted $CH_3.CH(OH).CO_2H$ as the correct formula for both lactic acids, proposed that they differed somehow in spatial arrangement and proposed the term "geometrical isomerism" later modified to "stereoisomerism" by Victor Meyer. The 1874 van 't Hoff/Le Bel theory solved this conundrum by explaining that optically inactive lactic acid was an equal mixture of the dextro- and levorotatory isomers ("enantiomers"), while "paralactic acid" was simply the dextrorotatory enantiomer of lactic acid. Acceptance of the van 't Hoff/Le Bel theory was not immediately universal, and many modern introductory organic chemistry textbooks include the biting, sarcastic words of the eminent organic chemist Hermann Kolbe. In a brief 1877 paper titled, in English translation, "Signs of the times", Kolbe states in part:

> A Dr. J. H. van 't Hoff, of the Veterinary School of Utrecht, finds, it seems, no exact taste for chemical research. He has considered it more convenient to mount Pegasus (apparently loaned by the veterinary school), and to proclaim in his "La chimie dans l'espace", how during his bold flight to the top of the chemical Parnassus, the atoms appeared to him to be arranged in cosmic space.

On March 18, 2024, at the National American Chemical Society Meeting in New Orleans, a symposium was organized to honor the sesquicentennial of the breakthrough by van 't Hoff and Le Bel. There were fifteen talks of which twelve are presented as chapters in this book. Three presenters were not able to provide chapters. An additional (13) chapter was contributed by co-authors who did not participate in the symposium. A specific list is included in Jeffrey Seeman's Chap. 12.

Chapter 1, by Daniel Rabinovich, reaches into his extensive, colorful "chemical philately" collection to provide a brief, selective historical outline of stereochemistry through worldwide postage stamps. Predictably, many of the stamps depict Pasteur. There is a surprising array of stamp-issuing nations represented, and some amusing anomalies are noted. Chapter 2, by Peter Ramberg, outlines and updates his comprehensive book *Chemical Structure, Spatial Arrangement. The Early History of Stereochemistry, 1874–1914.* He provides some tantalizing suggestions for future studies. In Chap. 3, Pierre Laszlo provides valuable insight into the very diverse interests of Joseph Achille Le Bel, independently wealthy and little involved with the French academicians. He was never admitted to the

French Academy of Sciences nor is there a stamp honoring his discovery. Marcellin Berthelot emerges as an opponent of Le Bel. In Chap. 4, Seth Rasmussen discusses the emergence of Werner's coordination theory from the earlier theories such as the chain theory of Blomstrand and Jorgensen. This chapter summarizes Werner's career, and of his gradual transformation from an organic chemist, who received his training in stereochemistry from Arthur Hantzsch, working with the stereochemistry of oximes and similar compounds, to an inorganic chemist and eventual Nobel laureate. In Chap. 5, Arthur Greenberg compares the research of William Jackson Pope and Kurt Mislow performed over one-half century apart. They both contributed to the fundamental understanding of stereochemistry—Pope's more static structural chemistry, and Mislow's more dynamic, depending upon the investigative tools of the period. They each made fundamental contributions toward understanding chiral central atoms beyond carbon and nitrogen. In Chap. 6, David Lewis discusses the early contributions of the Russian, Aleksandr Mikhailovich Butlerov to the structure of organic compounds, and his early involvement in the question of whether the four affinities of the carbon atom were identical or different. This led to him proposing a model for the carbon atom that was an irregular tetrahedron, but his focus on obtaining support for his structural theory meant that he missed out on this opportunity to advance the science. In Chap. 7, Gary Patterson traces the development and understanding of crystallography from Ancient through more modern times. René Just Haüy provides the critical connection between chemistry and crystallography and sets the table for the stereochemical advances by Biot and others during the next few decades and anticipating the research of Pasteur which explicitly connected the optical activities of crystals to the individual molecules that comprise them. Chap. 8, co-authored by Bart Kahr and Miriam Rossi, includes the first English translation of Primo Levi's undergraduate thesis at the University of Turin. As a Jew, Levi was prohibited by Italian race laws from performing experimental research, his thesis was a review of the work of the (Baltic German) Latvian chemist Paul Walden who discovered the S_N2 reaction. Levi, having suffered at the hands of the Nazis and Italian Fascists, survived to become one of the most renowned poets and novelists of the twentieth century. Ironically, Walden was a well-known Nazi sympathizer while denying being antisemitic. In Chap. 9, David Lewis discusses the series of experiments and the logic by which Emil Fischer was able to deduce the relative stereochemistry of glucose and other monosaccharides, including aldohexoses, aldopentoses and the ketohexose, fructose. Chapter 10, by Richard Johnson, provides a history of the development of the olefin linkage as well as cumulenes beginning with allene. Interestingly, he finds that van 't Hoff explored cumulene structures including the realization, using fused tetrahedra models, that suitably substituted, odd-numbered cumulenes are dissymmetric while even-numbered cumulenes are not. Using experimental and modern quantum-chemical calculations, he describes the properties of distorted cyclic alkynes, cumulenes, and other highly strained molecules. Chapter 11, co-authored by Kathleen Morgan and Kenneth Wiberg, examines some highly strained, even counter-intuitive organic molecules, including discussions of chemical reactions, calorimetry, and computational chemistry.

Bicyclo[1.1.0]butane possesses about 10 kcal/mol higher strain energy than the two (separated) cyclopropane rings that form the fused ring system. Among the [m.n.o]propellanes, compounds with "inverted tetrahedral carbons", [1.1.1]propellane, three-fused cyclopropane rings with decidedly inverted bridgehead carbons is surprisingly stable since cleaving the central bond leads to another highly-strained species-the bicyclo[1.1.1]pentyl structure. Although discussions of stereochemistry typically do not include quantum chemistry, Jeffrey Seeman (Chap. 12) makes explicit the necessity of incorporating quantum chemistry into the very definition of stereochemistry. He illustrates this point by demonstrating the core concepts that form the foundation of the mechanisms of pericyclic reactions and their stereochemical consequences. Chapter 13, by Martin Pomerantz, Isiah M. Borkowski, Kathleen E. Edwards, Zelek S. Herman, Shayan Khan, and Joel F. Liebman, is provocatively titled "Although sp^3 hybridization is not found for the carbon in methane nor in isoelectronically-related species, the concept of hybridization is still useful and powerful." Among the interesting illustrations of this argument are discussions of hybridization in hypothetically stretched and "scrunched" tetrahedral methane. This chapter concludes with an extensive mathematical addendum, optional for most readers.

The editors gratefully acknowledge the contributors to the New Orleans symposium and the present volume as well as the reviewers of contributed chapters. We are grateful for the support of the symposium by the History of Chemistry Division of the American Chemical Society.

Durham, USA — Arthur Greenberg
Eau Claire, USA — David E. Lewis

Contents

About the Editors

Arthur Greenberg Professor Emeritus at the University of New Hampshire, is an organic chemist interested in unusual (high-energy, distorted) molecules. Prior to joining UNH and serving as dean of Engineering and Physical Sciences (2000–2005), he was the chemistry department chair at the University of North Carolina at Charlotte (1994–2000) and prior to that director of the graduate program in environmental sciences at Rutgers University (1989–1994). He earned his Ph.D. from Princeton University in 1971. He is Fellow of the ACS and Fellow of the HIST Division. Most recently he served as Chair of the HIST Division in 2023 and 2024.

He has published more than 140 refereed articles and chapters, authored four books, co-authored one book, and co-edited eighteen books. He is the founding co-editor of the journal Structural Chemistry. Professor Greenberg's most recent book, *From Alchemy to Chemistry in Picture and Story*, was published by John Wiley and Sons.

David E. Lewis Professor Emeritus at the University of Wisconsin-Eau Claire, is an organic chemist whose interests are in organic synthesis and the history of organic chemistry in Russia. He was educated (B.Sc., Ph.D., D.Sc.) at the University of Adelaide, South Australia. He built his independent career in the USA retiring in 2022. He is the author of over 110 refereed journal articles and chapters, 12 open-access columns in Synform, and seven books; he served as the English-language editor of the English translation of a compendium published in Russia: Beloglazkina, E.; Beletskaya, I.; Lewis, D.; Nenajdenko, V. (Eds.) History of Organic Chemistry at Russian Universities. From Origins to the Present Day; NGB Publishing House: Moscow, 2022. His most recent books are *The Wolff-Kishner Reduction and Related Reactions Organic Name Reactions. Biographies, Discovery and Development* and *Addition, Elimination and Substitution: Markovnikov, Hofmann, Zaitsev and Walden. Discovery and Development*, both published by Elsevier. He is the American Associate Editor for *Synthetic Communications.*

He is Fellow of the Royal Society of Chemistry and of the Royal Australian Chemical Institute, a HIST Fellow, and former Chair of HIST. He received the 2018 Joseph B. Lambert HIST Award for Outstanding Achievement in the History of Chemistry and the 2019 Markovnikov Medal for Outstanding Achievement in Organic Chemistry (presented by Lomonosov Moscow State University, Russia).

Part I

Introductory Chapters

Stamps are Planar, Carbon is Not: A Philatelic History of Stereochemistry

1

Daniel Rabinovich

Abstract

Stereochemistry, the branch of chemistry that deals with the three-dimensional arrangement of atoms in space, and how such layout affects chemical reactivity, is of critical importance in biological systems and the pharmaceutical industry. This chapter involves the use of postage stamps to illustrate some milestones in the history of stereochemistry, including the pioneering work of Louis Pasteur, the seminal contributions of van't Hoff and Le Bel, and the contributions of several other key players in the field of stereochemistry in the nineteenth and twentieth centuries.

1.1 Introduction

Stereochemistry is the branch of chemistry that deals with the spatial arrangement of atoms within molecules, and how such three-dimensional layout affects chemical reactivity [1]. Significantly, stereoisomers have the same molecular formula and atom connectivity, but differ in the geometric positioning of the atoms in space, and this often leads to their distinct physical properties, chemical behavior, or biological activity [2].

Stereochemistry is certainly essential in biological systems, from protein folding and the structure of DNA to enzymatic catalysis. Not surprisingly, stereochemistry also plays a fundamental role in medicine and the pharmaceutical industry, where the pharmacological efficacy and pharmacokinetic action of drugs are both critically important [3–5]. In this regard, it has been estimated that more

D. Rabinovich (✉)
Department of Nanoscience, Joint School of Nanoscience and Nanoengineering, University of North Carolina at Greensboro, Greensboro, NC, USA
e-mail: Dan.Rabinovich@uncg.edu

A. Greenberg and D. E. Lewis (eds.), *The Birth of the 3rd Dimension in Chemistry*, Perspectives on the History of Chemistry, https://doi.org/10.1007/978-3-031-97743-5_1

than half of all drugs currently on the market are chiral compounds (i.e., have at least one asymmetric center and exhibit non-superimposable mirror images), even though nearly 90% of them are administered as racemates (i.e., equimolar mixtures of both enantiomers) [6, 7].

The history of stereochemistry is both enthralling and idiosyncratic, from the seminal work of Arago and Biot to understand the polarization of light early in the nineteenth century to the growing impact of asymmetric catalysis towards the end of the twentieth century [8, 9]. This account relies on the use of postage stamps to illustrate some milestones and key contributors in the development of stereochemistry (Fig. 1.1). While the scope of the article is limited by the availability of relevant stamps, the author hopes that it inspires the readers to access the other, insightful chapters in this volume. The use of postage stamps as tools in science communication and teaching, including many aspects of the history of chemistry, is well documented [10–17].

1.2 Let There Be (Polarized) Light

1.2.1 François Arago

The French physicist, astronomer, and mathematician François Arago (1786–1853) discovered optical rotation in 1811 when he observed that quartz crystals rotated the plane of "plane-polarized" (i.e., circularly polarized) light [18–20]. A French stamp issued in 1986 (Fig. 1.2) marks the bicentennial of Arago's birthday and is part of a set of five stamps celebrating significant discoveries by French scientists and engineers, including the isolation of fluorine by Henri Moissan in 1886 and the centennial of Paul Héroult's pioneering contributions to the industrial production of aluminum (i.e., the Hall-Héroult process). Arago also worked on electromagnetism with the renowned French physicist and mathematician André-Marie Ampère (1775–1836). In a demonstration before the members of the French Academy of Sciences in July 1820, the deflection of a magnetic compass needle was observed upon holding in proximity two electrically conductive metallic keys, an experiment that the Danish physicist Hans Christian Oersted (1777–1851) had conducted earlier that year [21]. Arago and Ampère are portrayed together on a 1949 French stamp (Fig. 1.3).

1.2.2 Jean-Baptiste Biot

The French physicist, astronomer, and mathematician Jean-Baptiste Biot (17741862) reported in 1815 that some non-crystalline substances (liquids, gases, and solutions) also rotated polarized light [22]. Biot's report in the *Bulletin of the Philomatic Society of Paris* included observations made with pure samples of

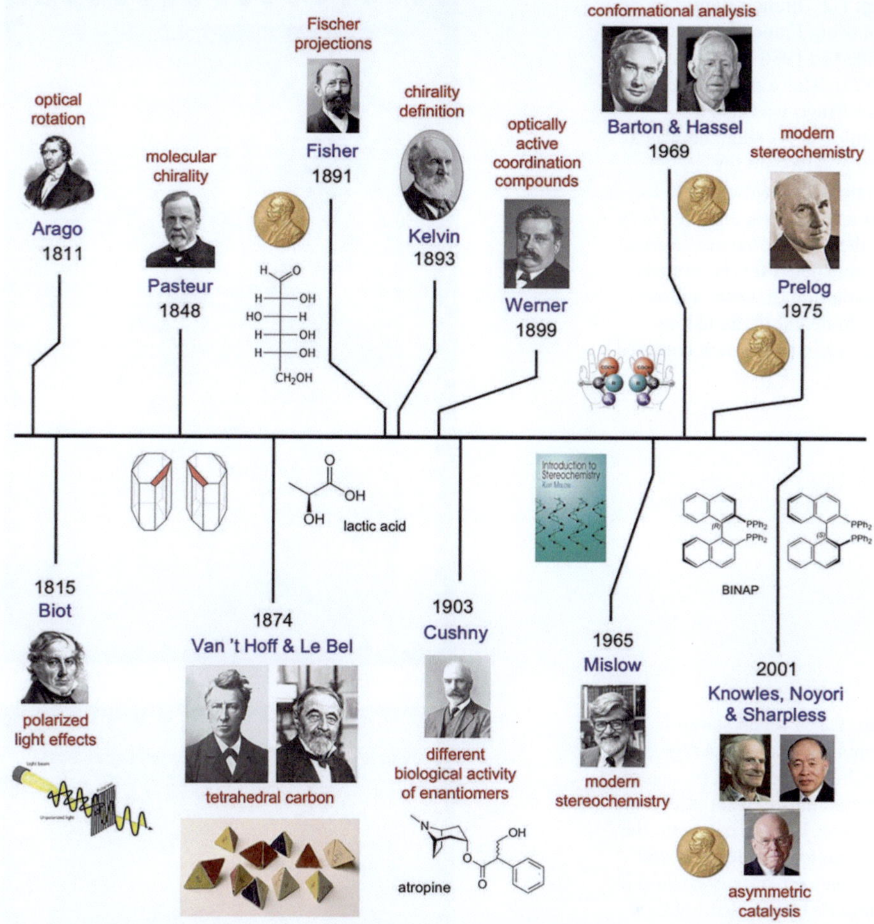

Fig. 1.1 Timeline featuring key milestones in the history of stereochemistry

turpentine and other volatile essential oils (laurel, lemon), and also with an alcoholic solution of camphor.[1] In addition to his work on polarized light, Biot also studied electromagnetism (the Biot-Savart law is an equation that describes the magnetic field generated by a constant electric current) and provided evidence for the extraterrestrial origin of meteorites [23]. Although there are no postage stamps depicting Biot, a subgroup of dark-colored phyllosilicate minerals from the mica group, referred to as biotite and with approximate chemical composition $K(Mg, Fe)_3AlSi_3O_{10}(F, OH)_2$, was named after him (Fig. 1.4).

[1] A philomatic society is an association of individuals who love the process of learning and acquiring new knowledge, especially in the sciences (from the Greek words *philos* and *manthanein*, which mean "loving" and "to learn").

Fig. 1.2 French stamp honoring François Arago, issued in 1986 (Scott No. B575). The stamp indicates that Arago was also a politician. As such, he was a strong advocate for republican ideals, served as Minister of War and the Navy in the provisional government established after the French Revolution of 1848, and was instrumental in abolishing slavery in the French colonies

Fig. 1.3 French stamp honoring François Arago and André-Marie Ampère, one of four stamps in a set issued on 13 June 1949 to celebrate the International Telegraph and Telephone Conference held in Paris that year (Scott No. 626)

Fig. 1.4 Stamp from the St. Thomas and Prince Islands issued in 2016, showing a specimen of biotite and (for no obvious reason) the Popocatépetl volcano in Mexico (Scott No. 3176c). The name biotite was proposed in 1847 by the German mineralogist Johann Friedrich Ludwig Hausmann (1782–1859) to honor Biot, and he is also credited with naming the minerals pyromorphite and rhodochrosite

1.3 The Discovery of Molecular Chirality (1848)

Louis Pasteur, the eminent French chemist and microbiologist (1822–1895), is well known for his research on the principles of vaccination, his contributions to understanding microbial fermentation, and the development of the process of food preservation named after him, i.e., pasteurization [24, 25]. But it was the discovery of molecular chirality, which preceded his work as a bacteriologist, that had an everlasting impact in the field of chemistry (Fig. 1.5).

Born on 27 December 1822 in the small town of Dole in the Franche-Comté region of France, Pasteur was not a particularly accomplished student in his formative years but managed to receive a *licencié ès sciences* (B.S.) degree in 1845 from the *École Normale Supérieure* (Teachers College) in Paris. He subsequently earned a doctorate from the University of Paris (1847) with two dissertations (chemistry and physics) and was briefly appointed professor of physics at the *Dijon Lycée* before moving to Strasbourg.

In April 1848, at the University of Strasbourg, Pasteur conducted his landmark experiments with sodium ammonium salts of the so-called paratartaric acid, which was a racemic mixture of the two enantiomers of tartaric acid. He manually separated the right-handed crystals, which were dextrorotatory in solution from the left-handed crystals, which were levorotatory (Fig. 1.6). Significantly, he concluded that the optical activity of each set of crystals was related to their shape and

Fig. 1.5 Louis Pasteur has been honored with more than 100 stamps from different countries, including Gabon in 1995 (Scott No. 811A), Transkei in 1983 (Scott No. 103), and the Republic of China (Taiwan) in 1995 (Scott No. 3026)

Fig. 1.6 Stamps from France (1995, Scott No. 2453) and Afars and Issas[2] (1972, Scott No. C70) honoring Pasteur and depicting pairs of enantiomorphous crystals of sodium ammonium tartrate

to the asymmetric internal arrangement of the molecules present in each compound [26–28]. Furthermore, he demonstrated that a 50:50 mixture of the two crystalline forms canceled each other's effect and did not rotate polarized light. Some scientists consider Pasteur's discovery of molecular chirality as his "most profound and most original contribution to science" and his "greatest scientific discovery" [29].

[2] French Somaliland, a French colony in the Horn of Africa (1884–1967) became the French Territory of the Afars and the Issas in 1967 and, in 1977, became the independent country of Djibouti.

On 22 May 1848, Pasteur presented for the first time the results of his experiments on molecular chirality to the *Académie des Sciences* and published a preliminary account shortly thereafter [30, 31]. A more detailed paper was subsequently published in October of the same year [32]. He continued working on crystals and molecular chirality for about eight years, and his discoveries included the thermal interconversion of enantiomers (i.e., the process of racemization).

Pasteur was appointed Professor of Chemistry and Dean of the new Faculty of Sciences at the University of Lille in 1854, and it was during his inauguration address, on 7 December 1854, that he declared "*Dans les champs de l'observation, le hazard ne favorise que les esprits préparés*" [In the fields of observation, chance favors only those minds that have been prepared]. In 1856, still at Lille, Pasteur abruptly began investigating fermentation processes, discovered the role of microorganisms in the spoiling of wine, beer, and milk, and invented (ca. 1864) the process of pasteurization.

Starting around 1877, Pasteur began studying infectious diseases and developed vaccines against chicken cholera, anthrax, and rabies. Later in life, in 1887, he founded the Pasteur Institute, which was inaugurated in Paris the following year and is still today a non-profit biomedical institution dedicated to the investigation of infectious diseases, molecular biology, and the development of vaccines. In addition, since 1891, new branches of the Institute have been opened in several countries, and currently there are institutes operating in 28 countries around the world, including Canada, Uruguay, Korea, Senegal, Algeria, Tunisia, Romania, Bulgaria, Italy, Greece, and two locations in France (Fig. 1.7).

Fig. 1.7 Stamps from Djibouti (1987, Scott No. 624) and New Caledonia (1988, Scott No. 601) commemorating the centennial of the Pasteur Institute in Paris

1.4 Tetrahedral Carbon (1874)

The Dutch physical chemist Jacobus Henricus van't Hoff (1852–1911) and the French chemist Joseph Achille Le Bel (1847–1930), working independently, proposed in 1874 a tetrahedral geometry to explain the optical activity of a carbon-centered compound bearing four different substituents [33–37]. In what could be considered a case of simultaneous discovery [38, 39], van't Hoff's and Le Bel's seminal reports on tetrahedral carbon were published only two months apart. Van't Hoff's self-published 12-page "pamphlet" (as it is usually referred to) was dated 5 September [40] and, since it was written in Dutch, a French translation with a shorter title and a new set of diagrams was published within a few weeks [41]. In turn, Le Bel's manuscript was published in the November 5th issue of a more established French scientific journal [42].

The two aforementioned 1874 publications suggested that the physical properties of organic compounds, especially their optical activity, could be derived from the presence of a tetrahedral carbon center linked to four different substituents. It is also worth noting that while van't Hoff's papers focused on the tetrahedron, Le Bel considered that chiral molecules with other molecular geometries would be viable sources of optical activity.

Although the initial reaction to van't Hoff's paper by the scientific community was lukewarm at best [43], the publication of an expanded version of the pamphlet as a small book in 1875 [44], and his pioneering use of cardboard models [45] (Fig. 1.8) to further explain his ideas to recipients of copies of the book, improved the general acceptance of his work [46]. Several organic chemists started to embrace van't Hoff's theory, most enthusiastically the German chemist Johannes Wislicenus (1835–1902), who had studied for several years the problem of "excess isomers" in lactic acid [47]. However, there were also some detractors, and the renowned German chemist Hermann Kolbe (1818–1884) was particularly critical [48].

As for Le Bel's paper, it never received the attention it deserved, perhaps because it was written in a more abstract style, and without any visual aids. Despite the facts that Le Bel never held an academic appointment or worked with any students, he had a variety of interests and wrote extensively on a range of subjects, including the chemistry of petroleum, the origin of the universe, the stereochemistry of nitrogen, and fermentation of gelatin. The chapter by Pierre Laszlo later in this volume further explores in detail Le Bel's relationship with the French scientific establishment.

Some five years after the publication of his influential 1874 paper, van't Hoff completely moved away from organic chemistry and embarked on a distinguished career studying chemical kinetics and thermodynamics. In 1884, he published *Études de Dynamique Chimique* [Studies in Chemical Dynamics], in which he used key concepts of thermodynamics to devise a mathematical model for the rates of chemical reactions based on the changes in the concentration of reactants over

Fig. 1.8 Personalized[3] Dutch stamp released in December 2008, featuring a three-dimensional representation of lactic acid and a picture of one of the surviving sets of van't Hoff's cardboard models, currently on display at the Museum Boerhaave in Leiden [49]. The red text in Dutch states "J. H. van't Hoff discovers spatial carbon chemistry"

time, a method that is still in used today. After serving as a professor of chemistry, mineralogy, and geology at the University of Amsterdam for 18 years, he moved to the University of Berlin in 1896, where he remained for the rest of his career. He received the first Nobel Prize in Chemistry (1901) "in recognition of the extraordinary services he has rendered by the discovery of the laws of chemical dynamics and osmotic pressure in solutions".

While Le Bel has unfortunately never been philatelically honored, there are several stamps celebrating the life and work of van't Hoff (Fig. 1.9).

1.5 Chirality in the 1890s

1.5.1 Fischer Projections (1891)

The German chemist Emil Fischer (1852–1919) made an incredible array of pioneering contributions to organic and carbohydrate chemistry towards the end of the nineteenth century [50, 51]. In rapid succession, he introduced in 1891 a simple method to draw in a plane organic molecules that have multiple chiral centers (i.e., Fischer projections) [52–54], proposed in 1894 the celebrated "lock and key" mechanism for enzymatic reactions [55, 56], and reported the following year both the total synthesis of caffeine [57] and the esterification reaction that now bears his name [58]. He also made insightful contributions to understand the stereochemistry of sugars, discovered that proteins are polypeptides, and conducted fundamental research in the synthesis of purines. He was honored with the 1902 Nobel Prize

[3] Personalized stamps are privately printed but valid for postage, as authorized by the appropriate government postal agency. One or more pictures or photographs provided by the customer are typically added to the stamp's design.

Fig. 1.9 Stamps honoring van't Hoff, issued by Grenada in 2001 to commemorate the centennial of the Nobel Prizes (Scott No. 3204) and by Mozambique in 2011 to mark the 160th anniversary of his birth (Scott No. 2539b)

in Chemistry "in recognition of the extraordinary services he has rendered by his work on sugar and purine syntheses" (Fig. 1.10).

Fig. 1.10 Stamps from Sweden (1962, Scott No. 618) and Ghana (1995, Scott No. 1826d) honoring Emil Fischer. The Swedish stamp also depicts Hendrik Antoon Lorentz and Pieter Zeeman, who shared the 1902 Nobel Prize in Physics "in recognition of the extraordinary service they rendered by their researches into the influence of magnetism upon radiation phenomena"

Fig. 1.11 Postage stamps from Guinea-Bissau (1983, Scott No. 540) and Serbia (2007, Scott No. 394) portraying Lord Kelvin

1.5.2 Lord Kelvin's Definition of Chirality (1893)

William Thomson (Lord Kelvin), the renowned British mathematician, physicist, and engineer (1824–1907), was born in Belfast and spent most of his career (53 years!) as a professor at the University of Glasgow (1846–1899). Among his many contributions are extensive work on the 1st and 2nd laws of thermodynamics, the implementation of transatlantic telegraphy, and the development of the absolute temperature scale (Fig. 1.11).

Lord Kelvin was also quite interested in the thermodynamic properties and structure of crystalline materials. On 16 May 1893, he presented the 2nd Robert Boyle Lecture to the Oxford University Junior Scientific Club, which was published the following year as a short book (~60 pages) with a peculiar title ("The Molecular Tactics of a Crystal"). It was in this lecture that he first introduced definitions for the terms "chiral" and "chirality", both derived from the Ancient Greek word for "hand" [59]:

> I call any geometrical figure, or group of points, *chiral*, and say that it has chirality, if its image in a plane mirror, ideally realized, cannot be brought to coincide with itself.

1.5.3 The Walden Inversion (1896)

Paul Walden, one of the best-known Latvian chemists (1863–1957), made important contributions to the study of stereochemistry and the electrical conductivity of molten salts (Fig. 1.12). In April 1887, Walden joined the Russian Physico-chemical Society and started collaborating with Wilhelm Ostwald, who was also born in Latvia, on the relationship between the electrical conductivity of aqueous solutions of salts and their corresponding molecular weights (Fig. 1.12). Walden is also credited with the synthesis in 1914 of the first room-temperature ionic liquid, i.e., ethylammonium nitrate, $[EtNH_3]NO_3$ (mp $\approx$ 12 °C).

Fig. 1.12 Souvenir sheet from Latvia issued in 2013 for the sesquicentennial of Walden's birth (Scott No. 832)

In September 1894, Walden became a professor of analytical and physical chemistry at the Riga Technical University, where he stayed until 1911. It was during that time that he made what is probably his most important discovery, namely the inversion of a stereogenic center in a chiral molecule during the course of a chemical reaction (e.g., the inversion of a tetrahedral carbon center in an S_N2 reaction) [60]. The phenomenon, now universally known as a Walden inversion, converts the configuration of a chiral molecule from one enantiomer to the other.

1.5.4 Optically Active Coordination Compounds (1899)

The Swiss chemist Alfred Werner (1866–1919), a professor at the University of Zurich, proposed the octahedral geometry of transition metal complexes in 1893 [61, 62] and is broadly considered the father of coordination chemistry [63, 64]. The 1913 Nobel Prize in Chemistry was awarded to Alfred Werner "in recognition of his work on the linkage of atoms in molecules by which he has thrown new light on earlier investigations and opened up new fields of research especially in inorganic chemistry" (Fig. 1.13).

Werner first mentioned asymmetric (optically active) complexes in 1899, in the 17th article in a series dealing with the constitution of inorganic compounds [65, 66]. After studying a series of bis(ethylenediamine)oxalatocobalt (III) salts of general formula $[Co(C_2O_4)(en)_2]X$, Werner concluded, using language akin to that of van't Hoff, that "it [the *cis* isomer] can be constructed in two spatial arrangements that are related as image and mirror image, and cannot be made congruent". Werner also recognized that because the two [enantiomeric] ethylenediamine compounds were identical, the origin of the asymmetry was molecular, and not derived by the inorganic equivalent of an "asymmetric" carbon atom (i.e., the cobalt center). Although Werner was uncertain about how the two isomers could be differentiated, he speculated that they may have distinct crystal forms.

Fig. 1.13 Stamps from Sweden (1973, Scott No. 1028) and Antigua and Barbuda (2001, Scott No. 2517f) honoring Alfred Werner. The Swedish stamp also features Heike Kamerlingh-Onnes, who received the 1913 Nobel Prize in Physics "for his investigations on the properties of matter at low temperatures which led, *inter alia*, to the production of liquid helium"

1.6 Conformational Analysis

Conformational analysis, the study of the energies associated with the different shapes a molecule takes due to rotation around one or more single bonds, is important to understand the stability of isomers and to predict and explain the outcomes, mechanisms, and rates of chemical reactions. The conformations of cyclohexane and its derivatives were systematically investigated by Hassel in the 1940s [67] and Barton in the 1950s [68, 69], both of whom are considered among the pioneers of modern stereochemistry (Fig. 1.14) [70].

Derek Barton, an English organic chemist (1918–1998), and Odd Hassel, a Norwegian physical chemist (1897–1981), shared the Nobel Prize in Chemistry in 1969 "for their contributions to the development of the concept of conformation and its application in chemistry".

1.7 Woodward–Hoffmann Rules (1965)

In 1965, while conducting research in the Department of Chemistry at Harvard University, Robert B. Woodward (1917–1979) and Roald Hoffmann (b. 1937) jointly developed a set of rules to predict and explain the outcome, activation energy, and stereochemistry of pericyclic organic reactions, which are those in which the transition state leading to the product has a cyclic structure [71–74]. Furthermore, in electrocyclic reactions, a subset of pericyclic reactions, the net result is the conversion of one pi bond into a sigma bond, or vice versa. Such

Fig. 1.14 Stamps from Norway (2004, Scott No. 1414) and Great Britain (1977, Scott No. 806) honoring Odd Hassel and Derek Barton, respectively. The British stamp explicitly highlights Barton's research on the conformational analysis of steroids

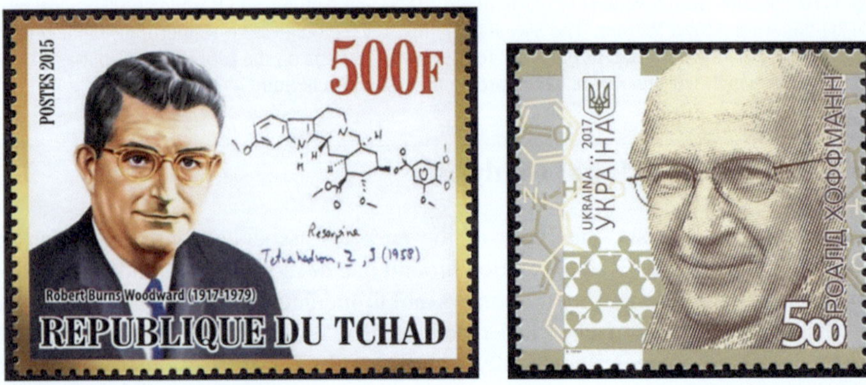

Fig. 1.15 Stamps from Chad (2015, Scott unlisted) and Ukraine (2017, Scott No. 1108) honoring R. B. Woodward and R. Hoffmann, respectively

reactions can be driven thermally or photochemically, and they can be either ring-opening or ring-closing in nature. Woodward received the 1965 Nobel Prize in Chemistry "for his outstanding achievements in the art of organic synthesis", and Hoffmann shared his with Kenichi Fukui (1918–1998) in 1981 "for their theories, developed independently, concerning the course of chemical reactions" (Fig. 1.15).

1.8 Cahn–Ingold–Prelog (CIP) Rules (1966)

The Croatian-Swiss organic chemist Vladimir Prelog (1906–1998), born in Sarajevo, the capital and largest city in present-day Bosnia and Herzegovina, developed together with British chemists Robert S. Cahn (1899–1981) and Christopher K. Ingold (1893–1970) a set of rules to unequivocally name a stereoisomer of a

Fig. 1.16 Stamps from Croatia (2001, Scott No. 475), Bosnia and Herzegovina (2006, Scott No. 564), and Serbia (2024) honoring Vladimir Prelog

molecule [75, 76]. More specifically, the CIP system involves the assignment of an *R* or *S* descriptor to each stereocenter and an *E* or *Z* descriptor to each double bond, so that the configuration of the entire molecule can be specified by including such labels in its systematic name.

The 1975 Nobel Prize in Chemistry was shared by Prelog "for his research into the stereochemistry of organic molecules and reactions" and the Australian-British chemist John Cornforth (1917–2013) "for his work on the stereochemistry of enzyme-catalyzed reactions". Of the CIP trio, only Prelog has been honored with the release of a few postage stamps (Fig. 1.16).

1.9 Asymmetric Catalysis (1968)

Enantioselective catalysis, known traditionally as "asymmetric catalysis", is a method used in organic chemistry to produce chiral molecules in an enantiomerically enriched form [77]. It is a key tool in the manufacturing of pharmaceuticals and agrochemicals, and it is often performed using chiral metal catalysts, which are typically chiral coordination complexes, obtained in turn from chiral supporting ligands like N-heterocyclic carbenes (NHCs), bis(oxazolines), or diphosphines such as BINAP[4] [78].

Metal-catalyzed enantioselective synthesis was pioneered by William S. Knowles, Ryoji Noyori, and K. Barry Sharpless back in the late 1960s. Significantly, half of the 2001 Nobel Prize in Chemistry was awarded to Knowles and Noyori "for their work on chirally-catalyzed hydrogenation reactions" [79, 80], and the other half was bestowed on Sharpless "for his work on chirally-catalyzed oxidation reactions" [81, 82] (Fig. 1.17). Knowles, working at Monsanto, was

[4] BINAP = 2,2′-bis(diphenylphosphino)-1,1′-binaphthyl.

Fig. 1.17 K. Barry Sharpless is the only individual among the 2001 Nobel Laureates in Chemistry to have been honored with a postage stamp (Guinea-Bissau, 2022). Interestingly, the stamp was issued to recognize the 2022 Nobel Prize in Chemistry that Sharpless shared with Carolyn Bertozzi and Morten Meldal "for the development of click chemistry and bioorthogonal chemistry"

also the first to apply enantioselective metal catalysis on an industrial scale, as implemented for a key hydrogenation step in the production of L-DOPA.

Interest in asymmetric organocatalysis, the use of small organic molecules to catalyze organic transformations, has also increased steadily since the early 2000s, as evidenced by the exponential growth of publications in the field [83–86] and the award of the 2021 Nobel Prize in Chemistry to Benjamin List and David W. C. MacMillan "for the development of asymmetric organocatalysis" [87].

1.10 Is Tetracoordinate Carbon Always Tetrahedral?

In 1968, Monkhorst suggested for the first time a *planar* tetracoordinate carbon *intermediate* in the interconversion of substituted methanes [88]. Several researchers, including Roald Hoffmann, Gerhard Erker, Walter Siebert, Gabriel Merino, and others, have subsequently studied computationally, and sometimes isolated, *bona fide* examples of compounds containing planar tetracoordinate carbon centers [89–92]. The interplay of insightful calculations and skillful experimental work to attain such results is a wonderful tribute to the power of synthesis 150 years after the birth of the third dimension in chemistry.

Acknowledgements I sincerely thank Art Greenberg and Dave Lewis for organizing (and inviting me to partake in) *The Birth of the 3rd Dimension in Chemistry* symposium held at the Spring 2024 national ACS meeting in New Orleans, Louisiana, USA. More importantly, their passion for the history of chemistry (and chemical storytelling) has been (and will continue to be) a source of inspiration for many years.

References

1. Eliel EL, Wilen SH, Doyle MP (2001) Basic organic stereochemistry. Wiley Interscience, New York
2. Patočka J, Dvořák A (2004) Biomedical aspects of chiral molecules. J Appl Biomed 2:95100
3. Nguyen LA, He H, Pham-Huy C (2006) Chiral drugs: an overview. Int J Biomed Sci 2:85–100
4. Brooks WH, Guida WC, Daniel KG (2011) The significance of chirality in drug design and development. Curr Top Med Chem 11:760–770
5. Singh K, Shakya P, Kumar A, Alok S, Kamal M, Singh SP (2014) Stereochemistry and its role in drug design. Int J Pharm Sci Res 5:4644–4659
6. Ceramella J, Franchini A, De Luca M, Saturnino C, Andreu I, Sinicropi MS, Catalano A (2022) A look at the importance of chirality in drug activity: some significative examples. Appl Sci 12:10909
7. Senkuttuvan N, Komarasamy B, Krishnamoorthy R, Sarkar S, Dhanasekarand S, Anaikutti P (2024) The significance of chirality in contemporary drug discovery—a mini review. RSC Adv 14:33429–33448
8. Gal J (2013) Molecular chirality in chemistry and biology: historical milestones. Helv Chim Acta 96:1617–1657
9. Hargittai B, Hargittai I (2024) 150 years of stereochemistry-selected examples. Struct Chem 35:383–392
10. Heilbronner E, Miller FA (1998) A philatelic ramble through chemistry. Wiley-VCH, Weinheim
11. Yardley CB (2015) The representation of science and scientists on postage stamps. ANU Press, Canberra, Australia
12. Habashi F, Hendricker D, Gignac C (1999) Mining and metallurgy on postage stamps. Métallurgie Extractive Québec, Quebec City, Canada
13. Habashi F (2019) Postage stamps: metallurgy, art, history part 2. Métallurgie Extractive Québec, Quebec City, Canada
14. Miller FA (1983) The history of spectroscopy as illustrated on stamps. Appl Spectrosc 37:219–225
15. Miller FA (1986) A postage stamp history of chemistry. Appl Spectrosc 40:911–924
16. Schreck JO (1986) Postage stamps as a teaching tool in chemistry. J Chem Educ 63:283–287
17. Rappoport Z (1992) Chemistry on stamps (chemophilately). Acc Chem Res 25:24–31
18. Arago JDF (1811) Sur une modification remarquable qu'éprouvent les rayons lumineux dans leur passage à travers certains corps diaphanes, et sur quelques autres nouveaux phénomènes d'optique [Memoir on a remarkable modification that light rays experience during their passage through certain translucent substances and on some other new optical phenomena]. Mém Inst 1:93–134
19. Kahr B, Arteaga O (2012) Arago's best paper. ChemPhysChem Educ 13:79–88
20. Lequeux J (2008) François Arago. Springer International Publishing, Switzerland
21. Magnusson CE (1920) Centennial of the discoveries of Oersted, Arago, and Ampère. J Am Inst Elec Eng 39:1031–1033
22. Biot JB (1815) Phénomènes de polarization successive, observés dans des fluids homogénes [Phenomenon of successive polarization, observed in homogeneous fluids]. Bull Soc Philomath Paris 190–192
23. Frankel E (1977) J. B. Biot and the mathematization of experimental physics in Napoleonic France. Hist Stud Phys Sci 8:33–72
24. Smith KA (2012) Louis Pasteur, the father of immunology? Front Immunol 3:68
25. Brogren C-H (2023) Louis Pasteur-the life of a controversial scientist with a prepared mind, driven by curiosity, motivation, and competition. APMIS 132:7–30
26. Kauffman GB, Myers RD (1998) Pasteur's resolution of racemic acid: a sesquicentennial retrospect and a new translation. Chem Educ 3:1–18

27. Vantomme G, Crassous J (2021) Pasteur and chirality: a story of how serendipity favors the prepared minds. Chirality 33:597–601
28. Derewenda Z (2008) On wine, chirality and crystallography. Acta Cryst A64:246–258
29. Flack HD (2009) Louis Pasteur's discovery of molecular chirality and spontaneous resolution in 1848, together with a complete review of his crystallographic and chemical work. Acta Cryst A65:371–389
30. Pasteur L (1848) Mémoire sur la relation qui peut exister entre la forme cristalline et la composition chimique, et sur la cause de la polarisation rotatoire [Memoir on the relation that may exist between crystal form and chemical composition, and on the cause of rotatory polarization]. C R Séances Acad Sci 26:535–538
31. Gal J (2013) Citation for chemical breakthrough awards: choosing Pasteur's award winning publication. Bull Hist Chem 38:7–12
32. Pasteur L (1848) Recherches sur les relations qui peuvent exister entre la forme crystalline, la composition chimique et le sens de la polarisation rotatoire [On the relations that can exist between crystalline form, and chemical composition, and the sense of rotary polarization]. Anal Chim Phys 24:442–459
33. Grossman RB (1989) Van't Hoff, Le Bel, and the development of stereochemistry: a reassessment. J Chem Educ 68:30–33
34. Meijer EW (2001) Jacobus Henricus van 't Hoff: hundred years of impact on stereochemistry in the Netherlands. Angew Chem Int Ed 40:3783–3789
35. Cintas P (2002) On the origin of tetrahedral carbon: a case for philosophy of chemistry? Found Chem 4:149–161
36. Wisniak J (2002) Joseph Achille Le Bel. His life and works. Rev CENIC, Cienc Quim 33:35–43
37. Fechete I (2016) Joseph Achille Le Bel, *Lux Aeterna* of stereochemistry. C R Chimie 19:3–6
38. Seeman JI (2018) From "Multiple Simultaneous Independent Discoveries" to the theory of "Multiple Simultaneous Independent Errors": a conduit in science. Found Chem 20:219–249
39. Ramberg PJ (2003) The tetrahedral carbon atom, 1874–1877. In: Ramberg PJ (ed) Chemical structure, spatial arrangement: the early history of stereochemistry, 1874–1914. Ashgate Publishing Company, Burlington, Vermont, Ch 3, pp 53–86
40. Van't Hoff JH (1874) Voorstel tot uitbreiding der tegenwoordig in de scheikunde gebruikte structuur-formules in de ruimte; benevens een daarmee samenhangende opmerkung omtrent het verband tusschen optisch actief vermogen en chemische constitutie van organische verbindingen [A proposal for extending the currently employed structural formulae in chemistry into space; together with a related remark on the relationship between optical activating power and chemical constitution of organic compounds]. Greven, Utrecht, The Netherlands
41. Van't Hoff JH (1874) Sur les Formules de Structure dans l'Espace. Arch Neerl Sci Exactes Nat 9:445–454
42. Le Bel JA (1874) Sur des Relations qui existent entre les Formules Atomiques des Corps Organiques et le Pouvoir Rotatoire de leurs Dissolutions [On the relationships that exist between the atomic formulas of organic bodies and the rotatory power of their solutions]. Bull Soc Chim Fr 22:337–347
43. Ramberg PJ (2003) Initial reception of the tetrahedron, 1874–1887. In: Ramberg PJ (ed) Chemical structure, spatial arrangement: the early history of stereochemistry, 1874–1914. Ashgate Publishing Company, Burlington, Vermont, Ch 4, pp 87–109
44. Van't Hoff JH (1875) La Chimie dans l'Espace. Rotterdam, The Netherlands
45. Van der Spek TM (2006) Selling a theory: the role of molecular models in J. H. van't Hoff's stereochemistry theory. Ann Sci 63:157–177
46. Snelders HAM (1974) The reception of J. H. van't Hoff's theory of the asymmetric carbon atom. J Chem Educ 51:2–7
47. Ramberg PJ (1994) Johannes Wislicenus, atomism, and the philosophy of chemistry. Bull Hist Chem 15/16:45–51
48. Kolbe H (1877) Zeichen der Zeit [Signs of the times]. J Prakt Chem 15:473–477

49. Rabinovich D (2009) Van't Hoff personalized stamps from the Netherlands. Phil Chim Phys 31:172–173
50. Lichtenthaler FW (2002) Emil Fischer, his personality, his achievements, and his scientific progeny. Eur J Org Chem 4095–4122
51. Kunz H (2002) Emil Fischer-unequalled classicist, master of organic chemistry research and inspired trailblazer of biological chemistry. Angew Chem Int Ed 41:44394451
52. Fischer E (1891) Ueber die Configuration des Traubenzuckers und seiner Isomeren. II. Ber Dtsch Chem Ges 24:2683–2687
53. Ramberg PJ (2003) Emil Fischer and carbohydrate chemistry, 1884–1891. In: Ramberg PJ (ed) Chemical structure, spatial arrangement: the early history of stereochemistry, 1874–1914. Ashgate Publishing Company, Burlington, Vermont, Ch 8, pp 243–276
54. Magner JA (2004) Emil Fischer (1852–1919), the stereochemical nature of sugars. The Endocrinologist 14:239–244
55. Fischer E (1894) Einfluss der Configuration auf die Wirkung der Enzyme. Ber Dtsch Chem Ges 27:2985–2993
56. Cramer F (1995) Biochemical correctness: Emil Fischer's lock and key hypothesis, a hundred years after–an essay. Pharm Acta Helv 69:193–203
57. Fischer E, Ach L (1895) Synthese des Caffeïns [Synthesis of caffeine]. Ber Dtsch Chem Ges 28:3135–3143
58. Fischer E, Speier A (1895) Darstellung der Ester. Ber Dtsch Chem Ges 28:3252–3258
59. Kelvin L (1894) The molecular tactics of a crystal. Clarendon Press, Oxford, p 27
60. Walden P (1896) Ueber die gegenseitige Umwandlung optischer Antipoden. Ber Dtsch Chem Ges 29:133–138
61. Werner A (1893) Beitrag zur Konstitution anorganischer Verbindungen. Z Anorg Chem 2:267–330
62. Kauffman GB (1977) Early experimental studies of cobalt-ammines. Isis 68:392–403
63. Werner H (2013) Alfred Werner: a forerunner to modern inorganic chemistry. Angew Chem Int Ed 52:6146–6153
64. Ramberg PJ (2003) Alfred Werner and coordination chemistry, 1893–1914. In: Ramberg PJ (ed) Chemical structure, spatial arrangement: the early history of stereochemistry, 1874–1914. Ashgate Publishing Company, Burlington, Vermont, Ch 9, pp 277–319 65. Werner A, Vilmos A (1899) Beitrag zur Konstitution anorganischer Verbindungen. XVII: Über Oxalatodiäthylendiaminkobaltisalze. Z Anorg Chem 21:145–158
65. Werner A, Vilmos A (1899) Beitrag zur Konstitution anorganischer Verbindungen. XVII: Über Oxalatodiäthylendiaminkobaltisalze. Z Anorg Chem 21:145–158
66. Kauffman GB (1899) A stereochemical achievement of the first order: Alfred Werner's resolution of cobalt complexes, 85 years later. Bull Hist Chem 20:50–59
67. Hassel O (1943) Cyclohexane. Tidsskr Kjemi Bergves Metall 3:32–34
68. Barton DHR (1950) The conformation of the steroid nucleus. Experientia 6:316–320
69. Barton DHR (1970) The principles of conformational analysis. Science 169:539–544
70. Toyota S (2024) Discovery of internal rotation and conformers of 1,2-dichloromethane: the dawn of the concept of conformation. Proc Jpn Acad, Ser B 100:101–113
71. Vollmer JJ, Servis KL (1968) Woodward-Hoffmann rules: electrocyclic reactions. J Chem Educ 45:214–220
72. Woodward RB, Hoffmann R (1969) The conservation of orbital symmetry. Angew Chem Int Ed 8:781–853
73. Seeman JI (2022) The ways of science through the lens of the Woodward-Hoffmann rules. The stories begin. Chem Rec 22:e202100211
74. Zhou Q, Kukier G, Gordiy I, Hoffmann R, Seeman JI, Houk KN (2022) A 21st century view of allowed and forbidden electrocyclic reactions. J Org Chem 89:1018–1034
75. Cahn RS, Ingold C, Prelog V (1966) Specification of molecular chirality. Angew Chem Int Ed 5:385–415
76. Helmchen G (2016) The 50th anniversary of the Cahn–Ingold–Prelog specification of molecular chirality. Angew Chem Int Ed 55:6798–6799

77. Trost BM (2004) Asymmetric catalysis: an enabling science. Proc Natl Acad Sci USA 101:5348–5355
78. Pfaltz A, Drury WJ (2004) Design of chiral ligands for asymmetric catalysis: from C2 symmetric P,P- and N,N-ligands to sterically- and electronically-nonsymmetrical P,N ligands. Proc Natl Acad Sci USA 101:5723–5726
79. Knowles WS (2002) Asymmetric hydrogenations (Nobel lecture). Angew Chem Int Ed 41:1998–2007
80. Noyori R (2002) Asymmetric catalysis: science and opportunities (Nobel lecture). Angew Chem Int Ed 41:2008–2022
81. Katsuki T, Sharpless KB (1980) The first practical method for asymmetric epoxidation. J Am Chem Soc 102:5974–5976
82. Sharpless KB (2002) Searching for new reactivity (Nobel lecture). Angew Chem Int Ed 41:2024–2032
83. MacMillan DWC (2008) The Advent and development of organocatalysis. Nature 455:304–308
84. Xiang S-H, Tan B (2020) Advances in asymmetric organocatalysis over the last 10 years. Nat Commun 11:3786
85. Han B, He X-H, Liu Y-Q, He G, Peng C, Li J-L (2021) Asymmetric organocatalysis: an enabling technology for medicinal chemistry. Chem Soc Rev 50:1522–1586
86. Mancheño OG, Waser M (2023) Recent developments and trends in asymmetric organocatalysis. Eur J Org Chem 26:e202200950
87. Ault A (2002) The Nobel Prize in chemistry for 2021. J Chem Educ 79:572–577
88. Monkhorst HJ (1968) Activation energy for interconversion of enantiomers containing an asymmetric carbon atom without breaking bonds. Chem Commun 1111–1112
89. Hoffmann R, Alder RW, Wilcox CF Jr (1970) Planar tetracoordinate carbon. J Am Chem Soc 92:4992–4993
90. Erker G (1992) Planar-tetracoordinate carbon: making stable anti-van't Hoff/LeBel compounds. Comments Inorg Chem 13:111–131
91. Siebert W, Gunale A (1999) Compounds containing a planar-tetracoordinate carbon atom as analogues of planar methane. Chem Soc Rev 28:367–371
92. Crigger C, Wittmaack BK, Tawfik M, Merino G, Donald KJ (2012) Plane and simple: planar tetracoordinate carbon centers in small molecules. Phys Chem Chem Phys 14:14775–14783

The History of Stereochemistry: Progress and Prospects

2

Peter J. Ramberg

Abstract

In the fifty years since the centennial celebration of tetrahedral carbon atom, we have learned a great deal about the emergence of stereochemistry after J. H. van't Hoff and J. A. Le Bel first proposed the tetrahedral carbon atom. My own book on the subject was published in 2003. Since the appearance of that book there has been additional work on van't Hoff, Louis Pasteur, Emil Fischer and Albert Ladenburg, and other research on modeling and the general development of the visual language of chemistry. In this essay, I will recount the major arguments from my book, and reflect on the newer literature and its significance for understanding of the history of stereochemistry.

2.1 Introduction

Fifty years ago, the chemist-historian O. Bertrand Ramsay organized a symposium at the September 1974 Atlantic City ACS meeting to celebrate the centennial of the tetrahedral carbon atom. He gathered together historians of chemistry and chemist-historians to explore topics in the history of stereochemistry, including, among others, the origins of van't Hoff and Le Bel's hypothesis, comparisons of van't Hoff and Le Bel's ideas, early conformational analysis, the early use of molecular models, and the origins of configurational notation of stereochemical centers. Most of the papers presented at the symposium and several other essays were collected by Ramsay for a publication in the newly formed ACS Symposium series [1]. The resulting volume, albeit fragmented, was the first attempt at an overview of the origins and history of stereochemistry. In 1981, Ramsay published his own

P. J. Ramberg (✉)
School of Science and Mathematics, Truman State University, Kirksville, MO, USA
e-mail: ramberg@truman.edu

A. Greenberg and D. E. Lewis (eds.), *The Birth of the 3rd Dimension in Chemistry*, Perspectives on the History of Chemistry, https://doi.org/10.1007/978-3-031-97743-5_2

monograph of the history of stereochemistry, from the discovery of optical activity to the work of Derek Barton on conformational analysis in the 1950s, but was primarily written as a textbook, with little thematic development or references [2]. By the 1980s, there was still much room for a comprehensive treatment of the emergence of stereochemistry.

We have now reached the sesquicentennial of the tetrahedral carbon atom, and for better or worse, I have taken on Ramsay's mantle as the leading historian on the early history of stereochemistry. Many years ago, when I was searching for dissertation topics, I was originally interested in R. B. Woodward and organic synthesis. But I eventually slid backwards in time to the nineteenth century, intrigued by the use of what looked like familiar formulas in the nineteenth century chemical literature. What did nineteenth century chemists think these formulas represented? In other words, I was looking for a historical answer to the philosophical problem of the meaning of chemical formulas. Alan Rocke's book on chemical atomism had just been published in 1984, and I had become aware of Ramsay's centennial volume on stereochemistry [3]. I soon zeroed in on looking at a more thorough examination of the reception of the tetrahedral carbon atom in the late nineteenth century, and completed a dissertation in 1993 with the support of the faculty at Indiana University, despite their lack of knowledge of chemistry. Ten years later, after removing a chapter, adding three more, and extensively revising the rest of the dissertation, I managed to complete my early history of stereochemistry [4].

In the twenty plus years since my book appeared, there has been additional historical research that addresses the issues I wrote about, and fortunately, the number of publications is not too large to examine all of them here. In what follows, I will provide a synopsis of my book, and then survey the recent literature in light of what I argued earlier.

2.2 Chemical Structure, Spatial Arrangement

The major goal of the book was to put together a fuller context of the emergence of the tetrahedral carbon atom by van't Hoff and Le Bel in 1874, and follow the reception of their ideas through the remainder of the nineteenth century. I don't claim originality for many of the major themes and arguments in my book, but I certainly fleshed them out from existing suggestions in earlier literature. I quickly realized that I was not able to do a comprehensive history of stereochemistry in the period, and confined my study to German chemists. Even under those constraints, I was unable to examine all of the stereochemical topics in the German chemical literature. My hope was to collect and integrate the case studies that I did include into a more coherent framework for the incorporation of additional examples.

2.2.1 Van't Hoff and Le Bel

As was well known before I wrote the book, the origins of the tetrahedral carbon atom lay in the study of optical activity and its correlation to chemical composition. In 1848, Pasteur's study of the tartrates suggested that optical activity could be explained at the molecular level, although he was unable to offer anything specific at that time. The correlation was first offered by J. H. van't Hoff (1852–1911) and Joseph-Achille Le Bel (1847–1830) in 1874, when they independently proposed that a tetrahedral arrangement of groups around a carbon atom would produce the asymmetry required for optical activity [5, 6].

Although very similar, van't Hoff's and Le Bel's ideas were not the same. Both made a general correlation between optical activity and chemical structure, but the origin and resulting scope of both theories were very different. Van't Hoff introduced the tetrahedron as an expansion of traditional structural chemistry's power to explain isomers, and suggested a specific atomic form to do that. As van't Hoff himself recounted, the inspiration for his theory was the problem of the isomeric lactic acids established by Johannes Wislicenus between 1869 and 1873 [4, 7]. The resulting theory was highly deductive, and made specific predictions about whether compounds would be optically active, and predicted the existence of new kinds of isomerism. Le Bel, on the other hand, had come to the problem through the French tradition of crystallography, and looked for general rules of symmetry in separate groups of similar chemical formulas. He did not use any diagrams or introduce a picture of the carbon atom, and his approach is in general more inductive than van't Hoff's. Van't Hoff also took the unusual step of constructing model sets to accompany his 1875 French pamphlet. He sent these sets to prominent chemists in Germany, France, and Great Britain, and noted that these sets were also available on request.

The deductive and predictive character of van't Hoff's theory is what made it more influential by the 1880s. As van't Hoff presented his theory in the 1877 German edition, it was actually a cluster of related hypotheses based on the idea that all carbon atoms are tetrahedral, and so, unlike Le Bel, his theory went beyond simply explaining optical activity. The first hypothesis was the asymmetry of the tetrahedral carbon atom itself and the explanation of optical rotation. The second hypothesis consisted of a model for the carbon–carbon single bond, in which the carbon atoms had free rotation. The third hypothesis predicted a restricted rotation around the carbon–carbon double bond, and another form of spatial isomerism (*cis–trans*). This core of ideas provided a framework for later research programs into stereochemical problems.

Until my book, the prevailing historical conception about the early reception of the tetrahedral carbon atom was that chemists were either critical of it or indifferent, and Hermann Kolbe's famous violent diatribe against van't Hoff's theory has been taken as a proxy for general criticism. But surviving letters to van't Hoff following the publication of his 1875 work reveal Kolbe as an outlier. In general, chemists were supportive and cautiously optimistic toward the theory. There were certain problems—there were too few examples of optically active compounds with definitive structures, and the tetrahedral array of valences seemed to defy known physical forces—but overall, it seemed a promising theory that merited further investigation.

2.2.2 The Meaning of Structural Formulas

Throughout the 1860s, as chemists became increasingly drawn to visualization techniques, they began to rely on both physical hand-held molecular models and drawings of molecules using structural formulas [8]. Kekulé's famous tetrahedral models allowed him, for example, to easily depict single, double, and triple bonding. The now familiar drawings using Berzelian notation with dashes for bonds that we now take for granted were first introduced by Alexander Crum Brown in 1864. But these, and any other representations of molecules, had a particular meaning for chemists at the time. To use terms from semiotics, chemists regarded formulas as *symbolic*, that is, they were not intended to directly mimic the appearance of the molecule in the same way that eagles, for example, are meant to represent freedom. Many chemists of the time issued explicit warnings about the overinterpretation of these models. Crum Brown, for example, wrote in 1864 about his new form of representation:

> [These formulas are] used to express constitutional formulae, and by which, it is scarcely necessary to remark, I do not mean to indicate the physical, but merely the chemical position of the atoms. … and while it is no doubt liable, when not explained, to be mistaken for a representation of the physical position of the atoms, this misunderstanding can easily be prevented [9, p. 708].

In other words, chemists made a sharp epistemological distinction between *chemical* formulas and the *physical form* of the molecule. It may be that chemical formulas were a window into the microword, but they were based on chemical, and not physical data, and that precluded any such judgment.

In 1874, when van't Hoff suggested that the carbon atom was tetrahedral, he expanded the meaning of chemical formulas to include an *iconic* meaning–formulas *do* intentionally mimic the appearance of the molecule. The story of early stereochemistry is therefore the story of how chemical formulas were transformed from symbolic to iconic forms of representation. The details surrounding this transformation is one of the major themes of the book.

2.2.3 The Importance of Johannes Wislicenus

The third major argument I made was for the central importance of Johannes Wislicenus (1835–1902) in the emergence of stereochemistry. First, his work on the structure of lactic acid directly inspired van't Hoff to develop the tetrahedral carbon atom to explain optical activity. Second, on receiving his copy of *La chimie dans l'espace* in 1875, Wislicenus became one of van't Hoff's strongest advocates, sponsoring the translation and expansion of the theory into German. Third, in 1887, Wislicenus published an extensive and influential monograph on unsaturated compounds using van't Hoff's second and third hypotheses to organize information about the unsaturated acids and determine the genetic relationships between them to assign configurations [10]. Wislicenus and his students in Leipzig followed up on the general program outlined in the monograph with a series of publications describing specific methods for establishing the configuration (*cis* and *trans*) of various pairs of stereoisomeric unsaturated acids. His work on unsaturation reached a high point in 1892, when the critique of Wislicenus' methodology by the American Arthur Michael significantly undermined his approach to configurational assignments [11]. Although chemists did not accept all of Wislicenus' arguments, and one of his assumptions behind establishing configurations was shown to be faulty, as a result of his work, there seemed little doubt that the unsaturated acids were spatial isomers. Another way to emphasize Wislicenus' importance is to imagine an alternative history in which he did not, for example, encourage the translation of van't Hoff's pamphlet, or did not pursue the stereochemistry of the unsaturated acids. Under this scenario, the adoption of van't Hoff's ideas would have had a very different trajectory.

2.2.4 Victor Meyer

The remainder of the book discusses some of the other early approaches to stereochemistry. Victor Meyer (1848–1897) read Wislicenus' monograph on the unsaturated acids shortly after his call to Göttingen, and marveled at Wislicenus' skill in marshaling the evidence for van't Hoff's second and third hypotheses, even if he thought Wislicenus was a bit overenthusiastic. It seems likely that reading Wislicenus' work piqued Meyer's interest in van't Hoff's hypothesis, and in particular, modifying the second hypothesis to explain the three structurally identical but isomeric benzildioximes that he and Karl Auwers first identified in 1883. Meyer suggested that this isomerism could be explained by a restricted rotation around the carbon–carbon single bond in benzildioxime that created spatial isomers. Meyer's most enduring contribution was the creation of the word "stereochemistry" itself. He found stereochemistry appealing for two reasons—it provided a partial reduction of chemical phenomena to mechanics, and laid the ground work for understanding the concept of valence at a deeper level. He and Wislicenus both argued that stereochemistry could reveal the parts of atoms, just as the earlier stoichiometric laws had revealed the atoms themselves.

2.2.5 Arthur Hantzsch and the Stereochemistry of Nitrogen

Meyer's stereochemical theory of the benzildioximes would not last long, for in early 1890, Arthur Hantzsch (1857–1935) and his student Alfred Werner (1866–1919) presented an alternative theory, based on the stereochemistry of the nitrogen itself. According to this theory, actually Werner's doctoral dissertation, the carbon–nitrogen double bond in the benzildioximes was an analog to the carbon–carbon double bond, and could also create stereoisomers. Although Werner would soon move to inorganic chemistry, Hantzsch pursued the stereochemistry of oximes, hydrazones and diazo compounds in an innovative research program that relied on physical as well as chemical methods to establish configurations. Like Wislicenus, Hantzsch's approach was deductive, using the hypothesis of the stereochemical nitrogen atom to generate his research program. But unlike Wislicenus and Meyer, Hantzsch did not develop any sophisticated theory of valence, going so far as to say,

> at least in its present stage of development, stereochemistry requires no specific idea about the type and cause of intramolecular cohesion of atoms (the nature of chemical affinity), or concerning the type and cause of ratios in which different atoms combine (the nature of valence); at present it requires only the idea, proved by the existence of isomerism itself, that the atoms are not situated within the molecule in a chaotic state, but in a stable equilibrium position within certain limits (p. 239) [4].

2.2.6 Emil Fischer and Sugar Chemistry

While Wislicenus, Meyer and Hantzsch followed the implications of van't Hoff's second and third hypotheses, during the 1880s, Emil Fischer (1852–1919) gradually came to realize that van't Hoff's first hypothesis could help bring order to sugar chemistry. The key to Fischer's success was the discovery of phenylhydrazine in 1875. Used as a reagent, phenylhydrazine reacted with carbonyl groups to make highly crystalline compounds. In the mid 1880s, Fischer realized that sugars, usually syrupy, intractable substances, reacted with phenylhydrazine to make highly crystalline compounds that could easily be identified by melting point. Fischer followed a number of different experimental trails in sugar chemistry that eventually converged by 1890 into a coherent methodology for identifying and classifying sugars. Among these trails were the chemistry of sugars with phenylhydrazine, the formation of α- and β-acrose, the isolation of mannose, and the oxidation and reduction of monosaccharides. Optical activity and van't Hoff's hypothesis did not enter into Fischer's thought until 1889, when he and his student Josef Hirschberger realized that mannose and glucose were structurally identical sugars. At this point, Fischer introduced van't Hoff's theory to differentiate between them. These two sugars must possess a different symmetry at the second carbon atom, making mannose and glucose configurational isomers. In Fischer's subsequent research, van't Hoff's theory took on increasing importance for classifying the sugars, and he eventually established the configurations of all sixteen possible isomers of glucose. Fischer's work has justifiably become one of the most famous accomplishments in the history of chemistry, and earned him the 1902 Nobel Prize.

2.2.7 Alfred Werner and Inorganic Stereochemistry

After completing his dissertation under Hantzsch, Werner outlined an innovative theory of valence as an isotropic attraction across the entire surface of the atom, rather than directed forces akin to the balls and wires of hand-held models. Werner outlined this theory in his 1891 *Habilitationsschrift* and then applied it to the metal-amine complexes in his 1893 paper introducing his coordination theory [12, 13]. This 1893 article resembles in many ways van't Hoff's discussion of carbon stereochemistry. Werner did not introduce any new evidence for the coordination theory, but completely reinterpreted the known chemistry of the metal-ammonia compounds under his new theoretical framework of the octahedral or square planar arrangement of atoms around a central metal atom. Werner's theory relied heavily on "isomer-counting"—matching the number of known compounds with the number predicted by the formulas. But Werner's paper was also much broader in scope than anything van't Hoff wrote, in that it suggested a radical restructuring of the understanding of metal-ammines with a new conception of valence, and invented a new iconic visual language for the metal-amines. Werner and his students spent the next twenty years working out the implications of his new theory, isolating the many compounds it predicted, including *cis* and *trans* isomers, the first optically active metal complex, and the first optically active compound that did not contain carbon.

2.2.8 Major Themes in the Early History of Stereochemistry

After reading my book again after twenty years, I found it difficult to summarize adequately all of the various themes that emerged, but here are four brief points. First, although chemists were general receptive to the new theory, there did not seem to be much application or extension of the theory before 1885, even by van't Hoff himself, who eventually would leave organic chemistry altogether. It was not until Adolf von Baeyer's strain theory (1885) and Wislicenus' monograph on unsaturation (1887) that chemists became aware of the potential in van't Hoff's ideas. After 1887, the number of publications discussing stereochemical ideas increased considerably, with 1890 being a watershed year. Although chemists continually referred to the theory as the "Le Bel-van't Hoff theory," chemists generally pursued the much richer implications of van't Hoff's three hypotheses.

Second, the growing literature on stereochemistry all proceeded with relatively little controversy, as an application and expansion of van't Hoff's initial ideas to new groups of compounds and elements, but without discussion of any alternative theories. There were critiques—from Kolbe, Rudolph Fittig, Wilhelm Lossen,

Arthur Michael, and Rudolf Claus, and Sophus Mads Jørgensen—but they were isolated and addressed specific aspects of other chemists' work. There was no organized movement to offer alternative explanations for structurally identical compounds that did not rely on the older principles of structural chemistry. Van't Hoff's theory has remained famous, and rightly so, but the fundamental nature of the transformation in the meaning of chemical formulas has been forgotten or never noticed, even at the time. There was no commentary or criticism about the shift to an iconic meaning of chemical formulas, despite the numerous warnings and caveats by chemists about the meaning of structural formulas in the 1860s.

Third, the development of an iconic visual language containing the tetrahedral carbon atom implied a theoretical problem about how an atom could divide its valence into partial, directed, attractive forces. Van't Hoff, Fischer and Hantzsch remained pragmatic and ignored the problem, while Wislicenus, Meyer, and Werner attempted to reconcile the tetrahedral arrangement with known physical laws, with varying degrees of success. But in general, chemists constructed and used stereoformulas in a pragmatic way to give the explanation of isomers a graphic clarity. They used these models *as if* they were real, despite any physical shortcomings about valence they might raise (pp. 327–328).

Fourth, the emergence of stereochemistry also reflects various trends in nineteenth century chemical methodology. These include the continued use of various physical properties (now sometimes including optical activity) of compounds as "markers" for indicating chemical, structural, or stereochemical change, and the use of hand-held physical models. Stereochemistry also exemplifies the bifurcation of methodology between the inductive classificatory chemistry of the first half of the nineteenth century, and the deductive and predictive character of the second half. Stereochemists strongly emphasized the hypothetico-deductive nature of spatial arrangements, in that it made predictions of new compounds and made possible research programs to assign specific configurations to spatial isomers. Many of the criticisms of stereochemistry were primarily based on an objection to this speculative, deductive character of stereochemical theory. The stereochemists were also split between inductive and deductive approaches. Wislicenus, Hantzsch and Werner extensively relied on the predictive power of stereochemistry, whereas Meyer and Fischer were more inductive, making conclusions, cautiously adopting spatial arrangements as explanations.

2.3 Recent Historical Work on Stereochemistry

We can turn now to historical work done since I completed the book. These studies include studies of Louis Pasteur, the nature of early modeling, the growth of visualization techniques in organic chemistry, Albert Ladenburg's synthesis of coniine and a new study of Emil Fischer's work on the sugars. As we shall see below, to some degree, all of these studies have enhanced the story I presented in the book. Because I cannot read Dutch, I haven't been able to read Rob van den Berg's recent work on van't Hoff, including the first modern biography of van't Hoff, and an article on van't Hoff's unsuccessful call to Leipzig in 1887, and so I have not included them here [14, 15].

2.3.1 Louis Pasteur

Since 2003, Joseph Gal has published a number of articles on Louis Pasteur and the discovery of molecular asymmetry in 1848. Although too great to summarize in detail here, Gal has examined a number of important issues both in regard to Pasteur's work and about the emerging vocabulary of stereochemistry. First, Gal traced the origin of the word "enantiomorphous" to the Leipzig mineralogist Carl Friedrich Naumann (1797–1873) who coined the word in 1857. Pasteur did not use the word, nor did van't Hoff, until the 1877 German edition of his work when it began to see more regular usage. The initial usage in 1877 likely derived from the translator, Felix Herrmann, who may have been more aware of Naumann's work. Le Bel also did not use the term "enantiomorphous" in his original French article, but in the standard 1901 English translation of Le Bel's article, G. M. Richardson rendered Le Bel's phrase "not superposable on its [mirror] image" as "enantiomorphous" [16, 17]. This erroneous translation has found its way into the historical literature, including my own.

Gal has also worked to correct the historical record from the standard accounts of his work by Pasteur's son-in-law and grandson, both of which are hagiographic, and embellish on Pasteur's own attempts at crafting his public image. For example, Gal notes that historical accounts of Pasteur's discovery of biological enantioselectivity of racemic tartaric acid have been dominated by Pasteur's own reconstruction of the events, furthered by his son-in-law and grandson, that make it look like chance did not play a large role in his discovery [18]. Gal has also noted that Pasteur's initial presentation on the molecular chirality of the tartrates to the *Academié des science* occurred on May 22, 1848, one day after the death of his mother. This date is unambiguous in the printed record of the Academy, yet Pasteur's initial biographers consistently gave the date as May 15, 1848, perhaps in order to avoid any public judgment about Pasteur's possible insensitivity to his mother's death [19].

Gal has also noted Pasteur drew multiple portraits of friends as a youth and had considerable artistic ability. One of the techniques Pasteur used to create copies these portraits was lithography, which involves creating a mirror image of the portrait on a stone, such that during printing a correct image will appear. In a letter to his parents describing the process, he noted that he got the best results by using the image in a mirror to create the mirror image on the stone [20]. This ability to think about and draw mirror images perhaps sensitized him to think spatially and be familiar with the nature of mirror images.

Gal has also written about Pasteur's use of language, for example, why he chose the term "disymmetry" rather than "asymmetry" in his early works. Gal argues that "asymmetry" was a well-known term, but filled with ambiguity, while "dissymmetry" was lesser known. He was also possibly influenced by Jean-Baptiste Biot, who used the term. The term "chirality" was introduced by Lord Kelvin in 1884, but was not used in chemistry until 1962 by Kurt Mislow. Overall, Gal's work reminds us that historians, must pay close attention to the words that people choose, and also why they choose them, and how they are used, because the meaning of words can change, often dramatically, over time [21, 22].

Finally, Gal has presented a lengthy critique of the late historian Gerald Geison's treatment of Pasteur's work on tartaric acid in *The Private Science of Louis Pasteur* [23]. In a close examination of Geison's book and comparison to Pasteur's notebooks and published results, Gal argues that Geison was inaccurate and misleading in his analysis of Pasteur's work. Geison displayed some fundamental misunderstandings of numerous topics, including, but not limited to, the principles and language of chirality, the state of organic chemistry in the 1840s, Biot's understanding of optical activity, and lack of appreciation for the fundamental importance of Pasteur's insight into molecular chirality. He also notes many errors in translating and transcribing Pasteur's French. This critique is important, Gal notes, given Pasteur's own erudition as a refined speaker and writer of French [24]. While Geison's work on Pasteur certainly presents a healthy corrective to the hagiographic biographies from Pasteur's relatives, Gal has shown that there remains more work to be done to create a still more historically accurate depiction of Pasteur's research on tartaric acid.

2.3.2 Models

The second topic involves the origin and use of hand-held molecular models in stereochemistry. In 2005, Trieke van der Spek conducted a survey of the surviving cardboard tetrahedral models constructed by van't Hoff, and made a few observations that had eluded me about these models [25]. First, the cardboard models resemble geometric and crystallographic models more than the earlier "ball and stick" models created by August Hofmann and August Kekulé to illustrate valence and bonding [8, 26]. Second, van der Spek suggests a more specific timeline for when van't Hoff first created the models. Van't Hoff did not mention models at all in the initial 1874 pamphlet, a fact that I had overlooked. It seems that the idea of constructing these models appeared in the year between the 1874 pamphlet and the 1875 French expansion of the pamphlet in which he not only mentioned the use of models, but sent boxes of constructed models to prominent chemists throughout Europe. It may also be while playing with these models that van't Hoff found that allenes could be asymmetric without an asymmetric carbon atom, an idea that first appeared in the 1875 French pamphlet.

Third, the way in which van't Hoff used these models indicates that he didn't yet consider them fully iconic, because he did not think of the carbon atom as literally a tetrahedron. It was a simple way to illustrate the correlation between optical activity and molecular structure. As van der Spek points out, in his early correspondence, van't Hoff noted that the two different ways in which he constructed the tetrahedra—having the valences depicted on the faces of the tetrahedra versus the points–were functionally identical, because one could imagine the face-centered model as having an apex-centered model *inside* the tetrahedron. Finally, van der Spek argues more explicitly than I did that van't Hoff's models, as a form of representation, offered a novel means of "experimentation" and manipulation, and provided a form of "experimental proof" at a time when there was little or no experimental evidence in favor of the theory.

In 2016, Michael Friedman examined the relationship between the early folded models of the carbon atom made by van't Hoff and Hermann Sachse (1862–1893) and their mathematical foundations [27]. As we saw above, van't Hoff first started creating cardboard models before publishing his 1875 French pamphlet, and sent copies of these models to prominent chemists. By the 1877 German edition of his work, these cardboard models were described in the Appendix, where van't Hoff also provided templates of models that could be cut out of thick cardboard, scored with a knife, folded, and glued into tetrahedra. The first of these templates were entirely symmetrical tetrahedra, but van't Hoff then included templates with sides of non-uniform length, providing a mathematical derivation of these dimensions. But as Friedman notes, in subsequent stereochemical works by van't Hoff and others, these templates and the irregular tetrahedra and the mathematical descriptions disappeared from the literature, in favor of simple line drawings of regular tetrahedra.

The second example Friedman examines comes from Hermann Sachse's 1890 models of cyclohexane. These models are famous for two reasons, first, they suggested that the cyclohexane ring was not planar, as assumed by Adolph von Baeyer. Second, they predicted two spatially isomeric monosubstituted cyclohexanes (today called the axial and equatorial forms) [28]. But as Friedman notes, this 1890 paper was a qualitative description of Sachse's more sophisticated mathematical derivation of these configurations. Sachse published the full mathematical analysis in a lengthy paper (38 pages) published in the *Zeitschrift für physikalische Chemie* in 1892 [29]. Sachse followed this paper with another lengthy mathematical analysis of chemical affinity, also in the *Zeitschrift*, in 1893, in which he wrote that "the language of mechanics will ultimately dissolve the language of our science" [30, p. 185].

With hindsight, Sachse's folded models of substituted cyclohexane proved to be "correct," although it took nearly sixty years and the work of Derek Barton to vindicate the non-planar nature of cyclohexane. The reasons for this are complex, and well outlined by Colin Russell, although the full story remains to be told [31]. More important here is Friedman's observation about the fate of van't Hoff's and Sachse's attempt to use mathematics to derive the stereochemical relationship in molecules. In the end, chemists ignored the mathematical underpinnings of their models, but for different reasons. Van't Hoff attempted to give some mathematical basis for the tetrahedron, but his and other chemists' focus shifted to simply the tetrahedron itself. Sachse's mathematics, intended as a reductive replacement for the tetrahedron, was far beyond what any organic chemist could comprehend or use in the laboratory.

In 2014, I published an article that explored another aspect of modeling in early stereochemistry that involved a dispute between Johannes Thiele (1865–1918) and Emil Knoevenagel (1865–1921) [32]. In 1899, Thiele suggested that the reactivity of multiple bonds could be explained by what he called "partial valence" (*Partialvalenz*), without consideration of the tetrahedral nature of carbon, and based on the assumption that valences were directed to specific locations on the atom. In 1900, Knoevenagel developed a deeper explanatory model for partial valence with a three-dimensional model of bonding, by assuming an isotropic model of chemical affinity and using tetrahedral models with curved sides made of shellac that could easily be stuck together. Tetrahedra touching their faces in a single bond would have a "rocking motion" (*rollende Bewegung*) that allowed a residual attraction to appear as the valence became incompletely saturated. Knoevenagel expanded his models to explain addition to conjugated systems, conversion of tautomers, and the stability of aromatic compounds. Like his mentor Baeyer, Thiele was far more cautious theoretically, and was reluctant to add an additional level of mechanical speculation suggested by Knoevenagel.

Knoevenagel's use of models illustrates the tendency in the late nineteenth century for chemists to take the tetrahedral carbon atom beyond its original capacity to predict new isomers or optical activity, and to use it to imagine the dynamic properties of molecules for explaining chemical reactivity. Thiele had left the concept of partial valence unexplained, but Knoevenagel attempted to give it a deeper, mechanical explanation, and reduce it to the mechanical properties of the tetrahedron [32].

2.3.3 Visualization and Iconic Formulas

Visualization techniques are an essential aspect of modern chemistry that chemists created during the nineteenth century. This process began with the adoption of Berzelian formulas as theoretically neutral "paper tools," followed by the creation of heuristic structural formulas, and culminating in the appearance of stereochemistry and the shift to an iconic meaning of chemical formulas [33]. As I mentioned above, this transition occurred largely without notice, but in 2010, Alan Rocke argued in the book *Image and Reality* that although there was some reluctance to make the shift to thinking about formulas in an iconic way, chemists had already laid some significant groundwork for the shift during the 1860s [8].

First, one of the major themes throughout Rocke's book, too great to recount in detail here, is that chemists were *already* thinking in highly visual ways by the 1860s, such that inhibition toward an iconic view of molecules was already waning by the 1870s. Second, it is well-known that many chemists well before 1874 had speculated about three-dimensional properties of atoms and subatomic structure–chemists used the tetrahedron at least four times during the 1860s in connection with the carbon atom. But these earlier tetrahedral models assumed that the tetrahedron was *irregular*, which implied a difference in strength between the four affinity units of carbon. By 1864, this hierarchical view of carbon bonding had been removed in favor of four equal valences [8]. Van't Hoff adopted this non-hierarchal view of carbon's valence, and suggested that asymmetry was inherently a *molecular*, not atomic property, and the asymmetry relied on the rigidity of the four bonds surrounding the carbon atom.

This rigidity had not been an essential part of structural theory as it was originally conceived (most likely because of the intended symbolic nature of the representation), and Adolf von Baeyer later suggested that he and others had taken too seriously Kekulé's admonition that the bonds in the molecule were not rigid like the bonds in his tetrahedral models, and were always capable of switching places. Kekulé, in fact, held two contradictory visions of molecules—one in which atoms were vibrating, rotating and moving within the molecule, and one which required the atoms in the molecule to be fixed in position, in order to maintain a constant structure [8].

Finally, Rocke notes that both Stanislao Cannizzaro and Kekulé had read Rudolph Clausius and were familiar with the emerging conception of the microworld within physics (resulting in the kinetic theory of gases) that supported Avogadro's hypothesis and a realist account of molecules. As early as 1858, Kekulé directly cited Clausius' gas theory, and stated that "the chemical molecule is identical to the physical gas molecule" [8].

2.3.4 Albert Ladenburg (1842–1911) and the Synthesis of Coniine

The final historical study I will examine here is by Catherine Jackson, who has written much on the nature of nineteenth century organic chemistry, especially on the origins of organic synthesis. Here, I will address two of Jackson's works. The first is her book *Molecular World: Making Modern Chemistry* (2023), and the second is "Emil Fischer and the 'Art of Chemical Experimentation'" (2017), an earlier essay on Emil Fischer's sugar chemistry.

In the last chapter of her 2023 book, Jackson examines at length Albert Ladenburg's 1886 synthesis of the alkaloid coniine (α-propylpiperidine), which Ladenburg described as the "first total synthesis (*vollständige Synthese*) of an alkaloid." Jackson first recounts the early attempts to make coniine by Hugo Schiff (1834–1915) and August Hofmann (1818–1892). Schiff's attempt was unsuccessful, as was Hofmann's, but Hofmann did manage to conclude that coniine was an alkylated piperidine. Ladenburg began a study of alkaloids with atropine in 1881, but soon moved to coniine, shortly after Hofmann had determined its structure. As Jackson notes, one of Ladenburg's significant challenges was unequivocally establishing the identity of his synthetic α-propylpiperidine as coniine. After numerous unsuccessful attempts to alkylate pyridine, Ladenburg managed to synthesize α-propylpiperidine from α-picoline (2-methylpyridine) and obtained an optically inactive product. Because coniine was already well-known to be optically active, Ladenburg attempted to resolve the racemic product, at first unsuccessfully with the mold *Penicillium glaucum*, but then successfully by fractional crystallization using dextrorotatory tartaric acid (p. 2583) [34]. Ladenburg eventually identified his synthetic coniine on the basis of the melting point of its hydrochloride salt, its boiling point, its specific optical rotation, and its toxicity in rabbits (p. 83) [35]. At the end of the paper, Ladenburg announced that "Here I may point out that these experiments have led to the first total synthesis of an alkaloid."

In a separate one-page paper immediately following the paper describing the synthesis, Ladenburg noted that he could use his synthesis to test the validity of van't Hoff's hypothesis:

> The evidence presented in the preceding investigation, showing that α-propylpiperidine can be split into two optically active isomers, has inspired me to make use of the piperidine bases to test the Le Bel-van't Hoff hypothesis [36].

He noted that the structure had one asymmetric carbon atom, and as expected by van't Hoff and Le Bel, α-propylpiperidine did indeed exist in two enantiomeric forms. He also suggested that the theory predicted the corresponding β- and γ-propyl piperidines should not exist in enantiomeric forms. As Jackson rightly points out, however, Ladenburg was mistaken here, as only the γ-isomer would not have an asymmetric carbon atom [37]. Jackson is correct that Ladenburg's use of coniine's structural formula illustrates his ability to use it visually to validate the tetrahedral carbon atom, but that his error reveals that he was not yet completely familiar with this novel way of using structural formulas.

Jackson's account of Ladenburg's synthesis is a welcome addition to the discussion about the meaning of synthesis and the role of optical activity in demonstrating a compound's identity. She recounts the many details surrounding the nature of the process involved in elucidating coniine's structure by Schiff, Hofmann and Ladenburg. Like many chemists before them, they all struggled with what Jackson calls the chemical "identity crisis"—how can chemists unequivocally identify compounds as unique, or in the case of coniine, declare that the synthetic and natural products are identical. Ladenburg falls into the methodological tradition of making inductive conclusions, cautiously applying van't Hoff's theory only after completing his synthesis of coniine. In other words, at least as far as the published record implies, the tetrahedral carbon atom was not a guiding assumption in his synthesis. Ladenburg's mistake about possible stereoisomers of coniine illustrates the early difficulties in fully understanding or locating points of asymmetry in the molecule. I did not include Ladenburg's synthesis in my book, but it is clearly one of the early examples of support for van't Hoff's theory of the tetrahedral carbon atom as an explanation for optical activity, appearing between Baeyer's strain theory and Wislicenus' monograph.

2.3.5 Revisiting Emil Fischer's Work on the Sugars

We can now turn to Jackson's study of Emil Fischer's sugar chemistry. Jackson presents a detailed portrait of Fischer's early career, including his early difficulty learning chemistry in Kekulé's Bonn laboratory. He and his cousin Otto sought out another location and settled on Adolph von Baeyer's laboratory in Strasbourg, where Fischer flourished, becoming Baeyer's most significant student. Fischer spent ten years in Baeyer's Strasbourg and Munich laboratories, climbing the academic ranks before leaving Munich in 1882 for Erlangen. It was in Baeyer's laboratory, working with the compound fluorescein, that Fischer learned the "art of experimentation" or "laboratory reasoning," to use the phrase employed by Jackson, that entailed learning how to make solids crystallize, how to measure accurate melting and boiling points, and how to use standard reagents.

The key to Fischer's success with the sugars was his discovery of phenylhydrazine in 1875, which has usually been described as "serendipitous," but Jackson suggests otherwise. Jackson argues that Fischer's discovery was the result of the development of his bench skills acquired during his work on fluorescein and during the process that led to the discovery of phenylhydrazine itself. It was in the context of this work that Fischer became accustomed to his mentor Baeyer's more empirical style of small-scale chemical manipulation and the tasks involved in the identification of compounds.

As Jackson notes, Fischer's initial work on fluorescein derived from Baeyer's own interest in dye chemistry, and he hoped the chemistry of phenylhydrazine, published in a long series of articles between 1874 and 1888, would also provide entry into a new class of lucrative dyestuffs. His hopes proved unsuccessful, but in the 1880s, his interests moved toward the sugars, another group of potentially lucrative compounds. Many chemists had already looked at the chemistry of the sugars, but the tendency of sugars to form intractable syrups made them difficult to study. Fischer had become accustomed to creating compounds that were highly crystalline by including aromatic groups, and in 1886, spurred by Victor Meyer's use of hydroxylamine and its reactivity towards ketones and aldehydes, Fischer realized that phenylhydrazine had similar potential as a reagent for tackling the sugars, since by then the sugars were known to contain an aldehyde or ketone. Fischer found that they reacted with phenylhydrazine to make highly crystalline derivatives—osazones—that could be characterized easily by melting point. Jackson emphasizes, as I did not, the *practical* nature of Fischer's work on phenylhydrazine, its use as a "reagent" to identify specific carbonyl species, and Meyer's inspiration. Fischer moved primarily from the results of chemical reactions of the sugars to their chemical structures, and Jackson emphasizes that the structures themselves did not offer any suggestions as to how they could be made.

Jackson also argues that Fischer's research program was not a target-oriented synthesis of the sugars. Phenylhydrazine offered Fischer a means towards creating a series of what Jackson calls "synthetical experiments," using phenylhydrazine to make a series of easily characterizable compounds to elucidate chemical structure. This was a similar approach that August Hofmann had used in his study of coniine, and also thirty years earlier with amines (especially aniline) and the alkyl halides to derive the ammonia type [37, 38]. Jackson arrives at the same conclusion I did, that in 1889 Fischer and Hirschberger reached an impasse in differentiating between the osazones of mannose and glucose because they had identical melting points and it was only then that optical activity and van't Hoff's theory came into play. Jackson describes an important procedural detail that I missed; however, Fischer and Hirschberger realized that the osazones of mannose and glucose had identical melting points because they were able to modify their technique by measuring the melting points of the two osazones simultaneously with the same thermometer in the same apparatus, enabling a direct comparison of melting points. This technique bypassed any problems with melting point consistency in using different thermometers, different samples, or different apparatus.

Jackson's work as a whole is a significant contribution to the history of nineteenth century organic chemistry, but it is not without its problems. Her analysis of Ladenburg's papers omits important information about the properties of his synthetic coniine, and she makes claims about the significance of Ladenburg's synthesis and his use of optical activity that are not supported. A minor issue in the analysis of Fischer's work, is that after taking care to argue, I think successfully, that Fischer's discovery of phenylhydrazine was not serendipitous, Jackson unfortunately writes later in the article (p. 17), that "almost a decade after his *serendipitous* discovery of phenylhydrazine, Fischer had 'laid the foundation for [his] future sugar work'" (emphasis added) [39]. Jackson also claims that previous historical treatments of Fischer, including my own, have overemphasized the role of theory in enabling Fischer's synthetic work on the sugars [39]. In my book I argued that it was the strange structural identity of the osazones of mannose and glucose in February of 1889, the key event repeated by Jackson, that first led Fischer to use van't Hoff's hypothesis to differentiate them. I did not argue that van't Hoff's theory enabled Fischer's *synthetic* work, because, as Jackson emphasizes, and I have written elsewhere, it could not do this [40]. I suggested that van't Hoff's theory provided a useful role in classifying and predicting the existence of sugar isomers, a position that is not significantly different from what Jackson has argued.

2.4 Conclusion

All of these case studies have improved greatly our understanding of the early development of stereochemistry. The many contributions to this volume will no doubt also enhance our understanding of the development of stereochemistry.

Yet, at the sesquicentennial of the van't Hoff-Le Bel hypothesis, there is much work still to be done. I hope that my book laid the groundwork for additional studies on late nineteenth century stereochemistry. There remains a large number of topics to fill in the blanks, including more on optical activity and chemical structure, and the nature of racemization (Ladenburg, for example, wrote on this in the 1890s). We still do not know much about the context of the discovery of the Walden inversion in 1893, or about the early attempts to explain it. As Tonja

Koeppel wrote in the centennial volume, controversies over the structure of benzene spurred much stereochemical speculation, especially by Adolf von Baeyer [41, 42]. The development and use of models as representational devices are also understudied, but also somewhat difficult, because most of the models themselves have not survived. We also need a broader study of the national context of the reception of van't Hoff: we know a great deal about early stereochemistry in Germany, but not in Britain, France, or the U.S. Given Jackson's call for paying close attention to laboratory reasoning, a broad study of the different kinds of laboratory reasoning in different laboratories would also be welcome.

Moving to the twentieth century, there are even more topics that await historical study. Consider the emergence of conformational analysis in the 1950s, the rise of mechanistic organic chemistry and the subsequent explanation of the Walden inversion. The second half of the twentieth century saw the appearance of total synthesis of natural products with near total stereocontrol, and the appearance of asymmetric synthesis, making it possible to bypass the resolution of enantiomers to produce optically active compounds more efficiently. We do not yet know how stereochemical principles made their way into textbooks for students. Broader issues would include the history of the importance of asymmetry in biochemical processes, and the history of the study of the origins of asymmetry itself in the biological world.

We have made much progress in the fifty years since the centennial celebration, but there is far more to learn about the contours of the emergence, spread and application of the tetrahedral carbon atom. What will the next fifty years bring, both in the world of stereochemistry, and in the study of its history? I do not know, but hope that at the bicentennial celebration of the tetrahedral carbon atom, future chemists and historians will be able to reflect on what they have learned since the sesquicentennial.

References

1. Ramsay OB (ed) (1975) Van't Hoff–Le Bel centennial. American Chemical Society, Washington, DC
2. Ramsay OB (1981) Stereochemistry. Heyden, London
3. Rocke AJ (1984) Chemical atomism in the nineteenth century: from Dalton to Cannizarro. Ohio State University Press, Columbus
4. Ramberg PJ (2003) Chemical structure, spatial arrangement: the early history of stereochemistry, 1874–1914. Ashgate, Aldershot
5. Le Bel JA (1874) Sur des relations qui existent entre les formules atomiques des corps organiques et le pouvoir rotatoire de leurs dissolutions. Bull Soc Chim 22(337–347):161
6. van't Hoff JH (1874) Voorstel tot Uitbreiding der tegenwoordig in de scheikunde gebruikte Structuur-Formules in de ruimte; benevens een daarmeê samenhangende opmerkung omtrent het verband tusschen optisch actief Vermogen en Chemische Constitutie van Organische Verbindingen. Greven, Utrecht
7. Fisher NW (1974) Wislicenus and lactic acid: the chemical background to van't Hoff's hypothesis. In: Ramsey OB (ed) Van't Hoff–Le Bel centennial. Americal Chemical Society, Washington, DC, pp 33–54
8. Rocke AJ (2010) Image and reality: Kekulé, Kopp, and the scientific imagination. The University of Chicago Press, Chicago

9. Brown AC (1864) On the theory of isomeric compounds. Trans R Soc Edinburgh 23:707–719
10. Wislicenus J (1887) Ueber die raümliche Anordnung der Atome in organischen Molekülen und ihre Bestimmung in geometrische isomeren ungesättigen Verbindungen. Abhandlungen der Sächsischen Akademie der Wissenschaften zu Leipzig, Mathematisch-Naturwissenschaftliche Klasse 14:1–77
11. Ramberg PJ (1995) Arthur Michael's critique of stereochemistry, 1887–1900. Hist Stud Phys Biol Sci 26:89–138
12. Werner A (1891) Beiträge zur Theorie der Affinität und Valenz. Vierteljahrschrift der Zürcher Naturforschenden Gesellschaft 36:129–169
13. Werner A (1893) Beitrag zur konstitution anorganischer verbindungen. Zeitschrift für anorganische Chemie 3:267–330, 9
14. van den Berg R (2017) Van't Hoff en de roep uit Leipzig in 1887. Studium 10:173–191
15. van den Berg R (2021) Een gedreven buitenstaander: J. H. Van't Hoff de eerste nobelprijswinnaar voor scheikunde
16. Gal J (2007) Carl Friedrich Naumann and the introduction of enantio terminology: a review and analysis on the 150th anniversary. Chirality 19:89–98
17. Richardson GM (1901) The foundations of stereochemistry: memoirs by Pasteur, van't Hoff, Le Bel, and Wislicenus. American Book, New York
18. Gal J (2008) The discovery of biological enantioselectivity: Louis Pasteur and the fermentation of tartaric acid, 1857, a review and analysis 150 years later. Chirality 20:5–19
19. Gal J (2008) When did Louis Pasteur present his memoir on the discovery of molecular chirality to the Académie des sciences? Analysis of a discrepancy. Chirality 20:1072–1084
20. Gal J (2017) Pasteur and the art of chirality. Nat Chem 9:604–605
21. Gal J (2011) Louis Pasteur, language and molecular chirality. I. Background and dissymmetry. Chirality 23:1–16
22. Gal J (2019) Louis Pasteur, chemical linguist: founding the language of stereochemistry. Helv Chim Acta 102:e1900098
23. Geison G (1995) The private science of Louis Pasteur. Princeton University Press, Princeton
24. Gal J (2019) In defense of Louis Pasteur: critique of Gerald Geison's deconstruction of Pasteur's discovery of molecular chirality. Chirality 31:261–282
25. van der Spek TM (2006) Selling a theory: the role of molecular models in J. H. Van't Hoff's stereochemistry theory. Ann Sci 63:157–177
26. Meinel C (2004) Molecules and croquet balls. In: De Chadaverian S, Hopwood N (eds) Models: the third dimension of science. Stanford University Press, Stanford, CA, pp 242–275
27. Friedman M (2016) A failed encounter in mathematics and chemistry: the folded models of van't Hoff and Sachse. Theory Sci 359–386
28. Sachse H (1890) Ueber die geometrischen Isomerien der Hexamethylenederivate. Ber Dtsch Chem Ges 23:1363–1366
29. Sachse H (1892) Über die Konfigurationen der Polymethylenringe. Z Phys Chem 10:203–241
30. Sachse H (1893) Eine Deutung der Affinität. Z Phys Chem 11:185–219
31. Russell CA (1974) The origins of conformational analysis. In: Ramsey OB (ed) Van't Hoff–Le Bel centennial. American Chemical Society, Washington, D.C., pp 159–178
32. Ramberg PJ (2014) Partial valence, residual affinity, and early stereochemistry. In: Reinhardt C, Klein U (eds) Objects of chemical inquiry. Science History Publications, Sagamore Beach, MA, pp 141–170
33. Klein U (2003) Experiments, models, paper tools: cultures of organic chemistry in the nineteenth century. Stanford University Press, Stanford, CA
34. Ladenburg A (1886) Synthese der activen Coniine. Ber Dtsch Chem Ges 19:2578–2583
35. Ladenburg A (1888) Ueber Pyridin und Piperidinbasen. Annalen der Chemie und Pharmacie 247:1–98
36. Ladenburg A (1886) Ueber das spezifische Drehungsvermögen der Piperinbasen. Ber Dtsch Chem Ges 19:2584
37. Jackson C (2023) Molecular world: making modern chemistry. MIT Press, Cambridge, MA

38. Jackson CM (2014) Synthetical experiments and alkaloid analogues: Liebig, Hofmann, and the origins of organic synthesis. Hist Stud Nat Sci 44:319–363
39. Jackson CM (2017) Emil Fischer and the 'art of chemical experimentation.' Hist Sci 55:86–120
40. Ramberg PJ (2001) Paper tools and fictional worlds: prediction, synthesis and auxiliary hypotheses in chemistry. In: Klein U (ed) Tools and modes of representation in the laboratory sciences. Kluwer, Boston, pp 61–78
41. Koeppel TA (1973) Benzene structure controversies, 1865–1920. PhD, University of Pennsylvania
42. Koeppel TA (1975) Significance and limitation of stereochemical benzene models. In: Ramsey OB (ed) Van't Hoff–Le Bel centennial. American Chemical Society, Washington, D.C., pp 97–113

Part II

Stereochemical Pioneers

3 Joseph-Achille Le Bel: Polymath Extraordinary

Pierre Laszlo

Abstract

Joseph-Achille Le Bel (b 1847), co-discovered the asymmetric carbon atom, virtually simultaneously with Jacobus H. van't Hoff in 1874, both working in the laboratory of Adolphe Wurtz in Paris. Until just before his death, he was never elected to the French Academy of Sciences, due in part to the opposition of Marcellin Berthelot. This chapter outlines the lineage of the extremely wealthy Le Bel family, granted exclusive dealership throughout the Alsace of the Pechelbronn mined oil products. It focuses specifically on the influence of his uncle by marriage, Jean-Baptiste Boussingault, himself a distinguished chemist, on the young Joseph-Achille. His precocious intellect was discovered early, including his unusual mathematical aptitude, nurtured in the demanding Alsatian education system which facilitated his entry at seventeen into the École Polytechnique. Although expected by his family to manage the company, his father's death in 1867 offered enough flexibility to pursue a career in chemistry. Le Bel did not participate in the Franco-Prussian War of 1870 since he had resigned his commission as a polytechnicien cadet upon graduation in 1867. In 1869, Le Bel joined the research laboratory of the distinguished chemist Adolphe Wurtz, at the Medical School in Paris. Wurtz was renowned for training a generation of distinguished chemists from many countries. Even in Wurtz's laboratory, Le Bel kept faith with his deceased father's dictate and published some research on petroleum. Le Bel worked in Wurtz's laboratory through 1884, the year Wurtz died. He lost considerable research productivity due to the Franco-Prussian War of 1870 and its aftermath. Van't Hoff won the first Nobel Prize in Chemistry (1901) for his theoretical studies of solutions and

P. Laszlo (✉)
Université de Liège, Liège, Belgium
e-mail: clouds-rest@wanadoo.fr

École Polytechnique, Paris, France

A. Greenberg and D. E. Lewis (eds.), *The Birth of the 3rd Dimension in Chemistry*,
Perspectives on the History of Chemistry, https://doi.org/10.1007/978-3-031-97743-5_3

their similarities to the behavior of gases, not for the co-discovery of the asymmetric carbon. He was very much a part of the scientific establishment and remained better known than Le Bel. This chapter reminds readers that van't Hoff's model was flawed from logical inconsistency while Le Bel's was correct. The chapter reports other discoveries by Le Bel and follows him through an intellectually and socially rich life until his death in 1930.

3.1 Introduction

The 1874 joint discovery by van't Hoff and Le Bel [1], focusing on the contribution by the latter, still raises questions. Arguably, the least interesting is a comparison of both contributions: Snelders has done so, in the 1974 ACS volume, from the standpoint of Dutch chauvinism [2]. The questions I shall tackle here, about this important publication by the Frenchman are:

1. Who was he?
2. Why and when did he become an organic chemist?
3. Why did he espouse atomic theory and molecular structure?
4. Van't Hoff and Le Bel both worked in Wurtz's laboratory in Paris. Was it a factor in their joint and independent contributions?
5. Le Bel's explanation of rotatory power uses the language of polyhedral geometry: why?
6. Did Le Bel have an implicit agenda?
7. Can we guess at the reasons for his not having shared with van't Hoff the 1901 Nobel Prize in chemistry?

Fulfilling this task of writing a chapter about Joseph-Achille Le Bel, I wish to stress, from the outset, why I found it so challenging. His career as a scientist has met with a conspiracy of silence; as a rule, his publications have received very few citations.

The reasons for the relative obscurity shrouding his work are the following: owing to being personally wealthy, he was an independent scholar. He did not seek or obtain an academic appointment; hence he lacked the usual status. He did not devote himself to a single line of study, instead had multiple interests and followed them into different lines of work, from organic chemistry to prehistory. In his young years, he failed to apprentice himself, as was then customary, to an illustrious German chemist. At that time of the ascendency of German chemistry, he published only in the French language. For instance, he did not publish in *Berichte*. Moreover, he published in French exclusively. In addition, readers should not forget, he was trained in that uniquely French institution, the *École Polytechnique*; which gave him a sound and deeply ingrained mathematical outlook, that many a fellow chemist failed to understand or respect—not to say resented! Having taught there myself for a few years (1986–1999), I became personally familiar with the *polytechnicien* outlook and I shall strive to convey what such a polyscientific education entails. Hence the length of this chapter: I see it as my responsibility to try and rescue Le Bel from that undeserved and very much unfair oblivion.

3.2 The Le Bel Family [3]

Antoine Le Bel (ca. 1725–1788) was the first person in the family to own an oil production site and live in Pechelbronn. He was born about 1725 in a small village in the Aveyron, Saint Sernin-sur-Rance, about half-way between Millau and Albi. But he spent many of his adult years in Paris. He was an attorney and a socialite who interacted with many people, and a speculator. In 1759, Madame de la Sablonnière, whose late husband owned the bitumen and asphalt mine in Pechelbronn, in the Alsace, asked for his legal help in securing her husband's succession. To state it briefly, Antoine Le Bel, who had some money of his own, after a few years bought out Madame de la Sablonnière and became sole owner of that mine.

Antoine Le Bel visited his new property for the first time in October 1763. A year later, he arranged for a house in the nearby village. In February 1765 he married, in Paris, Anne-Catherine de Serre de Saint Roman (ca. 1735–1797), who also originated in the Aveyron—she was the daughter of a receiver general of the King's farms and *gabelles* (i.e., a taxation official) and treasurer of the Roussillon fortifications. In October of that year, Le Bel was granted the privilege of exclusive dealership throughout the Alsace of the Pechelbronn mine products.

What were the oil products used for? At that time, initially use by the military, sealing fortifications and lubricating cannon axles; subsequently, lubrication of the wheels of carts and carriages, and of mill gears; there was also an altogether distinct use, as an oil recommended for nervous conditions and for epidemic illnesses in cattle. In anticipation of enlarging the concession for oil production and sales, Antoine Le Bel purchased land and became both a farmer and a miner, using his own land in summer to employ workers, unable to go into the mining galleries due to the excessive heat at that time of year [4]. Once settled into Northern Alsace, the Le Bel family grew roots there [5]. The eldest daughter of Antoine Le Bel, Marie-Anne (1762–1848) assisted her brother Marie Joseph in launching the family business: by denouncing to the authorities as an *émigré*, a nearby competitor, the Baron de Bode, owner of the salt mine at Soultz-sous-Forêt; and secondly, by giving up her share of the heritage.

Monsieur Joseph: Marie Joseph Achille Le Bel (1772–1842), son of Antoine and brother of Marie-Anne, was known as Monsieur Joseph locally. As a teenager, he lived through the years of the French Revolution (1789–1792) and witnessed the rise of Napoleon. Indeed, he served afterward in the Army of the Republic during those years, returning to Pechelbronn in 1804. He wed Marie Salomé Krauss (1781–1844) in July 1806. She was a remarkable person. The daughter of linen merchants in nearby Wissembourg, who were Lutheran protestants, she had to cope with challenging times. She turned down emigration into Germany during the Revolutionary years. Nevertheless, she had to suffer persecution from Baron C. A. De Bode, German in origin and lord of nearby Soultz, a royalist who came with the army of Prince de Condé.

After this episode, the revolutionary regime confiscated her property, including the Pechelbronn asphalt and bitumen concession and seigneury. Jailed in nearby Brumath, she fell ill and never really recovered. As for her husband, in about 1814 he bought out the shares held by the other members of the family to run the Pechelbronn business on his own. He built a mansion (Château Le Bel) and its outbuildings; and devoted himself to farming and stockbreeding, leaving most of the management of the oil mine to his nephew Auguste Mabru (1780–1853), Antoine Le Bel's grandson and the great grandson of Charles Le Bel (1698–1773). Mabru held this position from 1812 until 1837. He went on to develop a printing ink made from bitumen lampblack. He set up a small manufacturing plant in Pechelbronn in 1824 and won his first award in 1827 at the *Exposition des produits de l'industrie française*, before moving to Haguenau. Supplier to major Alsatian printers such as Silbermann and Simon, he extended his trade in letterpress and lithographic inks throughout France. He also became Mayor of Lampertsloch, the village adjoining the Pechelbronn estate.

Louis Frédéric Achille Le Bel (J.-A. Le Bel's father, 1807–1867) [6]. After studying at the *École des Mines* in Saint-Étienne, he returned to Pechelbronn in 1829 to help his father, and in 1838 took over from Auguste Mabru as director of the mining plant. An outstanding engineer and industrialist, in 1857 he created the first distillery, where experiments were carried out that enabled his son, Joseph-Achille (the central figure in this chapter), to build a refinery. As early as 1865, the well—named "Georges"—made it possible to descend to a depth of 77 m. For the first time, heavy crude oil oozed from the galleries, marking the beginning of the mine's fortunes. From 1863 to 1866, the plant sold between 78 and 85 tons of asphalt grease annually [6].

The Le Bel family established a small private cemetery on their property in Lampertsloch. There lie in sequence: Marie Joseph Achille; his wife, Marie Salomé Kraus (1781–1844) (their three daughters are not buried with them); Joseph-Achille's father, Louis Frédéric Achille (1807–1867); his mother, Madeleine Le Bel née Martin (1819–1897); his sister Emma (1841–1926); his sister Adèle Herrenschmidt née Le Bel (1843–1919); her husband Dr. Ernest Herrenschmidt (Strasbourg 1834–Paris 1920); his other sister, Marie (1842–1934); and his nephew Louis Herrenschmidt (Strasbourg 1877–Paris 1897). Their private cemetery conveys a sense of the Le Bel family spirit. Conspicuous by their absence are, on one hand, Jean-Baptiste Boussingault (1802–1887), his uncle by marriage to Adèle and his wife Adèle, née Le Bel; he was buried at the Père Lachaise cemetery in Paris (Division 95) following his death in 1887; his aunt Adèle, who died in 1877, also lies there; and on the other hand, of course, Joseph-Achille Le Bel (1847–1930), solitary in his tomb, not at Père Lachaise, but in Bagneux, in a southern suburb of Paris. In death as in life, Le Bel thus followed Boussingault's example!

Jean-Baptiste Boussingault [7] was a deep-seated influence on J.A. Le Bel.

The scientist as rebellious, as an indomitable free thinker and totally unconventional: that is the lesson Le Bel received from his uncle by marriage, Jean-Baptiste Boussingault (1802–1887). Boussingault was a chemist. He studied at *École des Mines de Saint-Etienne*, a year after its creation. He then gave himself a career worthy of a novel: running a mine in the Alsace; then, spending 10 years in South America, following the example of Alexander von Humboldt, touring various countries, making mineralogical and geological observations, distinguishing himself in agricultural chemistry and being awarded the 1878 Copley Medal—the highest award of the Royal Society, climbing the tallest mountains, and also serving as a military aide to Simon Bolivar! What Le Bel got from Boussingault was both the self-imposed discipline of the experimental scientist and a deep-seated allergy to alignment with conventional wisdom, bureaucratic *diktats* and any impediment to freedom of speech and thought [7].

3.3 The Student

Le Bel's early education is likely to have been schooling in the family manor at Pechelbronn. In particular, the interaction with his uncle Jean-Baptiste Boussingault seems to have been deep and determining. Was he given in addition a private preceptor? I was unable to find out. In any case, aged 10 or so, he started attending school in nearby Haguenau, the third most populated city in the Alsace after Strasbourg and Colmar. He was brilliant and went through secondary school in record time. Following Haguenau, he was accepted, as a boarding student, at one of the best *lycées* in the entire country, the *Lycée Impérial Charlemagne* in Paris, to attend their preparatory classes for the entrance examination to *École Polytechnique*.

In 1863, aged 16, he was entered by his teachers in the countrywide *Concours Général des Lycées et Collèges*; and he earned himself a second prize in *Mécanique*—i.e., applied math at that time. Henry Léauté (1847–1916), who received the first prize, would also enter the *École Polytechnique*, in 1866, i.e., a year after Le Bel. Léauté became a professional mathematician, devoting himself to machine servocontrol and automation. He would become a professor of mathematics at *Polytechnique*, from 1895 to 1904. The following year (1864) Le Bel entered the class taught, at *Lycée Charlemagne*, by Simon Hauser (1813–1884). Hauser was a mathematician, a graduate of the *École Normale Supérieure* (class of 1833), an *agrégé de mathématiques* in 1836, i.e., a skilled teacher of the first rank. His students loved him. They had a nickname for him, *Tom Pouce* ("Tom Thumb"), because of his diminutive size. Simon Hauser was fond of geometry and admired the books published in analytical geometry by M. G. Salmon, an Irish mathematician. Hauser was so popular among students that his class at Charlemagne numbered 75 students in 1863–64: Le Bel had competition! Simon Hauser's son Adolphe (1846–1924) was in the same class at *Charlemagne* as Le Bel and also entered *Polytechnique* in the class of 1865. He would become important in the French Navy, as Head of Shipbuilding. Le Bel and Alphonse Hauser were both admitted to *Polytechnique* with an outstanding ranking, Le Bel ranked fourteenth (he was then only 17) and Hauser ranked twenty-fifth.

3.4 Preparing for *Polytechnique*

In the fall of 1862, Joseph-Achille Le Bel entered *Lycée Charlemagne* in Paris, in the *classe de rhétorique*, i.e., the last year of secondary school. His family was pleased by his admission into that school, then ranked second-best in the country after *Louis-le-Grand*, another Parisian *lycée* based on the results of its students in the *Concours Générale*. Le Bel was one of the youngest, one of the brightest too among the students. His teachers entered him, the next spring (1863), into *Concours Général,* a competitive examination pitting the most outstanding students against one another. Le Bel did very well, rising to the top in math and physics (*Mécanique* as already mentioned), where he came in second [8]. Two years later, in 1865, Le Bel again did extremely well in that competitive examination, then encompassing competitors from the entire country. He rose to gain two prizes: in mathematics, he was ranked sixth. In history, he was ranked third [9]. This happened as Le Bel completed the year of *mathématiques spéciales*, a drill that subjected the students to heavy work, centered on math and physics, preparing them for the competitive entrance examination to the two top academic institutions in the country, *École Normale Supérieure* and *École Polytechnique*. In 1852, this "special mathematics" class had been recently officially created. Only fifteen *lycées* in the entire country boasted it. It was set in order to prepare students for the entrance examinations to the *École Polytechnique* and the science section of the *École Normale Supérieure*. When in the early summer of 1865, Le Bel took the entrance examination to the *École Polytechnique*—the goal his family had set for him—in mathematics he tackled and elegantly solved the following problem in analytical geometry: given a circle centered at the focus of a parabola, what are the properties of their intersections? He did well in the other subjects too and was admitted to *Polytechnique*, ranked fourteenth in a class of 131 students. He was only 17, i.e., extremely young for his entry into higher education, in the *Polytechnique* format.

Continuing to be an outstanding student during his time at *Polytechnique*, Le Bel became doubly an heir: on one hand, through birth and wealth; on the other, through his educational profile, which only confirmed his membership of the French ruling elite [10]. In spite of his brilliance in history, mathematics, and physics, as recognized by his performance in the *Concours Général*, he had no choice since this young man was poised to follow the diktats from his family, his father and his uncle foremost. His mission, at the time of the Second Industrialization of France was to train in chemistry, to make the Péchelbronn plant competitive.

As it turned out, Le Bel was emancipated from such obedience by the death of his father in 1867, the same year he graduated from the *École Polytechnique*. Thus, he was allowed a middle course, a compromise between his duties as the male heir of the Pechelbronn estate and the taste he had acquired while at *Polytechnique*, i.e., doing research in chemistry. As the rest of this chapter will show, the latter took precedence during the period 1867–1884 and the former re-emerged to rule his professional life during the ensuing years 1884–1889, after which he returned from Pechelbronn to Paris and renewed doing research in chemistry (1890–1930). Returning briefly to the topic of his graduation from *École Polytechnique* in 1867, Le Bel also emancipated himself from that institution. Like his fellow-students, he was expected and trained to enter state service, first and foremost in the French Army—in the section of the *Génie*, the corps of the military engineers. He declined and resigned his commission, reimbursing the state for his tuition at *Polytechnique*—which his personal wealth enabled.

3.5 The *École Polytechnique*

Upon his entrance at the *École Polytechnique*[11], Le Bel was one of the youngest, if not the very youngest, in his class of 131 students [12]. He was only 17 and his classmates were, predominantly, 20 years old. We know what he looked like from the register in the *Polytechnique* archives: auburn hair, a large forehead, grey eyes, a small mouth, an aquiline nose and a round chin within an oval face. He was quite tall for his time, at 1.72 m (5 ft 7.7″), taller by a couple of inches than the average Frenchman, in the mid-1860s [13]. To some extent, he had a hard time as a student at *Polytechnique*: an epidemic of cholera—6,000 Parisians died in 1865–1866 [14]—made the Administration forbid leaving the buildings of the *École* for a few weeks; the military discipline translated into frequent punishments; and the food was not what he had been used to at home in Château Le Bel.

There were compensations, though: high-level and high-quality scientific classes, Le Bel favored those in mathematics [15] and chemistry. Friendships with fellow students. Their informality and their recourse to a slang, different from any at that time in France, the *argot de l'École polytechnique* [16]. France in 1865 [17] is a nation of 38 million inhabitants [18], 70% of whom are still farmers—in contrast with England, which benefited from an earlier industrialization. Confronted with Prussian laboratories, much better equipped, French scientists were crying foul in 1865. At the request of the government, Louis Pasteur agreed to study silkworms (1865–1869). He published his work on the microbial origin of diseases and patented pasteurization [19]. The first International Telegraph Conference was held in Paris in 1865 [20, 21]. Twenty European states—but not Prussia, which France, the inviting power, had excluded—signed the first International Convention, which laid the foundations for international telegraphy.

All in all, the Second Empire was a period of sustained French growth. From 1850 to 1870, the GDP rose from eleven billion to twenty-two billion francs, for an annual growth of four percent. The railroads network grew from four thousand kilometers at the end of the July Monarchy (1848) to twenty thousand at the end of the Empire (1870) [22, 23]. 1865 was a pivotal year in French science. Claude Bernard published his *Introduction to the Study of Experimental medicine* [24]: In it, Bernard proclaims the physicochemical foundations of physiology, thus amenable to the experimental method of Francis Bacon. In 1865, nearly 40% of the 62 members of the French Academy of the Sciences were *polytechniciens* [25]. This highly significant proportion gives an indication of the considerable weight of former students of the *École polytechnique* in the sciences of that time—which did not remain unnoticed by the young J. A. Le Bel.

In 1863, the publisher Germer Baillière (1837–1905) launched his *Revue des cours scientifiques de la France et de l'Étranger,* nicknamed *Revue rose* for its cover. It featured, in 1865, lectures by Alexandre Edmond Becquerel, Claude Bernard, Berthelot, Pasteur, Sainte-Claire Deville and Wurtz. In subsequent years, the *Revue rose* informed its readership of major advances abroad, such as Maxwell's laws and Clausius's formulation of the Second Principle of thermodynamics. Mendel formulated the laws of genetics in 1865. Le Bel would publish repeatedly in the *Revue rose*, for instance, reviewing in it the state of stereochemistry in 1891. Coming back to the period 1865–1867, when he was a student at *Polytechnique*, I see him as rebellious—contrary to the majority of students in his class, he turned down a career in the military or in the higher French administration, in favor of becoming a scientist—a *savant* in the terminology of his time [24].

3.6 Auguste Comte and Positivism

Under Le Bel's pen, the phrase "experimental verification" is a leitmotif. A sign of the times, partly. His British contemporaries also often used it [26]. Moreover, Pasteur, this genius of experimentation (Boussingault surely had directed his nephew towards this former student of his [27]) resorted to the Baconian style of research, the testing of hypotheses through experiment. And then, there was Auguste Comte and his Positivistic philosophy, movement and school—even religion, as well. Auguste Comte taught mathematics at the *École Polytechnique*, where he had also been a student. His interaction with *Polytechnique* was marked by his later expulsion as a teacher and mutual acrimony.

What was the place of chemistry in Comte's philosophy of science and classification of the sciences? Near the top: since of course, for Comte, mathematics was at the very top of the pyramid. Chemistry, though, he showed consideration for and termed it an "observational science."

In 1865, the year Le Bel started studying at *Polytechnique*, Claude Bernard published his *Introduction à l'étude de la médecine expérimentale*: his presentation of experimentation is very much in the spirit of positivism: "Experimental reasoning … has the same aim in all the sciences: the experimenter … seeks, by means of reasoning and experience, to relate natural phenomena to their conditions of existence … By this means, he arrives at the law that enables him to take control of the phenomenon. All natural philosophy can be summed up in this: to know the law of phenomena. The whole experimental problem boils down to this: predicting and directing phenomena" [24]. Le Bel, in his writings, seldom referred explicitly to positivism. There is one mention, though: in his 1891 presentation in *La Revue rose*, he terms in his conclusion stereochemistry "une science positive" [28]. During his attendance at *Polytechnique*, a lasting influence was indeed the influential yet then controversial positivism of Auguste Comte.

3.7 The Franco-Prussian War of 1870

Joseph-Achille Le Bel witnessed this war from outside the French army, since he had resigned his commission as a *polytechnicien* cadet upon graduation in 1867, as noted earlier, in order to pursue his career in chemistry. Nevertheless, he and his family in Lamperstloch (his mother and his three sisters) witnessed the events with a mix of passionate interest and despair—given that Lampertsloch, with its proximity to the German border was close to combat areas [29] and, in being part of the Alsace, would end up as part of Germany. Readers may need being reminded of the circumstances. Prussia, led by Bismarck, saw the conflict as the opportunity for uniting Germany. Hence, war was declared in mid-July 1870. France was headed by the ageing Napoléon III. Prussia had an army superior to the French in both numbers and firepower. By early September, with his army encircled in Sedan [30], the Emperor had no choice but capitulation [31, 32]. His abdication followed. The Third Republic was proclaimed and the French henceforth led by Gambetta resumed the fight against Germany. It was hopeless and the war ended in early 1871.

There were some heroic episodes on the part of the French, such as the resistance in the Belfort redoubt of Colonel Denfert-Rochereau (1823–1878, Polytechnique Class of 1842). This saved the Belfort Pocket from annexation into Germany. There also occurred the Paris Commune, from March to May 1871. The military arm of this popular insurrection by the Parisian downtrodden was led by another *polytechnicien,* Louis Rossel (1844–1871, Class of 1862). He was shot for his bravery. The peace treaty was signed in Frankfurt in May 1871. France suffered a crippling penalty, with payment to Germany of a huge settlement, 5.6 billion francs or—i.e., about 15% of the gross national product or about 300 billion current euros [33]. The peace treaty gave Alsatians the opportunity to go either way, by 1872 to become Germans together with annexation of their land into Germany or stay French, in which case they had to ultimately leave the Alsace. There were about 50,000 for both the Alsace and the Lorraine, who opted for French citizenship. Thus, they left the Alsace: for the Paris area, for the rest of France, for Algeria and even for Quebec, for some of them.

What about the Le Bel family, during and after this ordeal? It happened shortly after Joseph-Achille had entered Wurtz's laboratory. As the only son, he was family head and was expected to make decisions regarding the Pechelbronn family oil business. While bilingual, the Le Bels unmistakenly had their heart on the French side of the conflict—and they would all opt for staying French. However, they had a moral responsibility to their employees and their families, viz. maintenance of their jobs and livelihood. In this regard, the lifetime of Louis-Frédéric Achille, father to Joseph-Achille, had been most prosperous, it was a time of expansion for the family business—brought about by a combination of riding the Industrial Revolution, plus having a near-monopoly in France for supplying the asphalt at the time when Paris and major cities needed it for their sidewalks at first and roads secondly, when automobiles started being commercialized in the 1880s. Expansion of the Le Bel business can be gauged by the increase in population of Lampertsloch, 366 in 1800 to more than 500 by 1850—other nearby villages most probably showed similar increases by 25%. One can guess that the Le Bels had several hundred employees by the 1880s.

Despite the expansion of the family business and his responsibilities, Joseph-Achille had espoused chemistry as his call in life; he was not about to give it up. The compromise that was set was, indeed, for him to continue being a chemist, in Paris, while his mother and his sisters would continue to run the family business in the Alsace. This arrangement lasted more than a dozen years, from 1870 until 1884: Le Bel then returned to Lampertsloch and spent five years utterly modernizing the plant, bringing it up to snuff prior to selling out to a German conglomerate. Then and only then did the Le Bels leave Lampertsloch and resettle in Paris, following Joseph-Achille there.

3.8 Being Alsatian

His Alsatian roots were important to Le Bel: important enough that after he spent half of his professional career in Paris, at the top for a Frenchman, he returned home to his native village, Lampertsloch, in the Alsace; even though in the meantime it had become part of Germany.

A short digression is in order, to remind readers of the historical events and their resonance in French culture. In the 1870 War, recounted in an earlier segment of this chapter, Germany captured the two French provinces, Alsace and Lorraine. They were part of the German Empire for nearly a half-century until the end of World War I, in 1918. The loss was grievous to the French. I vividly recall a scene from my childhood (I was 9), our teacher in the lycée reading us Alphonse Daudet's "*La dernière classe*" (the last class) recounting the last meeting of the French language class in the Alsace, in 1871 just before its abolition by the Germans: he was so emotional that he was overcome by tears. As another evidence of the highly symbolical meaning of the loss to the French, to this day major streets and avenues in French cities, irrespective of their size, remain named "Alsace-Lorraine," for the two provinces annexed by Germany in 1871 and until 1918.

Of course, Le Bel's return to Lampertsloch in 1884, for a number of years, was motivated primarily by the need for modernizing the family business—oil extraction and refining—bringing it up and ahead of competition from foreign imports, such as American. He had sentimental motivations as well. To evoke them, I can do no better than sketching very briefly a few Alsatian particulars, they became engraved in the hearts of the French during the whole time of the loss to Germany. The first characteristic—it played a foremost role in both Le Bel's upbringing and intellectual brilliance—is the bilingualism. Alsatians were brought up in their own Germanic language (often termed a dialect). As is well known, additive bilingualism is an asset. Alsatian was and remains an authentic language, with poetry even published in it [34]. Related to this boost to the intelligence is the outstanding quality of Alsatian schooling—whether in Le Bel's time or to this day. It is due to the influence of the Protestant religion. Alsatian cities, such as the three major ones, viz. Strasbourg, Colmar and Haguenau, boasted in Le Bel's time high-quality secondary schools, *gymnasium* in German or *lycées* in French. There is no better evidence than the founding (1871–1874) by Alsatian exiles of the *École Alsacienne* in Paris [35]. In addition, throughout the second half of the nineteenth century the University of Strasbourg was a leading French (and German) institution; in spite of constant raids by the Sorbonne to steal many of its leading scholars, while it was French. Such excellence has remained a characteristic to this day [36].

Geographically, the Alsace is a wide plain in-between the Vosges mountains to the west and the Black Forest range on the east. The Rhine river goes through it. Hence, it is fertile from the alluvial deposits and it can readily conduct trade using the river, a major European thoroughfare for centuries. An interesting feature is the nurturing of industry by agriculture: Alsatian farmers work hard and, whenever poverty threatens, work even harder! [37] Over centuries, this feature has nurtured industrial development and it has made the province rather wealthy. Among the agricultural products, wine and spirits—including relatively unfamiliar ones such as holly brandy—can be singled out for their quality. Beer and breweries did not and do not lag behind. Mining products, exported for centuries, include petroleum products from Pechelbronn (asphalt and bitumen in particular, the very name Pechelbronn means "spring of bitumen") and potash. In terms of natural riches, there was also lumber from the Vosges forests and mining of rock salt. The nineteenth century industrial revolution developed the textile industry (cotton): it employed about 75,000 workers by the late 1860s [38].

To be Alsatian is to live a characteristic lifestyle, centered on homelife; a mere mention of traditional dishes will suffice: Cheesecake, known as *käsekuchen*; *spaetzle*, i.e., miniature pasta; *flammekueche*, a truly delicious tart: its thin pastry is filled with lardons, crème fraîche and onions, then baked over a flame; the *baeckeoffe*: it is made from three different meats: pork, beef and lamb. It takes 24 h to prepare: one bakes potatoes, cabbage, carrots, onions and meats in Alsatian white wine and spices; choucroute, *sauerkraut* in Alsatian: a mix of fermented cabbage, meat (usually pork), potatoes and cold cuts; the *Kougelhopf*, i.e., a cake baked by the local baker once a week, a huge cannelated (tube-shaped) production garnished with either sweet or salted goodies, consisting of a leavened brioche, baked in an earthenware mold with a characteristic high, fluted shape, a hollow center, etc. Alsatian gastronomy—not to forget goose *foie gras* —enjoys, justly, a well-deserved reputation [39].

3.9 To Be Alsatian

To be an Alsatian, as was Joseph-Achille Le Bel, is not only to partake of a culture, it is also and primarily to belong to an extended family. By interacting with fellow-Alsatians, whether sharing a glass of wine or beer in a *Weinstube* or a meal in a *brasserie*, is to share a set of values; that go well beyond the love of good food and drinks, they also include close friendships, conversing in the Alsatian language and the nostalgic evocation of places in the Alsace province. This was all the more true of Alsatian "exiles" and exiles, with or without quotation marks, in Paris, during the second half of the nineteenth century and the first quarter of the twentieth. Those were the times when quite a few Alsatians went to live in Paris—at first because they were attracted to the French capital or later on, after 1870, because they had opted to move out of the Alsace and remain French. But who were they for Le Bel, at the time a student at *Polytechnique,* and afterward, when he had espoused a career as a chemist? [40] His math teacher, Simon Hauser (1813–1884) at Lycée Charlemagne, while not an Alsatian proper, came from nearby Besançon, in Eastern France too. Amédée Hadamard (1828–1888), the father of the great mathematician Jacques Hadamard (1865–1963), also taught at Lycée Charlemagne, starting there about the time of Le Bel's graduation from *Polytechnique.* While still a student at the *École*, he discovered with gusto and alacrity the work of fellow-chemist Louis Pasteur (1822–1895), who also originated in Eastern France, in Dole, in the mountains of the Jura, not far from the Alsace.

Starting in 1869, when Le Bel was already engaged in a career as a research scientist in chemistry, he enjoyed the company, the guidance and the support of a number of fellow Alsatians, also in Paris at that time. Who were these chemists? [40] First and foremost, Adolphe Wurtz (1817–1884), whom Le Bel chose as a mentor (atomic theory [41]) and in whose lab he became at first a *préparateur* and where he stayed for 15 years, until Wurtz's death. But there were others, quite a few Alsatian chemists who had come to Paris [42] and shared with Le Bel similar human and scientific values [43]. Charles Friedel (1832–1899), best known for the Friedel–Crafts reaction, who was also the creator in 1871—note the date—of *École Alsacienne* in Paris. Friedel would become the first director of *École nationale supérieure de chimie*, built in the Latin Quarter near the *École normale supérieure* [44]. Achille Müntz (1846–1917), an almost exact contemporary of Le Bel and fellow-originator from the same village in Northern Alsace: we shall meet him in a little more detail when I recount Le Bel's first studies in Wurtz's laboratory. Charles Lauth (1836–1913), an outstanding scientist who, after the defeat of 1870, was very active in pushing for the creation of *École Municipale de Physique et de Chimie Industrielle de la Ville de Paris* [45]. It was started in 1882, with Paul Schützenberger (1829–1897), a fellow Alsatian. Lauth would succeed him in that position in 1897 [46]. Somewhat later, other Alsatian chemists, viz. the Schlumberger brothers, also settled in Paris, also of shining brilliancy, started their oil prospecting company, while a third brother was a star of his own in French letters: Conrad (1878–1936), Marcel (1884–1953), and Jean (1877–1968) [47]. But the deep influence on French chemistry of Alsatians by birth, who shone in both the University of Strasbourg [48] and in Paris, did not stop with them; another few deserve mention. There were two, who after the move in 1940 of the University of Strasbourg to Clermont-Ferrand, were captured by the Germans and deported to concentration camps: Albert Kirrmann (1900–1974), an organic chemist, who was deported to Buchenwald; and Charles Sadron (1902–1993), a polymer physicist, who was deported to Dora and was active there sabotaging manufacture of military rockets. After the end of World War II, Kirrmann became director of *École Normale Supérieure*, where he had been a student. Sadron was barred from an appointment in Paris, the Sorbonne or the *Collège de France*, by politically-influential polymer chemists, such as Georges Champetier (1905–1980), who lacked his talent or credentials [49].

To this roster of distinguished Alsatian chemists, having markedly enriched French science, one cannot fail to add Guy Ourisson (1926–2006), who followed Le Bel's example in studying the origins of petroleum, thus establishing the leading role of hopanoids from bacterial cell membranes [50]; he revitalized French organic chemistry in a number of ways; his student Jean-Marie Lehn (born 1939), who established supramolecular chemistry and won a Nobel Prize in 1981 [51]; and, in turn, Lehn's student, Jean-Pierre Sauvage (born in 1944) also a Nobel Prize winner (2016) for his design of molecular machines [52].

One ought to complement this highly distinguished roster with the names of others, also of Alsatian origin, Protestant or Jewish, who have elevated French academia and science: Georges Zink (1905–2003), a *normalien* who became a professor at the Sorbonne, in the Germanic languages; the Schwartz brothers, Laurent (1915–2002) and Bertrand (1919–2016), the former a *normalien*, Fields medalist in mathematics [53] and the latter a *polytechnicien* and a great educator too [54]. Marc Bloch (1886–1944), who renovated history in France and founded the *Annales* school and who had a heroic role in French Resistance during World War II, for which he was shot [55]. Alfred Kastler (1902–1984), a physicist whose contributions won him a Nobel Prize (1966) [56]. And the two siblings, Simone Weil (1909–1943), *normalienne*, a philosopher and a mystic [57]; and André Weil (1906–1998), *normalien* and a leading mathematician; one of the founders of the Bourbaki Movement [58].

3.10 Wurtz and His Laboratory

Why did Le Bel join Wurtz's group during the summer of 1869? What impelled him to make such a move? Wurtz was a fellow-Alsatian, which played a significant role. But there were additional factors and they led Le Bel to stay on for many years, until Wurtz's passing in 1884. Adolphe Wurtz, born in 1817, was remarkable as a professor of chemistry, within the Parisian School of Medicine and, in later years, the Sorbonne. What was remarkable about his teaching was his enthusiastic presentations, even his physical involvement in the classes he taught, during which he would walk about, literally, miles [59, 60]. He was joyful, friendly and congenial. Members of his research group would easily engage him in conversation. It was part of his attraction and his international reputation, that led scientists from outside France to come to Paris and join his research group. Jacobus H. van't Hoff (1852–1911) was one of these, attracted to Paris by the modernity of Wurtz's research and teaching [61]. Others included, over the years, Beilstein (1838–1906), Butlerov (1828–1886), Crafts (1839–1917), Ladenburg (1842–1911), Zaïtsev (1841–1910), as well as Auguste Cahours (1813–1891), who had taught organic chemistry at the *École Polytechnique* during Le Bel's scolarité there [62] and was a friend of Wurtz. They both had been members of Dumas's group. In the classes he taught at *Polytechnique*, Cahours used chemical formulas and equations, an excellent preparation for both Le Bel's work and his joining Wurtz's laboratory. Adolphe Wurtz had another endearing aspect; he was an activist in setting-up forward-looking reforms and institutions. Let me only mention entry of women students at the Medical School; the founding of *Société Chimique*—de Paris (1857), later to turn into *Société Chimique de France* (ca. 1884); and that of *École Alsacienne*, a precursor of the French *école laïque et républicaine* (secular and egalitarian a schooling). Balard, a co-founder with Wurtz of *Société Chimique*, hired Le Bel as his *préparateur* at Collège de France for two years (1867–1869) [63]. I have no doubt that he encouraged Le Bel to join Wurtz's group.

I have kept for the last, in this segment, Wurtz's pioneering role in the propagation of atomic theory in France. Outside France, the turning point had been the 1860 Karlsruhe conference. In France, atomic theory had powerful and distinguished opponents, chief among them, Marcellin Berthelot (1827–1907) but also Henri Sainte-Clair Deville (1818–1881), Edmond Frémy (1814–1894), Pierre Duhem (1861–1916), and Jules Pelouze (1807–1867). This feud, between Wurtz and the "equivalentists" had its culmination in a confrontation between Berthelot and Wurtz at the *Académie des Sciences* on May 7 1877 [64]. Earlier on, on September 20 1874—note the year!—Wurtz came out as an aggressive defender of atomic theory with a speech in Lille, during which he declared: "What fills space, i.e., matter, is not divisible at infinity, but is composed of a world of invisible, elusive particles, which nevertheless possess a real spatial extent and a determined weight. Those are the atoms" [65]. Le Bel, marked in Berthelot's eyes by his lengthy stay in Wurtz's laboratory, would suffer his hostility for many years, which helps to understand the many failed attempts by Le Bel to join the *Académie des Sciences*.

3.11 Topics Le Bel Started Pursuing

One may surmise that Le Bel, once in Wurtz's laboratory, started working out of fidelity to the wishes of his father, on the Pechelbronn petroleum. Indeed, he published in *Bulletin de la Société Chimique de Paris* a note on this topic during the first half of 1872 [66] and in *Comptes-Rendus de l'Académie des Sciences* a second note on this topic during the second half of 1872 [67]. Why only in 1872, even though he had started work in Wurtz's laboratory in August 1869? In-between intervened the Franco-Prussian War of 1870 and the annexation by Germany of the Alsace and the Lorraine in 1871, the abdication of the emperor Napoleon III, and the beginning of the Third Republic, after the Paris Commune. These momentous events kept Wurtz's laboratory closed for the duration. The very first piece of work Le Bel completed upon his return to the lab showed his devotion to his family and to both his father's and Boussingault's directions, in his becoming a chemist, viz. to investigate scientifically the Pechelbronn petroleum. Moreover, Le Bel did not carry it out entirely on his own; he secured the collaboration of Achille Müntz. Müntz (1846–1917) [68] was like a twin brother to Le Bel. He was born only a year earlier, in Soultz-sous-Forêt, the very village where Pechelbronn is located. He shared his first name, Achille, with Le Bel—and one may wonder at the meaning of this coincidence. Moreover, Boussingault took him under his wing and pushed him into chemistry; Müntz studied it in Paris, with Boussingault. These two Alsatians, Le Bel and Müntz extracted with ether the black substance responsible for the black color of both asphalt and bitumen [66, 67].

Le Bel felt so very much at ease and at home in Wurtz's laboratory that he became one of its longest research scientists. To him these were formative years, between 1869 and 1884—Wurtz died that very year 1884, an event surely influential in signaling Le Bel's return to heading the family oil production and refining business in Pechelbronn. What were the topics he worked on in Wurtz's laboratory? To go into them is to make out a little of Le Bel's personality, marked by an overwhelming and wide-ranging curiosity. As above stated, he started out, in line with his father's and uncle's wishes for him to become a petroleum scientist, in preparation for running the family business, by investigating fossil fuel hydrocarbons and possible mechanisms for their formation in the Earth's crust. Aware as he was of the contemporary Pasteur work on the relationship between hemihedral crystals and the presence of optical rotatory power in their dissolutions, his first step in that subdiscipline of organic chemistry was to prepare optically active amylic alcohol (2-methyl-butan-1-ol) [69]. He went on, jointly with Arthur Henninger (1850–1884), a fellow-member of Wurtz's cosmopolitan group, to devise a laboratory apparatus for fractional distillation. Arthur Henninger was both a medical doctor and a chemist. As he worked in Wurtz's laboratory, he came in daily interaction with Wurtz's second in command, his lieutenant so to say, Arthur Henninger (1850–1884). A Swiss, he was representative of the international group of young scientists who had joined Wurtz's group, from admiration for his personality and ideas, atomic theory foremost. Henninger was only 17 when he did so, with youthful enthusiasm [70]. He and Le Bel shared it, they were both young adults, Le Bel older than Henninger by only a couple of years. Wurtz had devised an apparatus for fractional distillation. To quote a nice description by Andrea Sella [71] in 1852 Wurtz published a short paper in which he isolated pure butyl alcohol by distilling potatoes. He commented that the process invariably resulted in a distillate consisting of two parts, one aqueous, and a second containing a mixture of alcohols that were a nightmare to separate. The solution he described was as concise as it was revolutionary: "To abridge these operations, which everyone knows are extremely time-consuming, I make use of a small ball tube (*tube à boules*) which I attach to the round distillation flask and which permits the less volatile vapors to condense and to fall back into the flask." The two young chemists, Henninger and Le Bel, decided to improve upon Wurtz's design, by adding an outer glass connection between the spherical bulbs, so that the vapors go unimpeded through the bulbs whereas liquid reflux can occur via this small outer tube or tubes, as the case may be [72] (Fig. 3.1). It was ingenious and practical and hence met with resounding success. Who actually was the glassblower chosen to build it? Who could it be, if not for Adrien Alvergniat, renowned throughout the Latin Quarter for his prowess! [73] After publication during the second half of 1874, this apparatus met with widespread success. Indeed, in terms of literature citation during the last quarter of the nineteenth century, it was markedly more successful than Le Bel's historical paper also published in 1874! Each and every year, judging from the *Bulletin de la Société Chimique*, there were at least a couple references to the Henninger-Le Bel apparatus. It was also featured in a number of laboratory manuals and catalogs published around the turn of the century [74].

His work on amylic alcohol led Le Bel to define what were the necessary and sufficient conditions for the presence of optical rotation in an organic molecule. Thus, he came to define these, most clearly and authoritatively—simultaneous with but independent from Jacobus H. van't Hoff's (1852–1911) work—in the

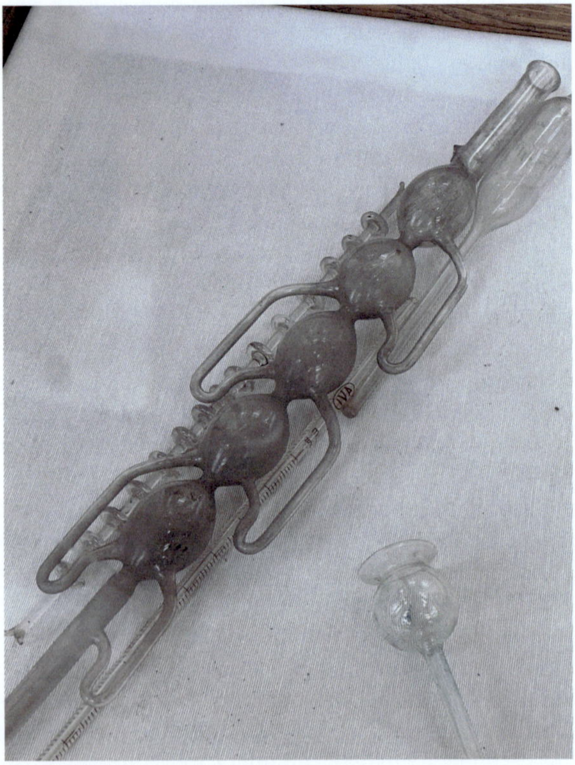

Fig. 3.1 Le Bel/Henninger condenser. Photograph taken at Museo de Geología de la UNAM in Mexico City (courtesy of Mariana Medrano)

autumn of 1874. He continued to investigate the origin of petroleum hydrocarbons and performed experiments on a number of olefins which he reacted with water. The following year, 1876, he returned to the question of the optical activity of amylic alcohol, the dextrorotatory enantiomer [75]. In the year 1876, Le Bel seized the opportunity of an attack by Bourgoin to defend atomic theory and the existing chemical formulas. In addition to the points he made, based on his widely encompassing knowledge of chemistry and familiarity with recent publications, he exclaimed thus: "Were we told that quinine does not cure all fevers and in some cases requires the simultaneous use of other remedies, would this be a reason to exclude it from therapeutics? The same applies to chemical formulas, which are indispensable for predicting and forming the innumerable chemical compounds" [76].

3.12 The 1874 Article

To this day, a teacher of chemistry might use it as a model of writing for his/her students! It is indeed totally clear and superbly organized. It numbers no fewer than 10 pages [77], at a time when Le Bel himself would limit most of his publications to in-between half-a-page, or just a couple. It is presented very much as a work in mathematics, starting by stating explicitly the premises and going on to their consequences. It is indeed organized like a demonstration: with thoroughness, a rhetoric of accumulation, and the pointing to the absence of exceptions to the general principles stated at the beginning. Accordingly, this paper is highly didactic. Its presentation resembles that of a class lecture—one is made aware that the author had been a student just a few years before; indeed, Le Bel had left the *École Polytechnique* only seven years before he wrote it.

The documentation is up-to-date. At the outset, Le Bel mentions explicitly his indebtedness to Louis Pasteur's pioneering work in that area [78]. In the body of the paper he refers by name to relevant work by contemporary organic chemists, three of them French, Pasteur, Luynes and Wurtz; the other three Germans including Johannes Wislicenus (1835–1902), then active in Würtzburg. After having relinquished his position in Zürich; from 1868 to 1872, he had discovered isomerism of the molecules of lactic acid [79]—the existence of the two enantiomers, in modern parlance. Lactic acid(s) would indeed turn into a major research interest for Le Bel. He refers explicitly to Adolphe Wurtz (1817–1884) for his work on caproïc/caprylic acid [80]. He cites Victor de Luynes (1828–1904) for his work on secondary butyl alcohol [81]. He devotes several paragraphs to the aromatic series and the work of Friedrich Kekulé (1829–1896), inserting explicitly in his text his hexagonal benzene formula [82]. The other chemist mentioned by name in Le Bel's article is F. L. Alphons Oppenheim (1833–1877), who in 1873 had become professor of chemistry in Berlin. Le Bel knew him because they both worked in Wurtz's laboratory. Oppenheim was a pioneer in the study of terpenes, devoting for instance an entire article to camphor [83]—then something of a rarity in Europe. Oppenheim would die, still relatively young, in 1877 by his own hands (cyanide?)

because he was devastated at the loss of his wife. Oppenheim, like Le Bel, was very much interested in molecular structure and theory. Very briefly stated, Le Bel's paper shines also by the quality of its up-to-date documentation.

The style of the article—its rhetorical approach in other words—is that of a demonstration: which clearly shows the domination of Le Bel's reasoning based upon math, even though he carefully abstains from being explicitly mathematical. It is thorough and honest, going through an accumulation of data; and noting carefully the absence of exceptions to the general principles stated at the outset. The title of Le Bel's paper, "On the relationships that exist between the atomic formulae of organic bodies and the rotatory power of their solutions," accurate and yet modestly discreet, does not hide the audacity of the subject matter, viz. establishing a correspondence between the two spheres—macro and micro, to use the modern shorthand. The article reports outstanding success in establishing correspondence between the two spheres, that of laboratory measurements and that of the postulated arrangements of atoms in the molecule—its actual geometry. Was such a daring transposition unique? Perhaps in terms of organic chemistry, even though van't Hoff's nearly synonymous contribution hints otherwise. However, I cannot help reminding the reader of a similar achievement, also dated 1874, i.e., Modest Mussorgski's "Pictures at an Exhibition," in which the Russian composer transfers to piano music his impressions—1874 is also the year, in Paris, for the epochal opening exhibition of the *Impressionistes*—from visiting an exhibition of paintings; from the visual sphere to the auditory sphere.

3.13 Van't Hoff's Blunder

As is well known, the 1874 discovery of stereochemistry was a multiple, in Merton's sense [84], occurring quasi-simultaneously for Le Bel and van't Hoff, moreover in the same location (Wurtz's laboratory in Paris), independently of one another, though. A consequence was and remains that these two scientists became linked together for many people, then and now. Accordingly, some of the features of their papers became—inadvertently—mixed up. This was the case for a bad mistake in van't Hoff's paper, it became attributed by some of their fellow-scientists to Le Bel as well, even though Le Bel had steered well clear of that loophole. van't Hoff considered the general case of a tetravalent carbon atom with four different substituents being *both* in the shape of a regular tetrahedron and asymmetric. Of course, this is absurd because it is self-contradictory, a bad mistake in logic and a major inconsistency. And yes, Le Bel got blamed for it even though he was totally innocent of that inner contradiction. An example is the 1892 article by Albert Colson (1852–1933) [85] in the same journal in which Le Bel had published his discovery in 1874, *Bulletin de la Société Chimique de Paris*: that paper thus opens, in English translation: "M. Colson and before him, a number of other authors admit that M. Le Bel bases his stereochemistry on the assumption that radicals attached to asymmetric carbon have their center of gravity at the apex of a regular tetrahedron with carbon at its center." Yet, Le Bel had pointed out,

Fig. 3.2 Joseph-Achille Le Bel at his polarimeter (courtesy the Wellcome Collection)

insistently, that he had not been guilty of such a major blunder. For instance, he wrote thus in 1890: "*Lorque j'ai publié, dans ce recueil, en 1874, mon mémoire sur 'les relations entre le pouvoir rotatoire et les formules atomiques …', j'ai mis lei plus grand soin, dans toutes mes démonstrations, à, éviter de m'appuyer sur l'hypothèse préalable que les composés du carbone de la fqrmule CR^4 ont la forme, d'un tétraèdre régulier*," [86] which translates as: "When, in 1874, I published in this journal my paper on 'the relationship between rotatory power and atomic formulas …' I took great care, in all my demonstrations, to avoid relying on the prior hypothesis that carbon compounds of the CR^4 formula have the shape of a regular tetrahedron" [86]. The published attack by Albert Colson started a polemic and there were other rejoinders by Le Bel, to the effect that he was totally innocent of that mistake. In spite of the many disclaimers by Le Bel, he had become immersed in the wash from that bad mistake by van't Hoff. Which is proof that the association of the two discoverers was both robust and indiscriminate. Figure 3.2 is a photograph of Le Bel at his polarimeter decades after his 1874 paper.

3.14 The Aftermath

As early as 1875, in his annual report on his laboratory, Wurtz wrote as follows, in English translation: "Mr. Le Bel has taken a completely new approach to the question of the rotatory powers of organic compounds. Everyone has said that rotatory power is due to molecular dissymmetry. Mr. Le Bel, together with a young and skilled Dutch chemist, Mr. van't Hoff, has tried to clarify this dissymmetry, by constructing spatial atomic formulae in which the individual atoms or groups of atoms are arranged around a central carbon atom. These are ingenious attempts which it will be very useful to pursue and extend. As of today, they give a certain and satisfactory result" [87]. Wurtz clearly understood the overwhelming importance of Le Bel's contribution. As he also wrote, in a brief but highly perceptive summary of Le Bel's contribution, "It is generally accepted that the rotatory properties of some chemical compounds are linked to the asymmetric structure of their molecules. The author's ingenious views on this subject are based on modern hypotheses concerning the atomic constitution of combinations. He is currently verifying the value and scope of his theoretical ideas with experiments that are as precise as they are delicate" [88]. Subsequently, Wurtz went on to enthusiastically support Le Bel's work. He thus wrote as follows in his laboratory report for the year 1879: "Mr. A. Le Bel is continuing his fine studies on the relationship between atomic structure and molecular rotatory power. Very familiar with the character and mode of action of figurative ferments, he cultivates certain microphytes in media containing organic compounds, whose dissymmetrical structure must determine, according to him, their rotatory power, but which, made up by a mixture in equal proportions of a dextrorotatory and a laevorotatory body, exert no deviation on the plane of polarization. The microphyte prefers to devour one of the active bodies; the other remains and can be isolated" [89]. The term "stereochemistry" was introduced in the years following the 1874 publications by Le Bel and van't Hoff by their contemporary, the German chemist Viktor Meyer (1848–1897), who would succeed Wislicenus in Zürich. The Hargittais gives 1890 as the date of its introduction [90]. As concluded in a perceptive article by Francis Robert Japp (1848–1925) in *Nature*, in December 1887: "The tridimensional formula of organic chemistry are thus an accomplished fact. The treatment of the subject is still, of course, only statical; but, taking care not to lose sight of the limitations thus imposed, the method is a perfectly legitimate one" [91]. The readers no doubt have noticed that, with of course the exception of Louis Pasteur, the chemists whose names have been quoted in this segment of the chapter—Le Bel, van't Hoff, Wislicenus, Meyer, Japp were all born in the 1840s: they constitute a generation—the young scientific generation of their time.

One is fortunate to be able to rely on Le Bel's own accurate and perceptive assessment of his 1874 contribution. As he wrote in 1891, at the beginning of a review article—meant, I believe, to support his obstinate candidacy to election at the *Académie des Sciences*—"In the developments that follow, I have deliberately left aside discussing the causes of the molecule's inner stability, i.e., the study of the forces that keep the atoms in their relative places in the molecule. Indeed, not only is this question entirely distinct, but it remains unresolved at the present time, and it is important to show that stereochemistry rests on experimental bases and reasoning that are independent of hypotheses concerning the forces that develop between atoms" [28].

3.15 Atomic Theory and Its Divisiveness

France periodically renews a debate, a controversy rather, between tradition and innovation. Science does not remain immune to it. In my time, during the second half of the twentieth century, French chemistry was thus divided between activists in the area of reaction mechanisms, set upon establishing in atomic detail the pathways between reactants and products, and more traditional chemists' intent upon synthetizing new molecules using well-established reactions. In Le Bel's time, such a heated debate also took place. It pitted partisans of the atomic theory against "equivalentists," the latter partisans of the notation of chemical changes using equivalents. Equivalentists claimed they were thus heeding positivistic ideas, following Auguste Comte in sticking to factual observation rather than engaging in speculations devoid of actual evidence [92–94].

Who were some of the foremost equivalentists? They were led by Marcellin Berthelot (1827–1907), who had been a student of Dumas and Balard, and who was a militant of a science unbacked by any ideology or religion. He was a friend of Victor Hugo, Joseph Bertrand and the Clémenceau family. During the War of 1870, he devoted himself to the manufacture of gunpowder for the French Army. Later, he gave support to Alfred Dreyfus. He was seconded by Henri Sainte Claire-Deville (1818–1881). He was an equivalentist and also an "energetist," espousing the view that the determining factors for chemical transformations were observables such as temperature, pressure, relative amounts of the reactants. He believed only in empirical observations and experimentation. Other anti-atomists included Émile Jungfleisch (1839–1916) who succeeded Berthelot in his chair at *Collège de France*; Louis-Joseph Troost (1825–1911), who had been a student at *École Normale Supérieure* (1848–1851); another *normalien*, Alfred Ditte (1843–1908), who had been a member in Sainte-Claire Deville research group; Clément Georges Lemoine (1841–1922), a *polytechnicien* (Class of 1858), who subsequently served for many years as a teacher at *École Polytechnique;* also the diabetologist Apollinaire Bouchardat (1806–1886).

The clan of the atomists, to which Le Bel clearly belonged, was led by Adolphe Wurtz (1817–1884). It included other Alsatians, such as, most notably, Charles Friedel (1832–1899) and Paul Schützenberger (1829–1897); another Alsatian chemist, Jules-Edmond Willm (1833–1910); other research workers in his (Wurtz's) laboratory including the spectroscopist Georges Salet (1814–1884); Edouard Grimaux (1835–1900), who taught at *École Polytechnique*, bravely stood for Dreyfus and was a gifted historian of science; Maurice Hanriot (1854–1933) who presented in 1880 a dissertation on "Hypothèses actuelles sur la constitution de la matière," with both Wurtz and Armand Gautier among the examiners; Armand Gautier (1837–1920) who would succeed Wurtz as head of his laboratory at the Medical School—a writing of his would inspire the novelist Joris-Karl Huysmans. Atomists versus equivalentists: a debate echoing another, about the epistemic status of chemistry. Is it a science worthy of the interest of philosophers [95], logicians [96] and linguists [97] or just a set of empirical recipes, akin to cooking?

3.16 Parisian Residences

In his long life (1847–1930), Le Bel spent most of his adult life in Paris, from 1863 to 1884 and from 1889 to his death, i.e., 62 years. I record here, as faithfully as I could establish it, in chronological order, his successive addresses; He dearly loved the French capital [98], from the time (1863) of his arrival, to study at the *Lycée Charlemagne* to prepare the entrance examination to the *École Polytechnique*. Both these establishments are in the Latin Quarter [99], i.e., the district of Paris that includes as well other prestigious centers of learning, such as the Sorbonne, the *École Normale Supérieure*, the *École de Physique et Chimie Industrielle* (started by Alsatian chemists during Le Bel's lifetime), and other prestigious lycées—Louis-le-Grand, Saint-Louis, Henri IV, Fénelon—with their *classes préparatoires* for entry into the *grandes écoles* [100]. It was first *Lycée Charlemagne* as a boarding student. *Lycée Charlemagne* is located in a splendid building in the historical Marais district, at 14 rue Charlemagne. It also faces the Saint-Paul subway station on rue Saint-Antoine, which of course postdated the years when Le Bel studied there [101]. He was admitted to that school on the basis of his outstanding record, earlier on, in Haguenau. At the time of the early 1860s, Charlemagne was considered one of the leading Parisian lycées, superseded only by the Lycée *Louis-le-Grand*, also located in the Latin Quarter. Once admitted into the *École Polytechnique*, Le Bel did not have to move very far: it is located on *Montagne Sainte-Geneviève*, near the Panthéon with its monumental architecture crowning it.

In August 1869, Le Bel joined Wurtz's laboratory at the École de Médecine. The Medical School was located in the Latin Quarter, in between Place de l'Odéon and Boulevard Saint-Michel. The *Société Chimique de Paris,* which Le Bel quickly joined, lists it as his address. Of course, the Franco-Prussian War of 1870 intervened and life in Paris was totally disrupted during the ensuing months [102]. In January 1877, the *Bulletin de la Société Chimique de Paris* lists Le Bel's address as 12 rue de l'Odéon—a five-minute walk from Wurtz's laboratory, which would have been very convenient; Le Bel, a gifted experimentalist, was very fond of his lab work; he strove, during the part of his career singly devoted to chemistry (the last third of the century) to live close to the laboratory he worked in at that time. Le Bel indeed retained that rue de l'Odéon address, so convenient for his love affair with chemistry, for several years until, in 1884, he returned to Pechelbronn to run the family business.

Rue de l'Odéon runs between Place de l'Odéon and the Odéon Theater of the *Comédie Française*. It is an attractive location. In Le Bel's time it featured a number of bookshops, which has remained the case—at least to some extent—to this day. One of these bookshops would gain celebrity: Adrienne Monnier set-up in 1915 "*La Maison des Amis des Livres*" at number 7 rue de l'Odéon. Sylvia Beach, who became her companion, set-up her own library, "Shakespeare and Company," at number 12, rue de l'Odéon [103], the very house in which Le Bel had lived, perhaps during and in any case after the War of 1870. A historical plaque on the façade reminds people of Sylvia Beach, also that James Joyce's *Ulysses* was published there in 1922. By the beginning of 1883, the *Bulletin de la Société*

Chimique de Paris gives Le Bel's address as 13, rue de l'Odéon: A mistake or had he moved across the street? In any case, this did not last very long, for in 1884—the year of the deaths of both Adolphe Wurtz his mentor and Arthur Henninger his friend and coworker—he moved back home to Pechelbronn in the Alsace (that had become German). The purpose was to modernize the plant in anticipation of selling it and moving away—back to a Parisian address.

Le Bel, this is obvious, loved Paris. Probably from the time (1865–1867) he was a student at the *École Polytechnique*. He developed at that time a fondness for the Latin Quarter. Indeed, the buildings of the *École Polytechnique* were embedded in it, to be more precise on the *Montagne Sainte-Geneviève*, the hilly district crowned by the Panthéon. This district had streets and squares bearing the names of illustrious polytechniciens, such as Monge, Laplace, and Lagrange; which pointed out even to a mere passerby the reverence for science and the deep, pervasive influence of the *polytechniciens* on French public life [104]. Rue Monge was indeed opened in the 1860s, contemporary with Le Bel's time at the *École*. Rue Lagrange was contemporary in its creation (1887–1890) with Le Bel's return from Pechelbronn to Paris. Rue Laplace, one of the oldest streets in Paris, the former rue des Amandiers, a stone-throw from the *École Polytechnique*, was thus renamed in the 1860s.

In any case, Le Bel during his time studying at the *École polytechnique*, as an obligatory boarder and under military discipline, whenever he could escape, would explore Paris; starting with the Montagne Sainte-Geneviève and the Latin Quarter. Hence, it is no surprise that upon his return to Paris in the late 1880s he chose to live on rue de Latran, at number 11. This small street, only 62 m (200 ft) long, had been opened in 1862 on land expropriated from the former enclosure of the Commanderie de Saint-Jean de Latran for the construction of the rue des Écoles. It was very close to both the *École Polytechnique* and the *Collège de France*, a building very familiar to Le Bel from his having worked there in the chemistry laboratories of both Thénard and Berthelot. Did he return there for his lab work? It may well be. In any case, during the 1890s and early 1900s Le Bel used his ample financial means to inhabit the fashionable 16th district. In January 1896 and again two years later, the *Bulletin de la Société Chimique de Paris* lists his address as 25 rue Franklin. This is a street, fashionable to this day, located between Place du Trocadéro, nowadays with its subway station and the nearby Passy station that overlooks the Seine. That address was that of his eldest sister Emma's Parisian address: she is already listed there in 1894, in *Bulletin de l'Association des Dames Françaises*. Emma Le Bel (1841–1926) was not married and, upon the sale of the Pechelbronn business and Le Bel estate there, had moved to Paris. Indeed, there is—if not proof an indication of the proximity of the two siblings—in Emma and Joseph-Achille having together attended in Nantes the Annual conference in 1875 of the *Association Française pour l'Avancement des Sciences*.

In terms of Le Bel's dwelling in the fashionable 16th district of Paris, there is evidence that he bought for himself at the very beginning of the twentieth century a chic house or *hôtel particulier* (mansion) on Place Possoz in 1901. Moreover, there is also an address of his given as the nearby 14 avenue du Trocadéro in 1900, in the list of the members of the *Société de Géographie*. As another evidence of Le Bel's time during the Belle Époque, around the turn of the century, as a member of the Proustian high society—not to forget, he belonged by birth to the HSP (the nickname for *Haute Société Protestante*)—he bought himself a lovely Panhard automobile [105], in which he would drive around, often jointly with Emma his sister (I surmise).

However, he could not resist the call of the Latin Quarter. At the turn of the century, in 1900, one finds him with two addresses in Paris: one in the 16th, at number 14, rue du Trocadéro; the other, in the 5th arrondissement, at number 7, rue Amyot. A former road in the thirteenth century, this street already appears on the 1588 plans of Paris, under the name "rue du Puits-Qui-Parle," i.e., street of the Well-That-Talks—archeologists have indeed found in that area ca. 20 wells dating back to Gallo-Roman times. The street was renamed for Amyot, a medieval bishop in Auxerre, in 1867, the year Le Bel graduated from *École Polytechnique*. Surely, Le Bel had gone through that street many times during his time at *École Polytechnique*. It flanks the laboratories of its sister institution and great rival on *Montagne Sainte-Geneviève*, the *École Normale Supérieure*. Le Bel was not the first scholar and scientist to choose to live on rue Amyot. A few years earlier, in 1892, his contemporary the botanist Gaston Bonnier (1853–1922)—a graduate of the *École Normale Supérieure*, the author of several treatises on the French flora [106]—lived there. Earlier on, Gay-Lussac (1776–1850) had also, not only lived but also had his personal laboratory on rue Amyot—was it the reason for Le Bel to follow suit? In any case, Le Bel loved living on rue Amyot; it was still recorded as his address in January 1902. Moreover, he was still recorded as living at 7 rue Amyot in October 1903. Was it in 1902 that he made the decision to move, from having built for himself a large new townhouse at 250 rue Saint-Jacques—the main North–South thoroughfare going across the Latin Quarter? When his new house was being planned, Le Bel had a small personal apartment and laboratories built to allow him to work there at his fancy. He lived subsequently at 250 rue Saint Jacques from 1903 until his death in 1930. During most of these decades, Le Bel was not alone in that mansion: he also generously offered an entire floor to the French Chemical Society and another floor to the French Prehistoric Society. Le Bel bequeathed the mansion [107] to the *Société Chimique de France*, which was responsible for maintaining his laboratories there. The laboratories at 250 rue Saint Jacques, however, closed after the Second World War. The entire building was refurbished in 1965. The *Société Chimique de France* set up its administrative offices there in 1966. Since then, the Société's head office has been located at 250 rue Saint Jacques in Paris.

3.17 Resolving Racemates

Amyl alcohol, i.e., 2-methylbutan-1-ol, with the formula below: to quote it directly from Le Bel's 1874 seminal paper to him was a favorite. He returned to this study time and again. In 1873, he reported its optical activity as ca. 20° in a column with a length of 20 cm [108]. In 1878, he reminded the readers of another paper on what he termed amylic alcohol that, to quote him directly in English translation, "Levorotatory amyl alcohol, marking—4°, 33′ for 10 cm of liquid column, can be rendered inactive by its transformation into sodium amylate and by the action of high heat on the latter body. I have described how dextrorotatory alcohol can be separated from inactive alcohol, by destroying with molds part of the left-hand alcohol" [109] (Scheme 3.1).

Interested as Le Bel was in those molds, he noted that "just as the mold is dying down, the vegetation starts up again. This initially consists of *Micoderma aceti*, followed by a white mildew of the *Oïdium* genus, and at the end a green *Penicillium* that seems to thrive in almost neutral liquors." And he noted, in a parallel note to the *Comptes Rendus*, that "In May, I sowed 16 L of this mixture, which contained 48 g of inactivated alcohol; the whole was contained in in four uncapped 8 L flasks. After a month, the green vegetation which at first was highly prosperous, seemed to wither away.

I then distilled the whole and removed a third of the alcohol used, which distilled at between 127 and 130°. This liquid, examined in a 22 cm column with the apparatus of Mr. Cornu, rotated the polarization plane of sodium yellow light by 2° 28′ to the right. It therefore represents a mixture already rich in dextrorotatory alcohol" [110]. In 1883, Le Bel described, in a note to the *Comptes Rendus*, isolation of amyl alcohol from alcoholic drinks: "I announced a while back that natural white wine contained quite significant proportions of amyl alcohol: 0.2 cubic centimeter per liter. Mr. Henninger has also found this alcohol in Bordeaux red wines. At this proportion of 0.2 cc, we should add the contents of the distillation tails, which pass through at around 85° and contain all the alcohols of fermentation and, in addition, water." He goes on to conclude in that Note that "Natural ferments must therefore contain more higher alcohols than that provided by sugar" [111, 112].

Here, to provide the flavor of his writing, I insert as an example in translation, one of Le Bel's papers:

"On active propyl glycol"

by J.-A. Le Bel, communicated by Mr. Wurtz. *Compt. Rend. Ac. Sci.*, 1881, 532–534.

Among molecules from the fatty series for which theory indicates optical rotarion, that we produce moreover only in the reactive state, Mr. Wurtz's propyl glycol

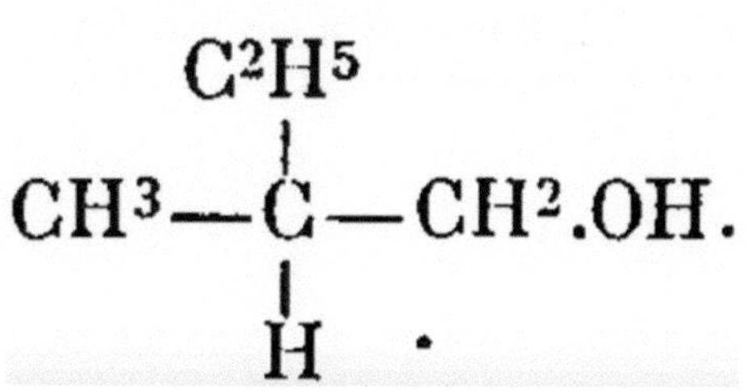

Scheme 3.1 Amyl alcohol (2-methylbutan-1-ol) as represented by Le Bel in 1874

had long attracted my attention. This compound has the formula, CH_3–HCOH–CH_2OH which shows the central carbon associated with four different radicals; moreover, we know that by oxidation it provides ordinary lactic acid; resolving (*dédoublement*) the propyl glycol must therefore lead to that of the acid. Preparation indicated by Mr. Belohoubeck, by means of soda glycerinate, allowed me to prepare almost 2 kg; but it was necessary to dilute with water and rectify three different times, to rid the glycol of very acrid empyreumes (*Editor's Note: strong-smelling substances derived from strong heating*) which initially greatly hampered my cultures.

Since I hoped to obtain a fermentation of propyl glycol similar to that of glycerin, I worked with 3% solutions, with mineral salts and precipitated calcium carbonate added. With glycols contaminated by empyreumes, I only obtained greenish molds, seemingly of the *Aspergillus* type. Once, I found them invaded by red spots, due to a very small bacterium, a parasite probably.[1] Another experiment was carried out in acidic medium, *Penicillium* grew then preferentially. In spite of the most diverse seedings, no fermentation per se took place, furthermore we were able to cultivate only two species from the Bacteridian family. On propyl glycol still very empyreumatic, that I had seeded with mud, numerous aerobic spirillums developed; as I still hoped to get a fermentation, I did not pursue this culture.

With well purified propyl glycol, we see whatever the seeding—developed *Bacterium termo*, the most widespread of microbes. However, I noted the existence of two *bacillus* similar to the butyric ferment. The bacteria form a thick skin on the surface; some, however, swim throughout the entire mass. By operating in an almost stoppered flask and tearing with jerks the skin formed on the surface, I obtain actually a vegetation, made of bacteria swimming in the mass. Those are the most favorable conditions for partial oxidation. These bacteria are very refractive and their cilia are easily visible using the X lens of Prazmowski. They show exactly the same appearance in glycerol or butylic alcohol, but do not thrive in the isobutyl alcohol of fermentation, saturated with zinc carbonate and brought to crystallization. The appearance of the crystals is that of ordinary lactate which requires 18.3% water: my crystallized salt contained 17.5. The mother liquor then deposited isolated crystals, similar to those of paralactate, but which we could not purify for lack of material. We prepared, from the active propyl glycol, an important derivative: it is propylene oxide, a compound with a closed chain whose asymmetry is also due to the central carbon; it is linked to two neighboring groups, which themselves have affinity relations. This compound, which boils at 35°, is the most volatile among known optically active compounds. It rotates by $+10\ 10'$ for 0^{10}, 22. It follows from this that double bonds and the reactions that introduce them do not seem to affect the existence of specific rotation (Scheme 3.2).

[1] These phenomena can almost certainly be obtained from cultivating molds on ammonia lactate.

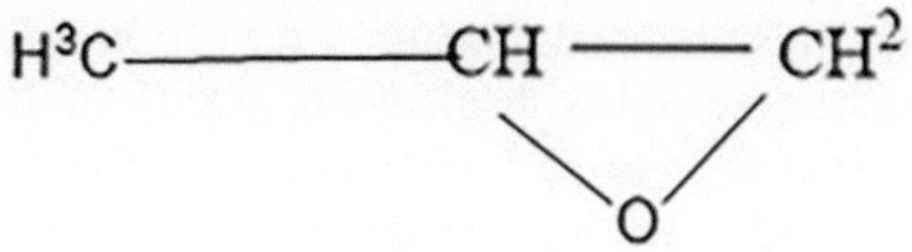

Scheme 3.2 Optically-active propylene glycol (Le Bel, 1881) redrawn here

3.18 Work on Petroleum

From his family ownership of the oil output at Pechelbronn, as noted earlier, Le Bel had a vested interest in petroleum, its components such as bitumen and asphalt, whose demand increased hugely at the end of the nineteenth century, due to their need for surfacing roads; a need made more acute by the appearance of the automobile. He was sensitized to the need for scientific study of the composition of petroleum, by both his father and Boussingault, as he moved to Paris and engineering studies at *École Polytechnique*. Upon entering Wurtz's laboratory in 1869, Le Bel almost immediately started studying the composition of the oil at Pechelbronn. In 1872, his first published article—co-authored with a fellow-Alsatian and fellow-Boussingault *protégé*, from Soultz-sous-Forêt, the village next to Pechelbronn, who also shared his given name, Achille Müntz [113], dealt with the black coloring matter in petroleum from Pechelbronn, concluding to a composition with 70.1% carbon and 7.7% hydrogen, with 11.2% of the components being mineral rather than organic. Le Bel returned, 16 years later (1888) to the study of the composition of asphaltene. He had procured, in the meanwhile, an Egyptian bitumen sample and asphaltenes from another Alsatian source (Lobsann) and from Germany and Crimea. What he found was silica- and iron-containing mineral residues. It seemed to him to give support to Mendeleev's theory (1877) on the origin of petroleum [114, 115], as resulting from the reaction of water with iron carbide, forming the Earth's core [116]. It remained a topic of intense interest to Le Bel. As he wrote in 1891, "From my early research into the pyrogenic carbides in petroleum, I was struck by the fact that unsaturated hydrocarbons combined with water to become soluble. This experience led me to admit the possibility of a pyrogenic origin of natural petroleum, and that the unsaturated part had been removed by groundwater." A few years earlier (1885), he published a little booklet within the *Bulletin de la Société d'Histoire Naturelle de Colmar* [117]. Le Bel notes in it, regarding the local geology, that "our district from Wissembourg to Gœrsdorf is the only one where the Tertiary terrain is in direct contact with the sandstone of the Vosges." He goes on to discuss the theories on the origin of petroleum, whether biogenic or abiotic [118]. Then, he relates precisely the history of the petroleum mining at Pechelbronn, providing a full chronological list of the 16 oil wells dug from 1745 until 1877. Le Bel concludes this booklet by comparing the composition of the Pechelbronn petroleum with those from nearby sources [116].

He studied a compound, amylane, present in petroleum of Alsatian origin. In his words, "I extracted this product from springs I'd just discovered in Alsace and, after fractionation, chlorinated the distillate at between 30 and 35 °C. The monochlorinated body thus obtained, having in turn been isolated from the more chlorinated products, was treated with alcoholic potash at 100 °C. Under these conditions, no isomerization is to be feared, and yet it has been recognized, by the cold hydrochloric acid reaction, that the amylene thus obtained (including 2-methylbut-2-ene) is a mixture of normal and *iso* hydrocarbons. As a result, petroleum itself contains both types of hydrocarbons" [117, 118]. Le Bel, in the mid-1880s, obtained samples of both Russian (Baku) and American oils. What he found, by chemical experiments, was a pronounced difference, viz. the likely presence of cyclohexane—in modern parlance—in petroleum of American origin, but not in that from Russia.

3.19 Return to Pechelbronn

The death of Wurtz in 1884 nudged Le Bel to immerse himself much more deeply into the family business in Pechelbronn. The ensuing years accordingly saw a spectacular modernization of the entire plant: drilling, coping with the different oil types, storage, commercialization. Le Bel showed that he was a skilled and accomplished—indeed of the first rank—*polytechnicien* engineer. This was the time, in France in particular, of the Second Industrialization; marked for instance by the building of a major railroads network. There was thus a need for new energy sources, oil being one of them. In the second half of the nineteenth century, Joseph-Achille Le Bel thus perfected the mining method. Thanks to a better geological understanding of the site (water injection, pumping), techniques became more efficient. Depths attained exceeded 200 m. Nothing on site was wasted: the collected gas was not flamed but stored in gasometers to supply industrial plants. Product diversification enabled commercialization of a wide range of products, from oil to naphtha to benzene. Simple heating systems were used to separate these different products. Around the oil field, the characteristic buildings and headframes stood out in the Alsatian countryside, with a complementarity between the miniature and large-scale activities very much in evidence, while rural activities around the plant remained in force.

A word about the features of the oil deposits at Pechelbronn. The Oligocene subsidence of the Rhine deposited a large mass of sediment, 1,200 to 1,800 m thick. Facies vary due to uneven subsidence: oil-bearing layers are in the Lower Rhine where Pechelbronn is located, potassic layers in the Upper Rhine. In the Lower Rhine, the oil mainly impregnates the sandy layers. From a depth of 90–150 m, hydrocarbons occur only as simple lenses containing a kind of thick, black thick, black grease. Deeper, from 150 to 600 m, the deposits reach 1 km in length and 200 m in width, with a thickness of 2.5–3 m, yielding a less heavy oil. Indeed, in that area, the oil-bearing sands are interbedded with Oligocene marl, which fills the Rhine plain over a thickness of more than 1,000 m, as above stated.

In 1879, J. A. Le Bel introduced the Fauvel drilling system to Pechelbronn. The attendant wooden towers were built by local carpenters: they were ca. 20 m high. In each, the drill bit, supported by hollow rods, would beat the clay with a repetitive vertical movement. The clay turned to mud under the effect of the water injected inside the rods, which rose back to the surface through the annular space between the hole wall and the drill rods. The wells reached depths of a few hundred meters. Auger drilling henceforth was only used for exploration or reconnaissance drilling. In Le Bel's words, "at No 146 the oil gushes out in a true flowing well, by more than 10,000 L a day. The violence of the flow is caused not only by the pressure of the gas contained in the oil, but also by the reduced charge resulting from the fact that the tube does not contain a column of oil, but real foam, whose density is much lower; a pressure of 3 atmospheres maintains itself at the top of the pipe." Whereas annual virgin fat production totaled only 4,500–5,000 t between 1812 and 1866, it rose to 14,500 t during the much shorter period 1867–1888. As early as 1880, drilling revealed new deposits at a depth of 130 m: a light oil gushed to the surface as a fountain. By 1886, seepage oil production had reached around 7,000 t. Mining using shafts and galleries was totally abandoned by 1889. That same year, J. A. Le Bel rendered operational pumping crude oil, when the oil was no longer gushing. To meet the new needs inherent in drilling and pumping, he created a complete refinery, including condensers and rectifiers for benzene and petroleum, equipment for distilling clear oils, separating kerosene from oils, distilling residues on pitch and coke, refining oils and creating new products.

Having thus upgraded the plant, Le Bel—fourth in the dynasty of Pechelbronn owners—opted for its sale. This occurred in 1889, purchasers were a group of Alsatians who, a few years later, sold out to a German consortium. The Le Bels, Joseph-Achille, his sisters, his aunt Adèle, and his uncle Boussingault left Pechelbronn for Paris. Indeed, after the defeat of France to Prussia in 1870, the Peace Treaty signed in Frankfurt (on May 10, 1871) had given the opportunity to Alsatians to choose either French or German citizenship. The Le Bels as a family opted for France: I document it here for both Joseph-Achille [120] and his aunt Adèle [121].

3.20 Frustrated by an Elusive Goal

With Le Bel back in Paris, he returned to research in chemistry, at first in his old laboratory, now headed by Armand Gautier since Wurtz's death. His main focus was stereochemistry. He intended to firm up his reputation as a founder and thus to parallel his earlier work on carbon with newer work on nitrogen—i.e., ammonium derivatives—using the same methodology that had earlier flourished in his hands. He was ultimately frustrated in that endeavor—not that he was unaware of possible difficulties on that path. As he wrote in 1891, "rotatory power was, in fact, to be observed on a body with four different radicals, on condition that these radicals were chosen higher in the series than those of trimethylisobutylammonium" [122]. Le Bel was indeed aware that some phenomenon interchanged the substituents, unless they were bulky enough. As he wrote, in the same 1891 paper, "Having prepared isobutylpropyl-ethylmethylammonium chloride (the radicals are named here in the order of their introduction into the molecule), I obtained, in fact, after culture with molds, two preparations which, as they were, already had powers of −3° and −25° and whose specific rotatory power, deduced from observations on concentrated solutions, exceeds −7° and −8°. These considerable figures leave no doubt as to the activity of the salt, whose chloroplatinate and chloraurate were analyzed after the action of molds." No doubt? Considerable doubt, in fact; at first from his co-discoverer of stereochemistry, van't Hoff [123], who, even though he did not base his skepticism upon original research, stated that the rotatory power was at best elusive. Later on, a German chemist, Marckwald, co-authored a paper to the effect that Le Bel work on ammonium salts was not reproducible [124]. Le Bel stood his ground though [125]. Later on, Pope and Read asserted that Le Bel could not have obtained methylethylpropylbutylammonium iodide, which nullified his claim of optical activity [126].

The first part of Arthur Greenberg's useful chapter in this book describes Pope's contribution to nitrogen stereochemistry—the topic that proved to be totally frustrating to Le Bel. And arguably the best description of Le Bel as a highly congenial friend is the obituary Pope published in 1930 [12]. What had defeated his efforts was nitrogen inversion, since ammonium compounds are in dynamic equilibrium with the corresponding trivalent molecule. This failure—whether one terms it a half-failure or a total failure is irrelevant—signed Le Bel's exit from chemistry. It was the very end of the nineteenth century. Le Bel was only 52. He decided to follow his wide-ranging curiosity elsewhere and effectively ended his career as an active research chemist.

3.21 The Non-academic, the Independent Scholar

Le Bel, during the last years of the nineteenth century, used a private laboratory in his apartment on rue Amyot, in Paris. From the beginning years of the twentieth century, he had moved it to the house built at 250 rue Saint-Jacques, also in Paris. In this personal lab, he would pursue his wide-ranging interests, whether in chemistry, microbiology, crystallography, geology, cosmology or prehistory. A word in passing about the latter: How come Le Bel, who was so fond of chemistry, turned into a prehistorian during the second half of his scientific career? It all came about because of the family name Hauser.

As described earlier in this chapter, at the Parisian *Lycée Charlemagne*, where Le Bel had studied and trained for the competitive entrance exam to the *École Polytechnique*, his leading teacher in mathematics was Simon Hauser. Once Le Bel was admitted to *École polytechnique*, he had for a classmate, in the same class of 1865 as he, Alphonse Hauser (1846–1924), the son of Simon Hauser. Alphonse Hauser, upon graduating from *Polytechnique*, became a naval engineer and had a distinguished career: he headed the French *Bureau des Constructions Navales* and was awarded the Legion of Honor for his contributions, including scholarly treatises in that field (*Cours de Construction Navale*, 1886. This shipbuilding course stands at the crossroads of two worlds, that of all-wood and all-metal construction:

the "Hauser" is part of the continuity of the great history of the French shipbuilding industry, and a reader more accustomed to older techniques will perceive the evolution in which metal gradually replaces wood).

One can imagine Le Bel's shock when he read, in 1910, in the Parisian daily newspaper *Le Matin* that a Swiss-German industrialist with the same family name as both his classmate friend and his teacher, one Otto Hauser (1874–1932), was devastating the renowned French prehistoric site at Les Eyzies, in the Dordogne area, attacking it with explosives following which he would sell and send out worldwide French prehistoric objects thus brutally acquired. Like many French persons, Le Bel was outraged. I figure that he promptly contacted Alphonse Hauser in order to ascertain whether this person was from the same family as he. Whatever Alphonse Hauser's answer, Le Bel, who was a kind of French Rockefeller, from his ownership and sale of the Pechelbronn oil well and petroleum-processing factory, proceeded to purchase the Laugerie Basse prehistoric shelter at Les Eyzies that Otto Hauser had already started brutalizing in 1898 and 1907. He then hired a trained archeologist, Maury, to conduct joint research work on this site [128].

I now return to Le Bel's absence of academic credentials. It is one of the reasons for his failure to be elected a member of the *Académie des Sciences.* Le Bel was a candidate in the years 1888, 1891, 1894, 1897, 1899, 1900, 1907, 1909—and failed to succeed on these numerous attempts [129]. He was finally elected to the *Académie* in 1929, when already too sick to attend sessions of the *Académie* and forced to stay at home. Why was he thus discriminated against? I will state here my conjectures. Jealousy on the part of his académiciens' colleagues, Le Bel being independently and very wealthy. That this Alsatian would continue to live in the Alsace after this French province was taken by Germany, during the period 1884–1889, surely was a factor—since his successful rival during the 1888 election was a fellow-Alsatian, but Paris-based, Paul Schützenberger (1829–1897). In the 1900 election, another fellow-Alsatian, also Paris-located, Albin Haller (1849–1925), was chosen. Likewise, also in 1909, when yet another Alsatian Emile Jungfleisch (1839–1916) was preferred to Le Bel. Possibly, the controversies Le Bel was embroiled with because of the assimilation of his and van't Hoff's 1874 contributions, van't Hoff's being tainted as I have mentioned earlier in this chapter, were another factor. Yet another was the hostility to atomic theory and to Wurtz's laboratory at the *École de Médecine* from the extremely influential Marcellin Berthelot [130]. Le Bel's uncle Boussingault had been a member of the *Académie des Sciences* for many years, since 1839. Boussingault had a sharp tongue; Le Bel too [131, 132] and assuredly this feature played a role in the discrimination against Le Bel.

The last but not least factor in the repeated non-election of Le Bel to the *Académie* is this all-important feature of French culture, known as *copinage*—best translated as buddy system" [133]. It is rooted in Meridional and even Italian influences. In any case, *copinage* has been and remains a significant factor in French life, whether economic, political or, even, scientific. Economic? As I write, the city of Paris prepares to host the 2024 Olympics. Major building projects are under way. *Copinage* was a factor in selecting the companies chosen to accomplish and complete them on time. Scientific? For many years, CNRS has been infected with *copinage* which, to some extent, accounts for the weakness and relative mediocrity of French chemistry on the international scene—the best French chemists nowadays work abroad, persons such as Jean Fréchet or Guy Bertrand, not to mention recent Nobel Prizes awarded to French chemists or biochemists also established outside France. To return to Le Bel, he did not belong to an in-group, hence his non-election to the *Académie des Science.*

3.22 Our Ease with Le Bel's Writing Style

What strikes this reader immediately in Le Bel's scientific writings is their modernity. No effort whatsoever is needed to read his prose, one-and-a half centuries after he set it down. He wrote in French, as one expects; only after World War II did English become the *lingua franca* of science. When he published his results, during the quarter century after the Franco-Prussian War of 1870, many a French chemist would have subscribed to Wurtz's proudly declaring " *la chimie est une science française.*" Indeed, at that time, it was not entirely without basis, obviously not for the whole science of chemistry, but assuredly for organic chemistry: its language [97], that of the *formules dévelopées,* had been arrived at and set primarily by French chemists, such as Dumas, Laurent, and Wurtz. This language thus has basically remained unchanged since Le Bel's time—to be more specific since his 1874 publication founding stereochemistry. That we find ourselves, modern readers in the first quarter of the twenty-first century, totally at ease with that language ought to be worrisome: does it mean that, ever since, organic chemistry has remained static, that its dominant paradigms have remained invariant? The answer is basically "yes," organic chemistry has hugely expanded in volume but has remained fundamentally unchanged. To quote Mitchell, "Modernity, we have said, seems to form a distinctive time–space, appearing in the homogenous shape of the West and characterized by an immediacy of presence that we recognize as the 'now' of history. This time and space are the products of an endlessly replicating system of representation" [134]. Or also, to quote Baudrillard, "Neither science nor technology are themselves modern," but the effects of science and technology are. Though founded on the historic emergence of science, modernity lives only at the level of the "myth of science" [135]. The myth of science? Indeed, Joseph-Achille Le Bel conforms very much to the familiar mythology of a nineteenth-century scientist, i.e., a rather visionary *savant,* a know-it-all as embodied by figures of scientists in the novels by Jules Verne.

We are at ease reading Le Bel's writings since we live indeed in the same mental space as he; with one major difference though, as expressed by Perry Anderson, the Nuclear Age: "To be modern is to find ourselves in an environment that promises us adventure, power, joy, growth, transformation of ourselves and the world—and, at the same time, that threatens to destroy everything we have, everything we know, everything we are" [136]. Yes, if not the science, at least the language of organic chemistry has remained at a virtual standstill. To go on with the comparison between Le Bel's writings and present-day chemistry, characteristic features of his are that most of his papers are concise, signed alone, written in the first person, and often entirely lacking bibliographic references; in contrast with a typical paper of today, with multiple authors who report their results as "we," numerous references as well and at least a dozen pages. We have gone from the Age of the Scientist (Le Bel's) to the Age of the Research Worker (ours). Another feature of Le Bel's publications is their occasional duplication, with the same results appearing the same year in both the *Bulletin de la Société Chimique* and the *Comptes-Rendus de l'Académie des Sciences*—which can be understood by invoking his most legitimate ambition—frustrated for many years—to enter the latter body, which his output obviously entitled him to. There is one marked difference between Le Bel's writing and today's publications though, his repeated reference to "experimental verification." Is it because our Western culture has been indelibly marked by the Age of Doubt? Or because of a takeover of history of science by philosophers and, later by sociologists? In any case, such beliefs are derided nowadays: one's research has Popperian falsification rather than verification as its epistemological goal.

References

1. See also: Laszlo P (2024) Découverte de la stéréochimie par J.-A. Le Bel. La Jaune et la Rouge (magazine of the École Polytechnique alumni) to appear in the October 2024 issue
2. Snelders HAM (1975) J. A. Le Bel's stereochemical ideas compared with those of J. H. van't Hoff. In: Ramsay (ed) Van't Hoff-Le Bel Centennial ACS Symposium Series, American Chemical Society, Washington, D.C., pp 66–73
3. De Chambrier P (1919) Historique de Pechelbronn (1498–1918) Paris-Neuchâtel
4. Streicher J-C (2006) Antoine Le Bel baron de Schoenenberg [Bas-Rhin]. L'Outre-Forêt 122:33–42
5. Streicher J-C (2011) Les pionniers de l'or noir de Pechelbronn: Jean Théophile Hoeffel, 1704–1781, Louis-Pierre Auzillon de La Sablonnière, 1740–1760, Antoine Le Bel, 1735–1788, Strasbourg, Hirlé
6. Le Bel L-F-A (1856) Rendement moyen des produits agricoles de la ferme de Pechelbronn, établi sur une observation de 11 années: prix moyen des denrées, moyenne de la chaleur et de l'eau tombée pendant la même période. J. M. Sauer, Wissembourg
7. (a) McCosh FWJ (1984) Boussingault, Peudel Publishing Company, Dordrecht/Lancaster; (b) Boulaine J (1994) Bousingault, Jean-Baptiste (1802–1887). Professeur d'Agriculture (1845–1848), de Chimie agricole (1851–1887). In: Les professeurs du Conservatoire National des Arts et Métiers. Dictionnaire Biographique 1794–1955. Tome 1: A.-K. Institut National de Recherche Pédagogique, Paris, pp 246–258; (c) Jonas S (2004) Jean-Baptiste Boussingault et l'Alsace. Revue des Sciences Sociales 32:156–161

8. Bulletin Administratif de l'Instruction Publique (1863) Concours général des lycées de Paris—Liste des élèves couronnés 14–164, 174–188
9. Bulletin Administratif de l'Instruction Publique (1865) Liste générale des lauréats des concours académiques pour 1865, 4–79, 405–445
10. (a) Bourdieu P, Passeron J-C (1977) Reproduction in education, society, and culture. Richard Nice transl; with a foreward by Tom Bottomore. Sage Publications, London, Beverly Hills; (b) Bourdieu P, Passeron J-C (1979) The inheritors: French students and their relations to culture, Richard Nice, transl. Chicago, University of Chicago Press; (c) Suleiman EN (1979) Elites in French society, the politics of survival. Princeton University Press. Princeton
11. Zeldin T (1967) Higher education in France 1848–1940. J Contemp Hist 2(3):53–80
12. Pass-Lanneau A (2015) Traditions et contradictions: un portrait de la promotion 1865. La Jaune et la Rouge 707:14–17
13. Tschouriloff M (1875) De l'accroissement de la taille en France. Journal de la Société de Statistique de Paris 16:5–8
14. (a) Loua T (1868) Notice statistique sur le choléra à Paris. Journal de la Société de Statistique de Paris 90: 142–145; (b) Bertillon J (1889) Cartogrammes et diagrammes relatifs à la population Parisienne et à la fréquence des principales maladies à Paris pendant la période 1865–1887: envoyés à l'Exposition Universelle de 1889, par le service statistique municipale de la ville de Paris. Masson, Paris
15. Grattan-Guinness I (2005) The Ecole Polytechnique, 1794–1850: differences over educational purpose and teaching practice. Am Math Mon 112(3):233–250
16. Lévy A, Pinet G (1894) L'argot de l'X. Émile Testard, Paris
17. Laszlo P (2015) La France en 1865, une embryogenèse silencieuse. La Jaune et la Rouge 707:8–9
18. Legoyt A (1871–1872) Mouvement de la population en France (mariages, naissances et décès) de 1861 à 1865. Journal de la Société de Statistique de Paris 12:113–124
19. Velmet A (2020) Pasteur's empire: bacteriology and politics in France, its colonies, and the world. Oxford University Press, New York
20. Hugill J (2015) Global communication electric: Business, news, and politics in the world of telegraphy. Hampf MM (ed.) Technol Cult 56(3):756–757
21. Sagnes J (2008) Napoléon III: le parcours d'un saint-simonien. Editions Singulières, Sète
22. Picon A (2007) French engineers and social thought 18–20th centuries: an archeology of technocratic ideals. Hist Technol 23(3):197–208
23. Chatzis K (2009) Coping with the second industrial revolution: fragmentation of the French engineering education system, 1870s to the present. Eng Stud 1(2):79–99
24. Bernard C (1865) Introduction à l'étude de la médecine expérimentale. Baillière, Paris
25. Fontanon C (2016) Les ingénieurs à l'Académie des Sciences de Paris (1840–1945). Quaderns d'Història de l'Enginyeria 15:295–310
26. A limited sampling of contemporary British scientists emphasizing "experimental justification": (a) Thomson J (1860) On recent theories and experiments regarding ice at or near its melting-point. Proc R Soc Lond 10:151–160; (b) Hofmann AW (1864) Researches on the colouring matters derived from coal-tar—No. IV. Phenyltolylamine. Proc R Soc London 13:485–494; (c) Maxwell JC (1865) A dynamical theory of the electromagnetic field. Philos Trans R Soc London 155:459–512; (d) Stewart B (1863) Reply to some remarks by G. Kirchhoff in his paper "On the History of Spectrum Analysis". Lond, Edinb, Dublin Philos Mag J Sci 25(169):354–360; (e) Perkin WH (1868) On the artificial production of coumarin and formation of its homologues. J Chem Soc 21:53–63; (f) Helmholtz H (1868) On discontinuous movements of fluids. Lond, Edinb, Dublin Philos Mag. J. Sci. 36(244): 337–346
27. Boulaine J (1994) Boussingault In Dictionnaire des Professeurs du Conservatoire des Arts et Métiers, 1. CNAM, Paris, pp 246–263
28. Le Bel J-A (1891) Les relations du pouvoir rotatoire avec la structure moléculaire. Revue des Cours Scientifiques de la France et de l'Étranger 48(20):609–617
29. Busser M (2006) Les événements de l'été 1870 en Alsace du Nord: Une histoire illustrée de la bataille dite de Reichshoffen. Calleva, Barr

30. Fermer D (2008) Sedan, 1870: the eclipse of France. Pen & Sword Military, Barnsley
31. Aronson T (1970) The fall of the third Napoleon. Bobbs-Merrill, Indianapolis
32. (a) Bond B (1983) War and society in Europe, 1870–1970. St. Martin's Press, New York; (b) Milner J (2000) Art, war and revolution in France, 1870–1871, myth, reportage and reality. Yale University Pres, New Haven
33. Giesberg RI (2015) Treaty of Frankfurt; A study in diplomatic history, September 1870–September 1918. University of Pennsylvania Press, Philadelphia
34. Zink G (2001) E Strss üs d'r Heimet: poèmes; trad., R. Matzen; ill., E. H. Cordier, Illkirch: le Verger éd
35. Allier J (1975) Les origines de l'Ecole Alsacienne. Bulletin de la Société de l'Histoire du Protestantisme Français 121:100–112
36. Laszlo P (2015) The University of Strasbourg and world wars. In: Sciences in the Universities of Europe. Nineteenth and twentieth centuries, Academic landscapes, Boston studies in the philosophy and history of science (BSPS), vol 309, pp 89–109
37. (a) Hau M (1987) Pauvreté rurale et dynamisme économique: le cas de l'Alsace au XIXe siècle. Histoire, Économie et Société 6(1):113–138; (b) Hau M (1989) Industrialisation et facteur culturel: le cas de l'Alsace. Revue Géographique de l'Est. 29(2):129–137
38. Stoskopf N (2011) Les industries textiles en Alsace vers 1840. Atlas historique d'Alsace Université de Haute Alsace. www.atlas.historique.alsace.uha.fr. Accessed 24 Oct 2024
39. Spoerlin M (2005) La Cuisinière Alsacienne, De Borée, Romagnat. Note: D L Spoerlin (1762–1852) is the author of one of the great classic books of Alsatian cuisine
40. Bachoffner P (1995) Chimistes et biochimistes en Alsace de 1850 à 1920. Rev Hist Pharm 83(307):380–383
41. Wurtz A (1989) La Théorie Atomique. Félix Alcan, Paris
42. Carneiro A, Pigeard N (1997) Chimistes Alsaciens à Paris au 19ème siècle: Un réseau, une école? Ann Sci 54(6):533–546
43. Fauque D (2005) Organisation des laboratoires de chimie à Paris sous le ministère Duruy (1863–1869): Cas des laboratoires de Fremy et de Wurtz. Ann Sci 62(4):501–531
44. (a) Curie P (1900) Charles Friedel. Bulletin de la Société Française de Minéralogie 23(6–7):171–190; (b) Fauque DM (2008) Charles Friedel (1832–1899) and the laboratory of practical chemistry in the Rue Michelet in Paris. In: Bertomeu Sanchez JR, Burns DT, van Tiggelen B (eds) Neighbours and territories, the evolving identity of chemistry. Mémosciences, Leuven
45. Wisniak J (2024) Charles Lauth: dyes and other compounds. Educación Química 35(2):157–167
46. Davis TLP (1929) Schützenberger J Chem Ed 6(9):1403
47. Gruner Schlumberger A (1982) The Schlumberger Adventure. ARCO Publishing Inc., New York
48. Stadler AM, Harrowfield J (2011) Places and chemistry: Strasbourg—a chemical crucible seen through historical personalities. Chem Soc Rev 40(5):2061–2108
49. Laszlo P (2020) The University of Strasbourg and World Press
50. Ourisson G, Rohmer M, Poralla K (1987) Prokaryotic hopanoids and other polyterpenoid sterol surrogates. Ann Rev Microbiol 41:301–333
51. Lehn JM (1985) Supramolecular chemistry: receptors, catalysts, and carriers. Science 227:949–956
52. Sauvage JP (1990) Interlacing molecular threads on transition metals: Catenands, catenates, and knots. Acc Chem Res 23(10):319–327
53. Schwartz L (1997) Un mathématicien aux prises avec le siècle. Odile Jacob, Paris
54. Schwartz B, Lambrichs LL (1995) Moderniser Sans Exclure. La Découverte, Paris
55. Bloch E (1997) Marc Bloch: 1886–1944, Une Biographie Impossible. Culture et Patrimoine en Limousin, Limoges
56. Kastler A (1976) Cette Étrange Matière. Stock G F, Paris
57. Pétrement S (1997) La Vie de Simone Weil. Fayard, Paris
58. Weil A (1992) The Apprenticeship of a Mathematician. Birkhaüser, Basel
59. Gault H (1921) Adolphe Wurtz, Sa Vie, Son œuvre. G. Fischbach, Strasbourg

60. Rocke AJ (2001) Nationalizing Science: Adolphe Wurtz and the Battle for French Chemistry. MIT Press, Cambridge, Mass., London
61. Pigeard-Micault N (2013) Charles-Adolphe Wurtz: un savant dans la tourmente: entre bouleversements politiques et revendications féministes. Hermann, Paris
62. (a) Fournier J (1994) L'enseignement de la chimie à l'École polytechnique de 1836 à 1880. Union des physiciens; (b) Fournier J (2006) Histoire des radicaux: contribution d'Auguste Cahours: (1813–1891). Revue d'Histoire de la Pharmacie 352:453–474
63. Charlot C, Flahaut J (2003) Antoine-Jérôme Balard. L'homme. Rev Hist Pharm 338:251–264
64. Pigeard-Micault N (2012) Wurtz et l'hypothèse atomique. Pour la Science 413:78–81
65. (a) Wurtz (1874) La théorie des atomes dans la conception générale du monde. Revue des Cours Scientifiques de la France et de l'Étranger (Revue rose) XIV, numéro du 22 août 1874:170–177; (b) Wurtz A (1875) La Théorie des Atomes dans la Conception Générale du Monde. J. Claye, Paris
66. Le Bel J-A, Müntz A (1872) Sur la matière colorante noire des bitumes naturels. Bull Soc Chim Paris 17:156–158
67. Le Bel J-A (1872) Sur les carbures pyrogénés de Pechelbronn (Bas-Rhin). Compt Rend Ac Sci 75:267–269
68. Girard AC (1917) Notice sur les travaux scientifiques de M. A. Müntz - Archives de l'Académie des Sciences, Paris, 1884 and 1893—Achille Müntz (1846–1917). Ann Sci Agron Obituaries: Bulletin de la Société d'encouragement pour l'industrie nationale, 1917-1, p 419; 1918-1, p 215
69. Le Bel J-A (1876) Sur l'amylène et quelques autres dérivés de l'alcool amylique actif. Bull Soc Chim Paris 25:545–547
70. Hanriot M (1884) Nécrologie. Arthur Henninger. Revue Scientifique 34:632–633
71. Sella A (2023) Wurtz's ball tube. Roy Soc Chem. https://www.chemistryworld.com/opinion/wurtzs-ball-tube/4016684.article. Accessed 30 Sept 2024
72. (a) Henninger A, Le Bel J-A (1874) Compt Rend Ac Sci 187:480–483; (b) Henninger A, Le Bel J-A (1874) Bull Soc Chim Paris 21:542
73. Van Berkel K, Homburg E (eds) (2023) The laboratory revolution and the creation of the modern university. Amsterdam University Press, Amsterdam
74. For example: (a) Joly A (1887) Cours Élémentaire de Chimie et de Manipulations Chimiques: Conforme aux Programmes de 1886. Hachette, Paris; (b) Docteur Sonnie-Moret (1896) Élémens d'Analyse Chimique Médicale Appliquée aux Recherches Cliniques. Société d'Édition scientifique, Paris; (c) Maquenne L (1897) Cours Complet d'Enseignement pour le Certificat d'Études des Sciences Physiques, Chimiques et Naturelles. Travaux Pratiques et Manipulations de Chimie. O. Doin, Paris; (d) Dupont J, Freundler P (1898) Manuel Opératoire de Chimie Organique. Hachette, Paris; (e) Barral E (1904) Précis d'Analyse Chimique Qualitative. E. Baillière, Paris
75. Could it be: Le Bel J-A (1876) Observations critiques sur le nouvel alcool dextrogyre de M. Baigues Bakhoven. Bull Soc Chim Paris 25:199–201; See Bakhoven GHB (1874) Annalen der Physik 229(S2):325–331
76. Le Bel J-A (1876) Réponse au mémoire de M E Bourgoin sur l'atomicité comme principe de classification. Bull Soc Chim France 25:540–545
77. Le Bel J-A (1874) Sur les relations qui existent entre les formules atomiques des çorps organiques et le pouvoir rotatoire de leurs dissolutions. Bull Soc Chim Paris 22:337–347
78. Pasteur L (1848) Mémoire sur la relation qui peut exister entre la forme crystalline et la composition chimique, et sur la cause de la polarisation rotatoire. Compt Rend Acad Sci 26:538–538
79. Wislicenus J (1869) Ueber die Einwirkung con Ammoniak auf sogcnannte wasserfreie Milchsii. Ann Chem Pharm 133:259–287
80. Wurtz CA (1864) Mémoire sur l'isomérie dans les alcools et dans les glycols. Ann Chim Phys 4(3):129–186
81. Luynes V (1864) Sur la réduction de l'érythrite par l'acide iodihydrique. Bull Soc Chim 2(1):11–13

82. (a) Kekulé A (1865) Sur la constitution des substances aromatiques. Bull Soc Chim Paris 2e série 3:98–110; (b) Rocke AJ (1985) Hypothesis and experiment in the early development of Kekulé's benzene theory. Ann Sci 42:355–381
83. Oppenheim A (1862) On the camphor of peppermint. J Chem Soc 15:24–29
84. Merton RK (1961) Singletons and multiples in scientific discovery: a chapter in the sociology of science. Proc Am Philos Soc 105(5):470–486
85. Colson A (1892) Sur la stéréochimie de l'acide diacétyltartrique: réponse à une communication de M. Le Bel. Bull Soc Chim 316–318. Ironically, Colson was a *polytechnicien* (Class of 1873) and became a chemistry professor at *Polytechnique* (1898–1923)
86. Le Bel J-A (1890) Sur les conditions d'équilibre des composés saturés du carbone. Bull Soc Chim 3rd series 3:788–796
87. Wurtz A (1875) Annuaires de l'École Pratique des Hautes Études, Paris, pp 37–42
88. Wurtz A (1873–1874) Laboratoire de Chimie de la Faculté de médecine. In: Rapport sur l'École Pratique des Hautes Études, Paris, pp 32–36
89. Wurtz A, Henninger A (1879–1880) Laboratoire de Chimie de la Faculté de Médecine. In: Rapport sur l'École Pratique des Hautes Études, Paris, pp 43–48
90. Hargittai B, Hargittai I (2024) 150 Years of stereochemistry. Struct Chem 36:383–392
91. Japp FR (1887) Tridimensional formulae in organic chemistry. Nature 37:121–123
92. Metz A (1963) La notation atomique et la théorie atomique en France à la fin du XIXe siècle. Revue d'Histoire des Sciences et de Leurs Applications 16(3):233–239
93. Bensaude Vincent B (1999) Atomism and positivism: a legend about French chemistry. Ann Sci 56:81–94
94. (a) Rocke AJ (2013) What did "theory" mean to nineteenth-century chemists? Found Chem 15:145–156; (b) Rocke AJ (2017) The "Indifferent Hypothesis" redux: the dilemmas of Pierre Duhem. Hist Stud Nat Sci 47(3):268–292
95. (a) McNulty MB (2014) Kant's philosophy of chemistry. PhD dissertation, University of California, Irvine; (b) von Engelhardt D (1976) Hegel und die Chemie, Studie zur Philosophie und Wissenschaft der Natur um 1880. Verlag Pressler, Wiesbaden
96. (a) Goodwin W (2003) Explanation in organic chemistry. Ann N Y Acad Sci 988:141–153; (b) Goodwin W (2010) How do structural formulas embody the theory of organic chemistry? Br J Philos Sci 61(3):621–633
97. Laszlo P (2011) Towards teaching chemistry as a language. Sci Educ 22:1669–1706
98. Charle C (2021) Paris, Capitales des XIXe Siècles. Le Seuil (Points), Paris
99. Berkeley-Smith F (1901) The real Latin Quarter. Funk & Wagnalls Co, New York
100. Dupont-Ferrier G (1913) Les Écoles, Lycées, Collèges, Bibliothèques; l'Enseignement Public à Paris. Renouard, Paris
101. de Ménorval E (1872) Les Jésuites de la rue Saint Antoine, l'église Saint-Paul-Saint Louis et le Lycée Charlemagne: Notice Historique. Aubry, Paris
102. Baconin J (2007) Paris 1870–1871: L'Année Terrible. A. Sutton, Saint-Cyr-sur-Loire
103. Mercer J (2005) Time was soft there: a Paris Sojourn at Shakespeare & Co. St. Martin's Press, New York
104. Dreyfuss B (1996) Le Guide du Promeneur: 5e Arrondissement. Parigramme, Paris
105. Vermeylen B, Le Lay Y (2005) Panhard & Levassor: Entre Tradition et Modernité. ETAI, Boulogne-Billancourt
106. Bonnier G (1907) Le Monde Végétal. E. Flammarion, Paris
107. Chebbi Y, Kazan M (2008) The Great Mansions of Paris. Ampersand, Paris
108. Le Bel J-A (1873) Procédé pour préparer l'alcool amylique actif. Compt Rend Ac Sci 77:1021–1023. At the end of that paper, acknowledging the work having been performed in Wurtz's laboratory, he spelled that name with an umlaut—an unusual mistake, I believe, probably made by someone else
109. Le Bel J-A (1878) Alcool amylique dextrogyre. Bull Soc Chim Paris 31:104–105
110. Le Bel J-A (1878) Recherches sur l'alcool amylique (suite): alcool dextrogyre. Compt Rend Ac Sci 87:213–215. At the very end, Wurtz's name is again misspelled with an umlaut

111. Le Bel J-A (1883) De l'alcool amylique produit accessoirement dans la fermentation alcoolique. Compt Rend Ac Sci 96:1368–1370
112. These phenomena can almost certainly be obtained from cultivating molds on ammonium lactate
113. Le Bel J-A, Müntz A (1872) Sur la matière colorante noir des bitumes naturels. Bull Soc Chim Paris 18:156–158
114. Mendeleev D (1877) L'Origine du pétrole. Revue Scientifique, 2e Ser. VIII: 409–416
115. Le Bel J-A (1888) Sur la matière minérale des pétroles naturels. Bull Soc Chim Paris 50:359–361
116. Le Bel J-A (1885) Notice sur les gisements de pétrole à Péchelbronn. Bull Soc Hist Nat de Colmar 24–26e Années: 446–459
117. Le Bel J-A, Greene M (1878) Sur le diméthyléthylène normal (divinyle). Bull Soc Chim 25:305–309
118. Le Bel J-A (1886) Sur les pétroles russes. Compt Rend Ac Sci 103:1017–1019
119. Le Bel J-A (1885) Notice sur les gisements de pétrole à Péchelbronn. Bulletin de la Société d'Histoire Naturelle de Colmar 24–26e Années: 446–459
120. Le Bel J-A (1872) Bulletin des Lois de la République Française, 1er juillet, p 3630, No 60
121. Le Bel J-A (1872) Bulletin des Lois de la République Française, 1er juillet, p 2632, No 418
122. Le Bel J-A (1891) Sur la dissymétrie et la création du pouvoir rotatoire dans les dérivés alcooliques du chlorure d'ammonium. Compt Rend Ac Sci 112:724–726
123. Van't Hoff JA (1899) Vorlesungen Über Theoretische und Physikalische Chemie, Zweite Heft. Die Chemische Statik. Druck und Verlag von Friedrich Viewig und Sohn, Braunschweig, pp 94–118. English translation: Lectures on theoretical and physical chemistry, translated by Lehfeldt RA. Part II: Chemical Statics. Edward Arnold, London, pp 90–125
124. Marckwald W, von Droste-Huelshoff AF (1899) Ueber die methyl-äthyl-propyl-isobutyl-ammoniumbase. Chem Ber 32(1):560–564
125. Le Bel J-A (1899) Sur la stéréochimie de l'azote. Compt Rend Ac Sci 129:548–550
126. Pope WJ, Read J (1912) Asymmetric quinquevalent nitrogen compounds of simple molecular constitution. J Chem Soc 101:519–529
127. Pope WJ (1930) Obituary notice, Joseph Achille Le Bel, born January 1847, died August 6th 1930. J Chem Soc 2789–2791
128. Maury J (1925) Laugerie Basse: the Excavations of M. J.-A. Le Bel, translated by Arthur Edmunds, Monnoyer, Le Mans
129. Emptoz G (2015) Achille Le Bel (1847–1930), Un chimiste innovant tenu à l'écart par ses pairs. Cahiers François Viète II-6/7: 121–135
130. Jacques J (1987) Berthelot, Autopsie d'Un Mythe. Belin, Paris
131. The Archives of the *Académie des Sciences* preserve a letter from Le Bel dated 1924 in which he depicts his uncle with equally great freedom of expression. "The idea of talking about Boussingault is very tempting, but the subject is a little vast, for the man is even more interesting than the chemist, and the most amusing is difficult to insert into an official speech." At the time of the appointment of a chemist they did not like, here's Boussingault's comment, reported by his nephew: "One more imbecile at the Sorbonne or the Academy in no way compromises France, whereas at the head of an army he can be very dangerous, the proof is Mac Mahon who sacrificed his army on the Alsace plain [in 1870] instead of defending the Vosges! […]". His conversation teemed with wit and good advice, and when he called me an imbecile I rejoiced in advance, knowing that he was going to tell me something amusing; once I told him that I had agreed to do a few articles for the *Dictionnaire de chimie* (Wurtz was its chief-editor). "I didn't think you were that stupid; Wurtz would have found some other feeble-minded person who would have done it as well! - But it's a very useful book! - No doubt, but the drainers are also very useful," "you're not going to pump the shit in their place." "I've kept it to myself [...]" (the articles in the dictionary confirm this)
132. Fournier J (2018) Les esprits libres: Jean-Baptiste Boussingault et Joseph-Achille Le Bel. Mediachimie. https://www.mediachimie.org/ressource/les-esprits-libres-jean-baptiste-boussingault-et-joseph-achille-lebel. Accessed 20 Oct 2024

133. (a) Médard J-F (2000) Clientélisme politique et corruption. Tiers-Monde 41(161):75–87; (b) Wionet C, Feng HJ (2015) Connivence pile et face: petit parcours historique du mot. Cahiers du GADGES 13:39–48
134. Mitchell T (2000) The stage of modernity. Quest Mod 11:134
135. Baudrillard J (1985) Modernite. In: Encyclopaedia Universalis, vol 12. Encyclopaedia Universalis, Paris, pp 424–426; Translated by Miller DJ, Purdue University (1987) Can J Polit Soc Theory/Revue canadienne de theorie politique et sociale XI(3):63–72
136. Anderson P (1984) Modernity and revolution. New Left Rev 144:96–113

Alfred Werner: Coordination Theory and the Stereochemistry of Metal Complexes

4

Seth C. Rasmussen

Abstract

In the early period of the theory of valency, difficulty was felt in explaining "molecular compounds" formed by the addition of species to molecules in which the valence was completely saturated. Common examples of these were produced through the reaction of metal salts with ammonia to give metal-ammine compounds. Attempts to understand such metal-ammines date back to the first half of the nineteenth century, with a leading theory introduced by Christian Blomstrand (1826–1897) in 1869. Blomstrand's chain theory was later extended and modified by Sophus Jorgensen (1837–1914) from 1878 onwards. Joining these efforts was Alfred Werner (1866–1919), who after studying the stereoisomerism of trivalent nitrogen in his 1890 doctoral dissertation, presented his coordination theory in 1893, a radical departure from the previous Blomstrand-Jorgensen chain theory. Not only did coordination theory better correlate the properties of compounds based on their composition and configuration, but Werner's proposed structures also introduced the first understanding of stereoisomers in metal complexes. While initially disputed by Jorgensen, Werner's theory finally displaced the older chain theory and resulted in the awarding of the Nobel Prize in Chemistry to Werner in 1913. The life and career of Werner is presented with a focus on his coordination theory and its introduction of stereoisomers in metal complexes.

S. C. Rasmussen (✉)
Department of Chemistry and Biochemistry, North Dakota State University, Fargo, ND, USA
e-mail: seth.rasmussen@ndsu.edu

A. Greenberg and D. E. Lewis (eds.), *The Birth of the 3rd Dimension in Chemistry*, Perspectives on the History of Chemistry, https://doi.org/10.1007/978-3-031-97743-5_4

4.1 Introduction

The second half of the nineteenth century brought the concepts of valence and structure to the forefront of chemical thought. The earliest idea of valence can be traced to Edward Frankland (1825–1899) in 1852, with the realization that elements tend to form compounds containing a set number of groups bound to the central atom, what he referred to as the *combining power* of the element [1]. This concept was then expanded upon in 1858 by both Archibald Scott Couper (1831–1892) [2], who referred to the *combining limit*, and Friedrich August Kekulé (1829–1896) [3], who coined the new term *atomicity* for this concept [4]. However, it was not until 1865 that the term *quantivalence* was introduced by August Wilhelm von Hofmann (1818–1892) to describe this combining power in his *An Introduction to Modern Chemistry* [5]. After that, the condensed terms *valence* or *valency* began to find use in 1866 [6], eventually replacing Kekulé's atomicity.

While valence provided the first indications of chemical structure in terms of atom connectivity, it wasn't until 1874 when Jacobus Henricus van't Hoff (1852–1911) [7] and Joseph-Achille Le Bel (1847–1930) [8] independently proposed the tetrahedral carbon atom in order to rationalize optical activity in organic compounds. Carbon's tetrahedral configuration then laid the foundation for the three-dimensional structure of chemical compounds. In addition, this led to the concept of *stereochemistry*, a term introduced by Viktor Meyer (1848–1897) in 1888 [9].

Models of valence at the time, however, failed to suitably explain what Kekulé called "*molecular compounds*," which were formed by the addition of species to molecules in which the valence was already completely saturated [4, 10–12]. Such molecular compounds were generally depicted using formulas in which the molecule and the added components were separated by a dot,[1] i.e., $NH_3{\cdot}HCl$ or $PCl_3{\cdot}Cl_2$ [11]. According to Kekulé, the forces acting between the components in molecular compounds were viewed to be weaker than those between atoms in conventional compounds, and thus molecular compounds should exhibit limited stability [10, 11]. Contrary to this view, however, was the large and ever-growing family of extremely stable metal-ammonia salts such as $CoCl_3{\cdot}6NH_3$, that had been known since the early 1850s [11, 13, 14]. Not only did these compounds not evolve ammonia with mild heating, but their reaction with acids also failed to produce ammonium salts [10, 11].

While various models and theories had been advanced in an attempt to explain metal-ammonia salts and their derivatives [11, 14–18], it was not until 1893 that Alfred Werner (1866–1919, Fig. 4.1) proposed a new theory that was able to consistently account for the constitution and configuration of these metal compounds [10–12, 14, 16–18]. Werner's coordination theory was revolutionary for its time, but it has survived the more than 130 years since with very little modification. As

[1] It should be highlighted that this notation is still commonly used for denoting the hydrates of salts and other compounds, i.e., $Na_2SO_4{\cdot}10H_2O$.

Fig. 4.1 Alfred Werner (1866–1919) (ETH-Bibliothek Zurich, Bildarchiv/Fotograf: Unbekannt/Portr_09965/ Public Domain Mark)

such, the following aims to review the life and career of Werner, with a focus on his development of coordination theory and its introduction of stereoisomers in metal complexes.

4.2 Early Life and Education

Werner was born on December 12, 1866, in the industrial town of Mulhouse [14, 18–23], located about 100 km south of Strasbourg in the Alsace region of France (Fig. 4.2). He was the fourth and last child of Jean-Adam Werner (1820–1893) and his second wife, Salomé Jeannette Werner (née Tesché) (1825–1903), with only Alfred and his older brother Adolf living to maturity [22, 23]. The family lived in modest circumstances on a family farm, with his father employed in Mulhouse as a foundry worker [18, 19, 23].

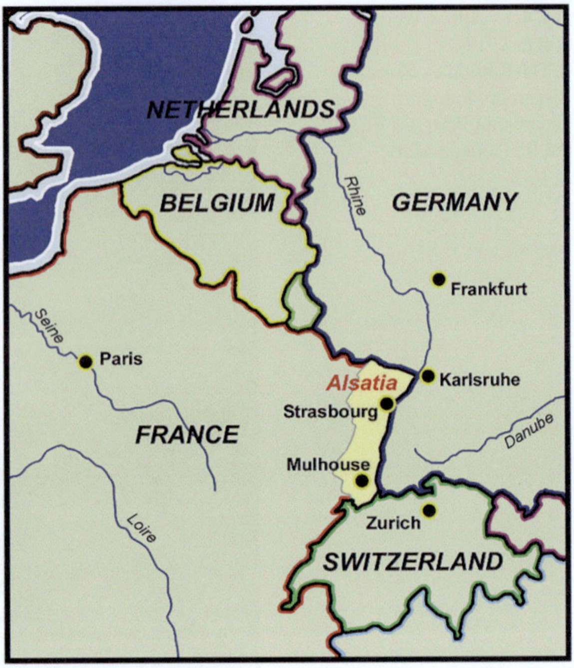

Fig. 4.2 Alace and the surrounding regions

Alsace was annexed by Germany in 1871 at the end of the Franco-Prussian War, becoming part of the German territory of Alsace-Lorraine[2] [18, 23]. As a result, the town of Mulhouse was commonly referred to as Mülhausen [18]. While more than 100,000 Alsatians chose to remain French citizens by emigrating to other regions of France, the Werner family elected to stay in Mulhouse, although their sympathies remained fully with France [23]. Thus, while German became the official language of the region, the family continued to speak French in the home. As a result, Werner was raised with fluency in both German and French.

Werner's education began at the age of six, when he was enrolled at the *École Libre des Frères* in Mulhouse [23]. Established in 1865 by a Catholic congregation of brothers, this free school was created with the aim of providing young people in the industrial city with quality teaching. As pointed out by Kauffman, however, this early training in Catholicism seemed to have little effect on Werner, as his later interest in religion was minimal [23]. Following his graduation from the École Libre des Frères in 1878, Werner then attended the *École Professionelle*, a technical school in Mulhouse, from 1878 to 1885 [18, 23]. It is here that it is believed that he first studied chemistry, which led to him setting up a home laboratory in the family barn [19, 23].

[2] After the conclusion of World War I, Germany ceded the region back to France under the Treaty of Versailles in 1919.

In October of 1885, Werner then began his one-year compulsory military service in the German army at Karlsruhe (Fig. 4.2) [18, 19, 21–24]. During his time in Karlsruhe, he also audited two organic chemistry courses at the *Technische Hochschule* (now the Karlsruhe Institute of Technology, KIT) during the 1885/1886 winter term [14, 18–24]. In July of 1886, Werner became a lance corporal and was considered suitable material for a non-commissioned officer. Still, he did not extend his military service and was officially discharged from the army in October of 1886 [23].

Within three weeks of leaving the army, Werner had set up residence in Zürich in order to attend the *Eidgenossisches Polytechnikum* (now ETH Zürich) [14, 18–24], enrolling in the *Chemisch-Technische Schule* for the 1886/1887 winter term [18, 23]. There, his teachers included Arthur Hantzsch (1857–1935),[3] Georg Lunge (1839–1923),[4] Heinrich Goldschmidt (1957–1937), Frederic P. Treadwell (1857–1918), Karl Heumann (1850–1893), Emil Constam, and Johannes Barbieri [14, 19–24]. In addition to his various required courses in chemistry, physics, mineralogy, and metallurgy, he also elected to take courses in Italian, the Franco-Prussian War, and Darwinism and Teleology [23]. In partial fulfillment of the requirements for his diploma, Werner submitted a 48-page *Diplom-Arbeit* supervised by Lunge and Hantzsch (Fig. 4.3) [18, 23]. He was then awarded the *Technischer Chemiker* (technical chemist) diploma on August 3, 1889 [19, 21–23].

That same year he became an unsalaried assistant in Lunge's Laboratory [12, 19, 21–24]. At the same time, he began carrying out doctoral research

[3] Arthur Rudolf Hantzsch was born in Dresden on March 7, 1857 [25, 26]. In 1875, he began the study of chemistry at the Polytechnikum in Dresden [25]. In 1879, he then went on to work under Johannes Wislicenus (1835–1902) at the Julius-Maximilians-Universität Würzburg, obtaining his doctor's degree in 1880 [25, 26]. Hantzsch spent the summer of 1880 working with August Hofmann (1818–1892) in Berlin, after which he became an assistant under Gustav Wiedemann (1826–1899) in Leipzig [25, 26]. He became *Privat-docent* in 1882 [25], after which he became Professor at the Eidgenossisches Polytechnikum in Zürich in 1885 [23, 25, 26]. In 1893, he then moved to Wurzburg to succeed Emil Fischer, who had moved to Berlin [23, 24, 26]. In 1903, after the death of Wislicenus, Hantzsch became his successor in Leipzig [25, 26]. He was most well-known for the self-named procedures to produce pyridine and pyrrole, as well as the development of a nomenclature to describe heterocyclic compounds. He formally retired in 1927, returning to Dresden, where he died on March 14, 1935 [25].

[4] Georg Lunge was born on September 15, 1839, in Breslau [27–29]. He received his doctor's degree from the University of Breslau in 1859 [27, 28], with a dissertation on alcoholic fermentation [26]. Lunge then studied under Robert Bunsen (1811–1899) at Heidelberg [27–29]. In 1860, Lunge became a chemist in a German fertilizer plant [27]. In 1862, he established his own plant for the production of potassium ferrocyanide, sal ammoniac, lead salts, and tartaric acid, but the plant did not operate as Lunge desired. As such, he decided to gain experience in the large-scale industry of England, where he initially worked at a plant where coal-tar products were made [27, 29]. He then moved to Zürich in 1876 to fill the position of professor of technical chemistry at the Eidgenossisches Polytechnikum, remaining there for 32 years to advance chemical technology and its influence on the development of chemical industry [27–29]. Lunge was forced to retire in 1907 due to poor health [26–28], after which his memory suffered greatly [27]. He passed away on January 3, 1923, at the age of 83 [29].

Fig. 4.3 Arthur Rudolf Hantzsch (1857–1935) (left) (ETH-Bibliothek Zürich, Bildarchiv/Fotograf: Grusche, Julius/Portr_16191-091-AL-FL/Public Domain Mark) and Georg Lunge (1839–1923) (right) (ETH-Bibliothek Zürich, Bildarchiv/Fotograf: Unbekannt/Portr_00184/Public Domain Mark)

under Hantzsch's direction [14, 18, 19, 21–24]. This resulted in the submission of Werner's first paper entitled "*Ueber räumliche Anordnung der Atome in stickstoffhaltigen Molekülen*" (*On the Spatial Arrangement of Atoms in Nitrogen-Containing Molecules*) in December of 1889 [23, 24, 30], which became the first of four sections of his doctoral thesis [23, 24]. Although this paper was jointly authored with Hantzsch, the paper included a closing note by Hantzsch that clearly assigned sole credit to Werner [30]:[5]

> Finally, the undersigned is only fulfilling a duty of justice by making the following declaration in the event that the above developments should be given any significance: that this theory, published jointly with Mr. A. Werner, is essentially the intellectual property of Mr. Werner.

Werner submitted his thesis at the University of Zürich, because the Polytechnikum did not receive the authorization to grant doctorates until 1909 [18]. His thesis was approved in July of 1890 [23], with Werner awarded the degree of Dr. Phil. (with special recognition of superior performance) on October 13, 1890 [18, 20–24].

[5] All quotes given in English have been translated from the original German by the author.

4.3 The Stereochemistry of Nitrogen

Werner began his presentation on the spatial arrangement of atoms in nitrogen species by pointing out "that all previous considerations of spatial isomerism refer solely to the carbon atom" [30]. Furthermore, he felt that previous treatments of carbon could be extended and applied to nitrogen species, stating [30]:

> In other words, it was necessary to investigate whether the hypothesis developed by van't Hoff and Wislicenus with regard to the carbon atom could not also be applied to the nitrogen atom, and ultimately perhaps even to other polyvalent atoms.

Thus, in analogy to related carbon species, Werner then proposed stereoisomers for three general classes of nitrogen compounds (Fig. 4.4): (I) Compounds containing a C=N unit; (II) Compounds containing an N=N unit; and (III) Ammonia-type compounds without double bonds.

To support the proposed stereochemistry for these nitrogen species, Werner then presented various known examples of isomeric pairs that could be explained through the above models. The greatest number of known isomeric pairs could be presented for class I, including various types of oximes (XYC=N–OH). Specific examples comprised benzaldoximes (X = Ph, Y = H), benzilmonoximes (X = PhCO, Y = Ph), and oximidosuccinic acid esters (X = EtOCO, Y = CH_2CO_2H) [30]. He also showed that the related dioximes could result in a set of three isomers, as illustrated by the known three isomeric benzildioximes [30].

For class II, Werner pointed out that the most important examples would be azo compounds such as azobenzene, even though isomeric pairs of azobenzene were not yet known. Of course, the difficulty in isolating isomeric pairs of azobenzene was due to the fact that this compound undergoes photoisomerization as

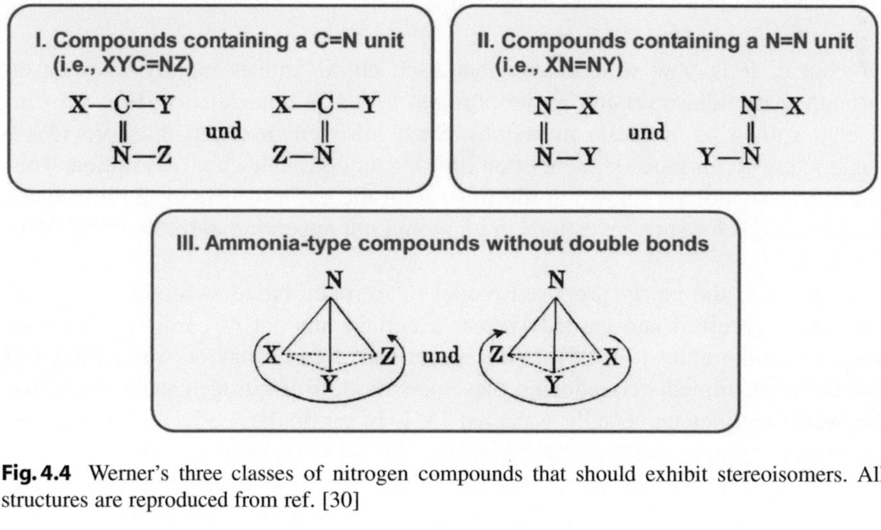

Fig. 4.4 Werner's three classes of nitrogen compounds that should exhibit stereoisomers. All structures are reproduced from ref. [30]

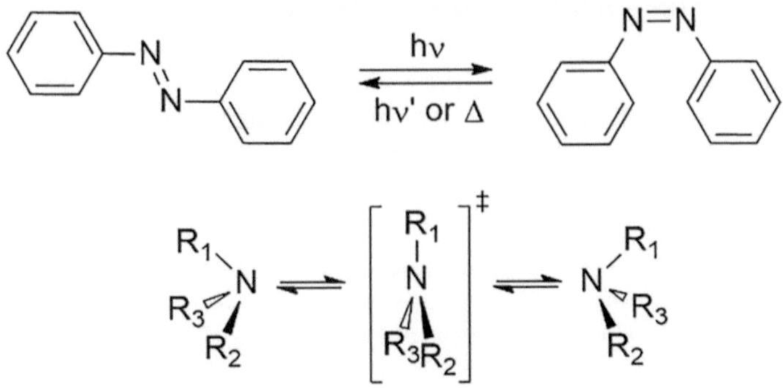

Fig. 4.5 Complicating isomerizations of nitrogen species that were unknown in 1890: the photoisomerization of azobenzene (top) and the thermal inversion of pyramidal nitrogen (bottom)

shown in Fig. 4.5. The reactivity of azobenzene under light, however, was not recognized until 1937, when it was first reported by G. S. Hartley, along with the isolation of the second azobenzene isomer [31]. As such, the only potential evidence that Werner could cite for isomeric pairs in this class was a single report of two trinitroazotoluenes that could be explained via his proposed stereoisomers [30].

For the final class, Werner admitted that such isomers have not yet been observed with certainty. However, he believed that this should be the focus of future research, stating [30]:

> we consider it to be one of the next tasks to determine by precise investigation whether there exist isomeric ammonia or hydrazine derivatives, perhaps only differing due to their opposite rotatory powers. In any case, it is important to decide whether or not there exists an "asymmetric nitrogen atom," i.e., a spatial arrangement of three monovalent radicals that is asymmetrical with respect to the nitrogen atom.

Of course, it is now well known that such chiral amines rapidly interconvert through pyramidal inversion at the nitrogen center as illustrated in Fig. 4.5 (this is also known as umbrella inversion). Such nitrogen inversion thus provides a facile pathway for racemization, often limiting any possible chiral resolution. This, however, was not yet known at the time, with the earliest theoretical and experimental studies for such inversions in ammonia not appearing until the early 1930s [32, 33].

In general, the newly proposed model of nitrogen-based stereochemistry was fairly well received and earned Werner a certain amount of cordiality from the scientific community [34, 35]. While prominent figures such as van't Hoff and Goldschmidt immediately adopted these new models of nitrogen stereochemistry, the work was not universally accepted [35]. In particular, early pushback came from Victor Meyer and Karl von Auwers (1863–1939), who had previously suggested that stereoisomerism in nitrogen species was possible, but attributed any isomerism to the carbon atoms, not the nitrogen [24, 35]. For isomeric oximes, which provided perhaps the strongest support for Werner's stereochemistry, Meyer and von Auwers instead viewed such isomers as resulting from different arrangements with respect to a pair of single bonded carbon atoms. Over time, however, the discovery of additional isomeric species by a number of researchers lent further support to Werner's model and Meyer eventually dropped his objections by 1896 [35]. In the end, Werner's nitrogen stereochemistry has withstood the test of time, with the theory assuming its rightful place alongside the Le Bel and van't Hoff concept of the tetrahedral carbon as one of the foundations of stereochemistry [24, 35].

4.4 Werner's Habilitation and Early Years as Privat-Dozent

Following the completion of his doctoral degree, Werner began work for his habilitation [24]. Over the period of 1890–1891, Werner had been carrying out independent research on benzhydroxamic acid derivatives at the Polytechnikum [22, 34], and these results served as the basis of the second part of his habilitation thesis (*Habilitationsschrift*) [34]. The first and primary part of Werner's thesis was entitled, "Beiträge zur Theorie der Affinität und Valenz" (Contributions to the theory of affinity and valence) [14, 18, 21, 24, 34, 36], which detailed his attempt to introduce a flexible view of valence as a force emanating outward from an atom equally in all directions [18, 24, 34, 36, 37]. Although this can be viewed to be a preamble to the more developed model of valence in his later coordination theory, this earlier work dealt exclusively with organic compounds [24, 34, 37].

Although he had not yet submitted his habilitation thesis, Hantzsch told Werner that he could begin lecturing in 1891. However, Hantzsch also advised him not to do so, but rather to go abroad for a semester as was custom. Werner thus planned to go to England, but Hantzsch thought that Paris would be a better choice as, in his mind, the old French chemistry, with its fine methods, was still practiced there [34]. Thus, after formally submitting his habilitation thesis on October 16, 1891 [18, 34], Werner departed for Paris to work under Marcelin Berthelot[6] (1827–1907) at the College de France. It should be highlighted that this was the furthest Werner had ever traveled relative to his hometown of Mulhouse (see Fig. 4.2). Nevertheless, he remained in Paris during the winter semester of 1891/1892 carrying out research in thermochemistry [18–22, 24, 34]. It has been suggested that the funding for this visit was generously provided by Lunge [12, 18].

Werner's habilitation thesis was accepted by the *Hohe Schweizerische Schulrat* on January 4, 1892 [18, 24, 34, 37], at which time he was still working in Paris

[6] His name appears as both 'Marcelin' or 'Marcellin' in the literature. On his first publication in 1850, Berthelot signed his name as 'Marcellin' and he used this spelling in most of his scientific writings. However, authorized documents were always signed as 'Marcelin Berthelot' [38].

with Bethelot [34]. In the process, Werner (at age 25) was also named a *Privat-Dozent* (unsalaried lecturer) of the Polytechnikum [18, 20, 22, 24, 34, 37]. Upon returning to Zurich, he thus taught atomic theory, comparative organic chemistry, selected topics in inorganic chemistry, and stereochemistry at the Polytechnikum beginning in April of 1892 [18, 24, 34]. As Werner was only a few years older than his students, he grew a mustache and beard in order to appear older and more authoritative. While the beard did not last long, the mustache remained a defining point of his appearance throughout his lifetime (as illustrated in Fig. 4.1) [34].

According to Paul Pfeiffer (1875–1951), a student of Werner during the late 1890s, it was while preparing one of his Polytechnikum lectures that Werner became absorbed in the theories of metal-ammonia species [20, 39]. In the process, he became convinced that the constitution of these compounds could not be suitably explained via conventional valence theory. Again, according to Pfeiffer, the solution came to Werner suddenly one night [21]:

> According to his own statement, the inspiration came to him like a flash. One morning at two o'clock he awoke with a start; the long-sought solution of this problem had lodged in his brain. He arose from his bed and by five o'clock in the afternoon the essential points of the coordination theory were achieved.

As highlighted by Kaufmann [34, 39], other students have given similar stories [12], such that this account is thus commonly accepted [14, 34, 39]. However, versions that have implied that he wrote the published paper over this initial 15 h period are unlikely, considering both its significant length (63 pages) and the amount of prior work cited. Still, as Kaufmann points out [34, 39], the total span of time from when he began lecturing to the submission of the paper in December of 1892 was less than 8 months, a phenomenally short period of time to develop this new theory. It was thus in early 1893 that his resulting landmark paper "Beitrag zur Konstitution anorganischer Verbindungen" (Contribution to the constitution of inorganic compounds) was published in *Zeitschrift für Anorganische Chemie* [40].

4.5 Blomstrand-Jorgensen Chain Theory

Before presenting Werner's new theory, it is necessary to briefly review the prevailing theory used to describe the structure and reactivity of metal-ammonia salts prior to 1893. This theory, known as chain theory, was introduced by Christian Blomstrand (1826–1897)[7] (Fig. 4.6) in 1869 and was an outgrowth of his interest

[7] Christian Wilhelm Blomstrand was born on October 20, 1826, in Växjö, Sweden. His father died when he was only four years old, and he was raised by his mother. Blomstrand entered Lund University on October 1844, where he initially studied the humanities, but soon turned to the sciences, in particular mineralogy. He studied chemistry under Nils Johannes Berlin (1812–1900), and obtained his master's and doctor's degrees in 1850. He was appointed leader of practical experimentation at the University's chemical laboratory in 1851, becoming assistant in 1853. He became

Fig. 4.6 Christian Wilhelm Blomstrand (1826–1897) (left) (Edgar Fahs Smith image collection, Kislak Center for special collections, rare books and manuscripts, University of Pennsylvania Libraries) and Sophus Mads Jorgensen (1837–1914) (right) (circa 1900)

in variable valence [41]. In its simplest sense, this theory postulated that ammonia molecules in metal-ammonia species could link together as $–NH_3–$ chains, analogous to $–CH_2–$ chains in hydrocarbons [16, 17, 41].

As applied to metal species, the number of ammonia molecules associated with the metal (i.e., the length of the chain), depended upon both the metal and its valence. Provisions were also made for different reactivities of associated groups in the compounds. Thus, halogens which could be precipitated readily by $AgNO_3$ were called 'farther' and were regarded as bonded through the ammonia chains, while those which could not be precipitated were called 'nearer' and were considered to be bonded directly to the metal [16, 17, 41].

Blomstrand's chain theory was then extended and modified by Sophus Mads Jorgensen (1837–1914)[8] (Fig. 4.6) beginning in 1878 [17, 41]. A complicating

docent in chemistry in 1854 and soon accepted a position as instructor of chemistry and physics at the recently established Technical School at Malmö. Blomstrand returned to Lund in 1856, were he was appointed assistant in chemistry and laboratory supervisor. After Berlin transferred to the medical faculty in 1862, Blomstrand was appointed Professor of Chemistry and Mineralogy, a position he occupied until his retirement in 1895. He died on November 5, 1897, at the age of 71 [41].

[8] Sophus Mads Jorgensen (or Jørgensen) was born July 4, 1837, in Slagelse, Denmark [16, 17]. He was initially educated in Slagelse and later at the Sorø Velvillie. Here, his interest in chemistry developed due to F. Johnstrup, who allowed older students at the academy to work in his own

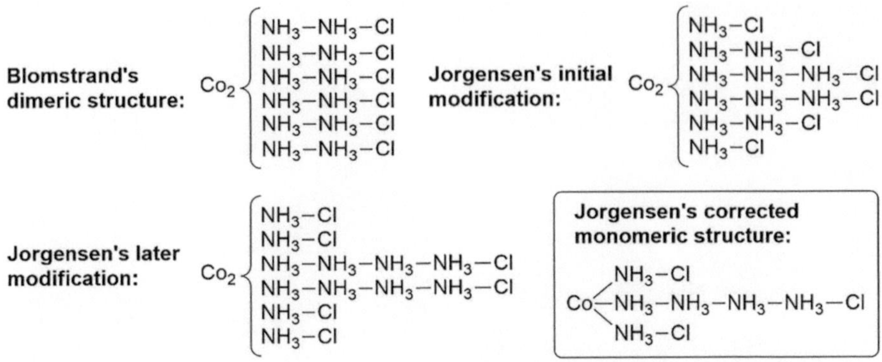

Fig. 4.7 Evolution of the $CoCl_3{\cdot}6NH_3$ structure as formulated by Blomstrand and Jorgensen

factor was that until 1882, there did not exist a reliable method for the determination of molecular weights for non-volatile compounds [16, 17]. As a result, $CoCl_3$ was originally thought to have the composition Co_2Cl_6 (analogous to the volatile Fe_2Cl_6) and thus cobalt-ammonia species were generally considered to be dimers (Fig. 4.7). In 1890, however, Jorgensen deduced evidence for monomeric molecular weights by freezing point and solution conductivity, thus resulting in Blomstrand's original formulas to be halved (i.e., Blomstrand's $Co_2Cl_6{\cdot}12NH_3$ was now correctly given as $CoCl_3{\cdot}6NH_3$) [16, 17].

Another modification introduced by Jorgensen was to address the fact that upon heating $CoCl_3{\cdot}6NH_3$, one-sixth of the ammonia is lost, after which only two of the three chlorides could then be precipitated by $AgNO_3$. Blomstrand's original formula consisted of uniform chains of two ammonia, capped by exterior chlorides. As this could not account for the reactivity upon heating, Jorgensen introduced ammonia chains of different lengths, with longer chains less reactive [16, 17]. As he later regarded four as the maximum chain length of ammonia molecules, which he regarded as a particularly stable arrangement, he again modified this formula to include chains of four ammonia in combination with shorter chains [16, 17]. The combination of these modifications thus resulted in the final structure of Fig. 4.7 by the 1890s.

private laboratory on Sundays, an opportunity that Jorgensen availed himself. Jorgensen entered the University at Copenhagen in 1857 [16, 17] and received his *Magisterskonferens i Kemi* in 1863 [17]. He became assistant to Edward Augustus Scharling in 1864 and became manager of the chemical laboratories and instructor at the *Polytekniske Lareanstalt* in 1867 [17]. He received his doctorate in 1869 with a dissertation on polyiodides of alkaloids [16, 17]. He then became *Lektor* at Copenhagen in 1871, and in 1887, Professor of Chemistry, a position which he held until his retirement in 1908. He died on April 1, 1914 [16, 17].

4.6 Werner's Coordination Theory

Werner presented his new ideas in a very methodical, logical, and comprehensive way, starting with the simple explanation of the metal-ammonia species that were the primary focus of his presentation [40]:

> metal ammonia salts are compounds that are formed from ammonia and metal salts after the same reaction that produces ammonium chloride from hydrochloric acid... and ammonia.

He then divided all known metal-ammonia species into three classes:

(1) Those that contain six NH_3 per metal and their derivatives—i.e., $Co(NH_3)_6Cl_3$.
(2) Those that contain four NH_3 per metal and their derivatives—i.e., $Cu(NH_3)_4(NO_3)_2$.
(3) Those in which the N:H ratio is no longer the same as NH_3—i.e., $H_2N–Hg–Cl$.

Of these, the final class was considered outside the scope of the current work and thus the focus would remain on just classes 1 and 2.

He then explained that the derivatives referenced were those compounds that contain the general nature of the pure metal-ammonia species of classes 1 and 2, but replace ammonia with other molecules that behave in the same way [40]:

> The most important thing to consider is the molecule of water, which behaves analogously to ammonia, but sulfides, alcohols, ethers, carbon monoxide, and ethylene gas also have a role in certain compounds that varies with their nature, but is basically completely analogous.

He went on to explain that further derivative species could also be produced via the loss of ammonia without the addition of other molecules—i.e., $Co(NH_3)_6X_3$; $Co(NH_3)_5X_3$; $Co(NH_3)_4X_3$; $Co(NH_3)_3X_3$.

Beginning with class 1, he then subdivided these again into various classes based on the valence of the metal: $M^{IV}(NH_3)_6X_4$; $M^{III}(NH_3)_6X_3$; $M^{II}(NH_3)_6X_2$. Of these, his initial focus was the $M^{III}(NH_3)_6X_3$ compounds, as these had been the best studied (M = Co, Cr, Rh) [40]. Thus, starting with $M^{III}(NH_3)_6X_3$ (commonly known as the *luteo* salt at the time),[9] he focused on the production of its metal-ammonia derivatives via the loss of ammonia.

[9] In 1852, Edmond Frémy (1814–1894) reported a series of cobalt-ammonia compounds and introduced a nomenclature for these species based on the colors of the compounds, including *luteo* (yellow) for $Co(NH_3)_6Cl_3$ and *roseo* (rose-red) for $Co(NH_3)_5(H_2O)Cl_3$ [42]. While such names were originally introduced for cobalt compounds, they were frequently used for other analogous species as well. As such *luteo* could refer to any species of the general formula $M^{III}(NH_3)_6X_3$. By the early 1890s, contributions by Frémy and others had grown this series to include $M(NH_3)_6X_3$ (*luteo*, yellow), $M(NH_3)_5X_3$ (*purpureo*, purple), $M(NH_3)_4X_3$ (*praseo*, green), $M(NH_3)_4X_3$ (*violeo*, violet), and $M(NH_3)_5(H_2O)X_3$ (*roseo*, rose-red) [13, 42, 43].

Beginning with the loss of one molecule of ammonia, the *luteo* salt is converted to the corresponding *purpureo* salt ($M(NH_3)_5X_3$). As Werner carefully points out, however, this is not just the simple loss of an ammonia molecule and is coupled with a change in function of an acid residue X [40]. For example, in $Co(NH_3)_6Cl_3$, all three chlorines behave as ions and can be precipitated by $AgNO_3$. This is not the case for the resulting *purpureo* salt $Co(NH_3)_5Cl_3$, however, in which only two chlorines behave an ions, with the third chlorine no longer acting as an ion (i.e., it does not react with either $AgNO_3$ or H_2SO_4). To emphasize this change in function, Werner used the notation $(M(NH_3)_5X)X_2$, where the unit X inside the parentheses no longer behaves as an ion [40]. Representing each member of this continuing series via this new notation, he was now able to compare his developing model with the Blomstrand-Jorgensen chain theory (Fig. 4.8).

As described above, Werner recognized that the loss of an ammonia to generate the *pupureo* salt results in one of the X units no longer being precipitated by $AgNO_3$, with this modified X unit now enclosed within the parentheses of his new notation. Jorgensen's model also accounts for this change in that the loss of one ammonia occurs from the shorter, more reactive chain, after which the X unit is directly bound to the metal and can no longer be precipitated.

The loss of another ammonia results in the production of another unreactive X unit, as indicated by the movement of a second X unit into the parentheses under Werner's model. Again, Jorgensen's model also accounts for this change with the loss of the second ammonia coming from the other shorter chain, again giving an

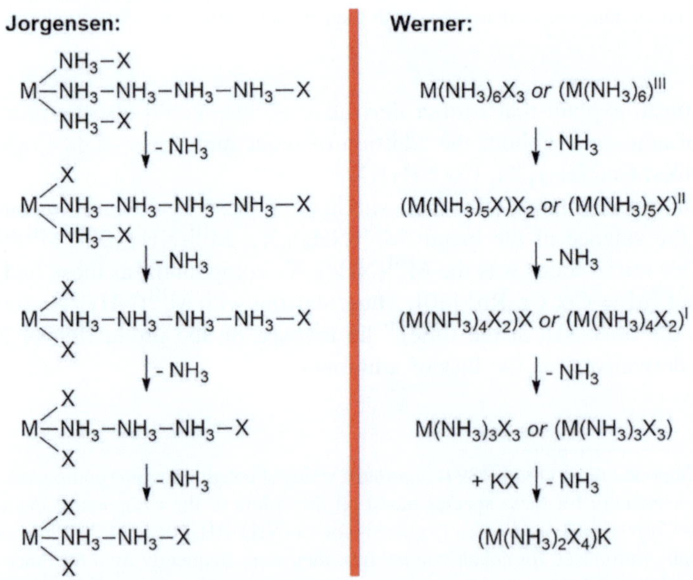

Fig. 4.8 Comparison of the Werner and Jorgensen models for the sequential generation of derivatives of $M^{III}(NH_3)_6X_3$ via loss of ammonia

X unit directly bound to the metal and a corresponding change in reactivity. As such, both models account for the experimentally determined properties of the first two derivatives.

The loss of third ammonia would now cause the final X unit to also become unreactive. Under Werner's model, this is reflected by the movement of the final X unit into the parentheses, leaving no remaining X units behaving as ions. For Jorgensen's model, the third ammonia lost would have to come from the remaining four-ammonia chain, shortening it to three. As the X unit capping this chain is still isolated from the metal center, however, it should still be precipitated by $AgNO_3$. As such, the loss of this third ammonia under Jorgensen's model would not predict any change in properties, which is not consistent with experimental results.

Werner then considered the loss of yet another ammonia. Under Jorgensen's model, this would again come from the remaining ammonia chain, shortening it to only two ammonia, as shown in Fig. 4.8. However, this again should not result in any change in properties and one X unit should still be precipitated by $AgNO_3$. As pointed out by Werner, the $M(NH_3)_2X_3$ species does not exist [40]. Rather, as the fourth ammonia is lost, it is again replaced with an X unit scavenged from the environment to give $M(NH_3)_2X_4$. Werner supported this claim with reference to known analogues such as $K(Co(NH_3)_2(NO_2)_4)$ and $NH_4(Cr(NH_3)_2(SCN)_4)$[40].

Thus, by the Werner model, the production of each of these derivatives in the above series does not really occur by ammonia loss, but rather by substitution of ammonia by an X unit. As stated by Werner [40]:

> in all developed transitions due to loss of ammonia, we are not just dealing with an escape of ammonia molecules, but with a real substitution of the ammonia molecules by acid residues, because at the moment when all the acid residues in the molecule have substituted, another acid residue enters from outside the molecular complex.

Furthermore, while Werner's model accurately accounts for the full series discussed above, the alternate chain theory could only account for the first two derivatives.

In all of the examples given so far, the compounds can be viewed as variations on a MR_6 core with, or without, associated ions. After illustrating that this same approach could be applied to a wide range of derivatives based on the MR_6 radical, Werner then turned to consideration of its spatial configuration [40]:

> If we think of the metal atom as the center of the entire system, we can most easily move the six molecules connected to it into the corners of an octahedron.

Werner's strongest evidence to support an octahedral structure (Fig. 4.9) was the fact that compounds of the form $(MA_4X_2)X$ should exist in two isomeric forms. As explained by Werner [40]:

> This conclusion finds confirmation in the facts. As we know from the beautiful work of Jörgensen, there are a set of two isomers $(CoA_4X_2)X$, where each A_2 is represented by an

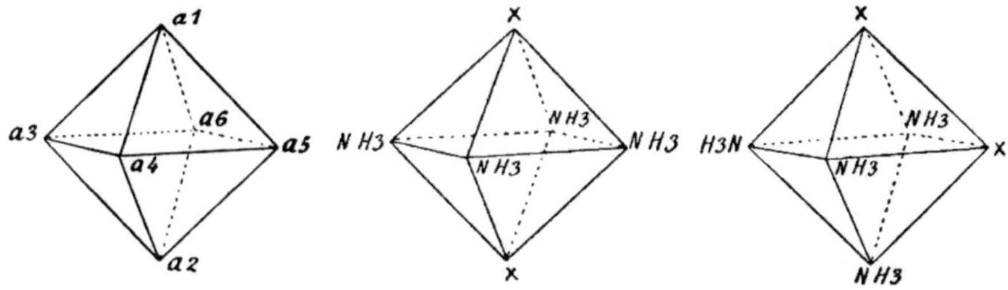

Fig. 4.9 Werner's depiction of his general MA_6 octahedral structure and his proposed structures for the two $M(NH_3)_4X_2$ isomers from ref. [40]

> ethylenediamine molecule. These isomers behave chemically in exactly the same way; of the three acid residues, only one behaves as an ion. However, the two differ in a very characteristic way due to their color. Some are green, the so-called praseo salts, the others are violet and are called violeo salts.

Furthermore, he highlights that this is also supported by the fact that two isomers are also known for $(Pt(NH_3)_2X_4)$. This model and its ability to account for known isomeric pairs thus marks the birth of stereochemistry in inorganic complexes [40]:

> The stereochemistry of carbon compounds and nitrogen compounds is thus joined by the stereochemistry of cobalt compounds and platinum compounds!

Werner's discussion then shifted to metal-ammonia compounds of class 2—those compounds based on the radical $M(NH_3)_4$. Here, he viewed the radical $M(NH_3)_4$ derived from the larger radical $M(NH_3)_6$ of class 1 by loss of two molecules of ammonia. To account for how this could occur without an ion entering the radical as described for the previous derivatives of class 1, he postulated that the positioning of the ions relative to the core radical was the key [40]. In his model, the ions interacted with the octahedral radical via planes consisting of $M(NH_3)_4$ units. If the loss of NH_3 from a $M(NH_3)_6$ radical can occur without disturbing the interaction of the ions with the core, then the ions would not necessarily enter the changing radical (Fig. 4.10).

Once two molecules of ammonia were lost, all that would remain would be a single $M(NH_3)_4$ plane as illustrated in Fig. 4.10. Any further loss would then react in the same way as the previous class 1 derivatives, disturbing the ion interaction and resulting in the formation of $(M(NH_3)_3X)X$. The loss of one more ammonia would then give the neutral $(M(NH_3)_2X_2)$, which can exist as a set of isomers as shown by the proposed structures given in Fig. 4.10. Here, Werner pointed to known isomeric pairs as supporting evidence of this model, such as the isomers of $Pt(NH_3)_2Cl_2$.

In developing his overall theory, Werner viewed valence alone insufficient to explain the nature of metal complexes [40]:

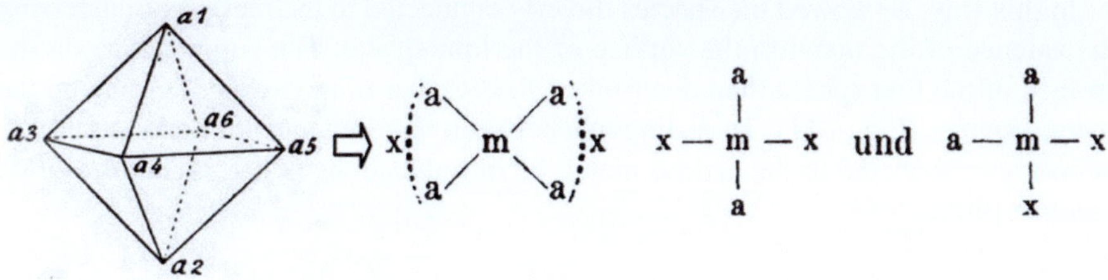

Fig. 4.10 Werner's modification of a MA_6 core to a MA_4 core and his proposed structures for the two MA_2X_2 isomers from ref. [40]

> It is clear that today's theory of valence, with the view of valence as a single force, does not achieve the goal, because it cannot explain how an atom that is divalent can be able to connect six species directly to itself.

To connect the idea of valence with his developing model, he began by describing the metal complex as a collection of concentric spheres surrounding a central spherical metal atom. Werner then considered not just the valence of the metal alone, but the sum valence of the complex making up the first sphere (Fig. 4.11) [40]:

> If we imagine the metal atom as a sphere, the six species that are directly connected to it as being in a first sphere described around it, and the other residues located outside this first sphere as being in a second sphere, we can do this for all the compounds under consideration to establish the general theorem: The valence of the radical (MA_6) formed by the metal atom and the six complexes connected to it in the first sphere is equal to the difference between the valence of the metal atom and the valence of the monovalent groups in the first sphere, completely independent of molecules present in the first sphere such as H_2O, NH_3, etc.

Fig. 4.11 Werner's overall model of concentric spheres

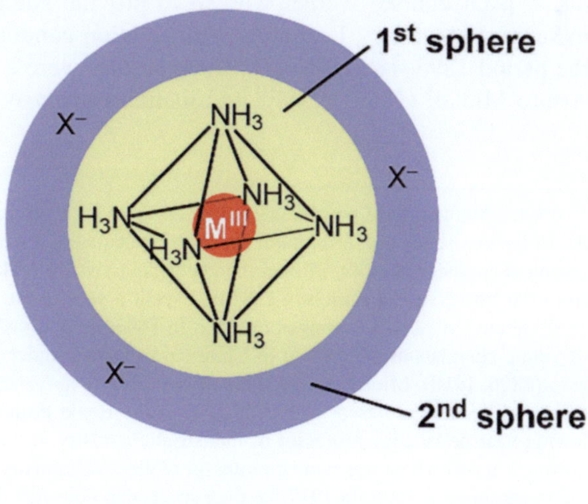

In this way, he viewed the species directly connected to the metal as transferring the valence of the metal to the surface of the first sphere. The sum valence on the surface of the first sphere then determined the number of monovalent groups in the second sphere (Fig. 4.11). To distinguish between this sum valence and the number of species connected to the central metal, he introduced the new term *coordination number* [40]:

> The number of atom groups with which an elementary atom coordinates in the manner discussed to form a compound radical is what I would like to call the coordination number of the atom in question. Thus, an atom that has the coordination number 6 then has the property to combine directly with six groups and to coordinate with six groups of atoms to form a compound radical.

In later papers, Werner refined the concepts of the sum valence and the coordination number into two types of valence: a primary or ionizable valence (*Hauptvalenz*), and a secondary or nonionizable valence (*Nebenvalenz*) [11].

4.7 Reception of Coordination Theory

Werner's theory brought him almost instant fame, as well as an appointment at the University of Zürich as professor extraordinarius (professor without a chair) in September of 1893 [18–22, 34]. His appointment at Zürich was the result of Victor Merz (1839–1904) retiring due to health reasons [34], and Werner formally succeeded Merz as Professor of Chemistry two years later [18–22]. Although offers came from a number of other universities [18, 20], Werner remained at Zürich for the rest of his career.

Werner's theory was not universally accepted, however. Jorgensen in particular viewed his theory as a break in the classical theory of valency and structure; nothing more than an ad hoc explanation insufficiently supported by experiment [17, 44, 45]. Of course, Werner worked to provide additional experimental evidence to support his theory, beginning with solution conductivity studies published over the period 1893–1896 [46–48]. These studies were a collaboration with his friend Arturo Miolati (1869–1956)[10] and included analysis of both cobalt and platinum

[10] Born in Mantua, Italy, on March 2, 1869, Arturo Miolati emigrated to Zurich, where he attended the Eidgenossisches Polytechnikum and became acquainted with Werner. Obtaining the degree of chemical engineer in 1889, Miolati then became assistant to Hantzsch. He received his doctorate from the University of Zürich in 1891 and spent a year as the assistant of Lothar Meyer (1830–1895) at the Universität Tübingen in 1892. In 1893, he returned to Italy in order to join the group of young chemists at the Istituto di Chimica, in Rome, under the direction of Stanislao Cannizzaro (1826–1910). Miolati received his *libera docente universitaria* at the Università di Roma in 1894 and became professor extraordinarius at the Reale Politecnico di Torino in 1902. The following year he became professor of industrial chemistry at the Museo di Chimica Industriale di Torino, after which he was named professor of electrochemistry in charge of physical chemistry at the Politecnico in 1906. In 1917, he then succeeded Giuseppe Bruni (1873–1946) as the chair of

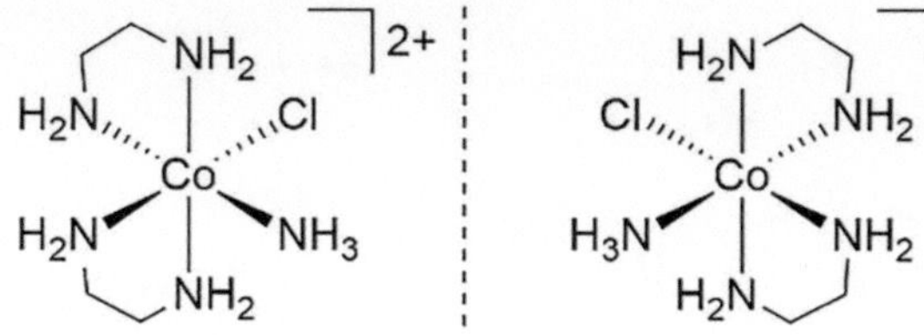

Fig. 4.12 Optical isomers of $[Co(NH_3)(en)_2Cl]^{2+}$

series that varied the number and type of ions in both Werner's first and second spheres. The experimental results confirmed that the conductivity was dependent on the number of ions in Werner's second sphere and that compounds without such ions did not act as electrolytes.

Another point of issue was that while Werner was able to cite isomeric species for the general formula $[MA_4X_2]$, his theory implied that $[Co(NH_3)_4Cl_2]Cl$[11] should also exist as an isomeric pair. As only a *praseo* (green) compound was known at the time, this seemed to contradict this theory. However, Werner was finally able to squash this critique when he, along with his assistant J. Dubsky, successfully synthesized the missing *violeo* (violet) isomer in 1907 [50].

That same year, Jorgensen's student S. P. L. Sorensen recounted discussing Werner's synthesis of the *violeo* isomer, during which Jorgensen concluded that "the strife is now decided" [44, 45]. While some, such as Kaufmann [16, 17, 51], have interpreted this to mean that Jorgensen had accepted the validity of Werner's theory, Kragh has argued that it is likely that Jorgensen never accepted the theory, but rather continued to privately reject it until his death [44]. As such, Jorgensen's statement could have been simply acknowledging the fact that the large majority of chemists would now likely accept Werner's theory [44, 45].

Werner realized that the most convincing evidence of his theory would be the resolution of metal compounds into optical isomers. After many years of attempts, success finally came in 1911 after one of his students, Viktor King (1886–1958), proposed increasing the dissimilarity of the isomers by using *d*-bromocamphorsulfonic acid as a salt-forming constituent [52]. Thus, using silver *d*-bromocamphorsulfonate, Werner, King, and Werner's assistant E. Scholze were able to resolve salts of $[Co(NH_3)(en)_2Cl]^{2+}$ (Fig. 4.12) and $[Co(NH_3)(en)_2Br]^{2+}$ in the summer of 1911 [53]. Following this initial success, additional resolutions quickly followed in the summer and fall of 1911, including salts of $[Co(en)_2(NO_2)_2]^+$ [54], $[Co(en)_2Cl(NO_2)]^+$ [55], and $[Co(en)_2Cl_2]^+$ [56]. Some still argued, however, that the optical activity was somehow due to the carbon content of the ethylenediamine (en) ligands. To remove all possible doubt, Werner finally reported the resolution of a carbon-free metal complex in 1914 (Fig. 4.13) [57].

general chemistry in charge of organic chemistry at the Universita di Padova. He retired in 1936 and died on February 23, 1956 [49].

[11] By 1907, Werner had switched to enclosing brackets rather than parentheses to denote the complex occupying the first sphere. This remains the standard notation still used in modern chemistry.

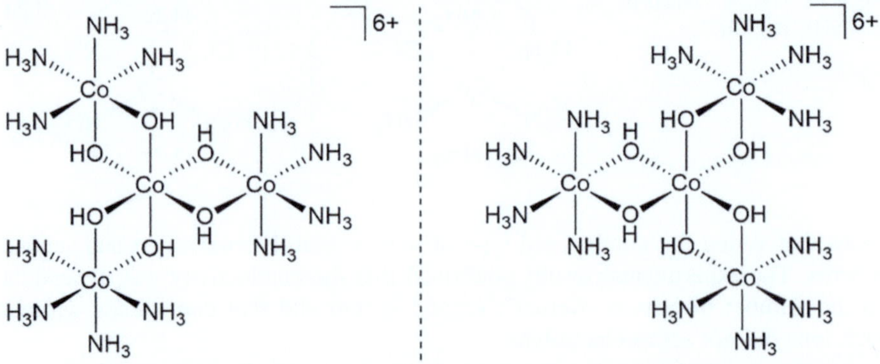

Fig. 4.13 Optical isomers of $[((NH_3)_4Co(\mu\text{-}OH)_2)_3Co]^{6+}$

4.8 Nobel Prize and Legacy

Werner's development of coordination theory led to his nomination for the Nobel in 1907 [58], the same year he synthesized the missing $[Co(NH_3)_4Cl_2]Cl$ isomer. This was followed by five nominations in 1908, three in 1911, four in 1912, and four in 1913, with him being finally selected for the 1913 Nobel Prize in Chemistry. His nominators in 1913 included Emil Fischer (1852–1919, Nobel 1902), Svante Arrhenius (1859–1927, Nobel 1903), Paul Sabatier (1854–1941, Nobel 1912), and his former mentor Arthur Hantzsch [58]. It is likely not a coincidence that his selection for the 1913 Nobel directly followed his successful resolving of metal enantiomeric complexes in 1911.

The wording of the 1913 Nobel Prize in Chemistry stated that it was awarded [59]:

> in recognition of his work on the linkage of atoms in molecules by which he has thrown new light on earlier investigations and opened up new fields of research especially in inorganic chemistry.

As such, this implies that his impact went beyond inorganic chemistry alone. Nevertheless, Werner was the first inorganic chemist to win the Nobel Prize. Furthermore, he was the only one so recognized prior to 1973, when the Nobel Prize in Chemistry was awarded to Ernst Otto Fischer (1918–2007) and Geoffrey Wilkinson (1921–1996).

Shortly after receiving the Nobel Prize, however, Werner began suffering from vascular disease such that by 1915, he felt that he could no longer deliver general lectures [14, 19–22, 60]. His health continued to decline forcing him to retire in October 1919. He died November 15, 1919, just shy of his 53rd birthday [14, 20–22, 60].

Coordination theory, with some refinement, has remained the prevailing theory for the structural description of metal complexes, with his predicted structures confirmed by modern X-ray techniques. Furthermore, many of the terms Werner introduced are still part of the modern inorganic lexicon. As such, it is not surprising that Werner is commonly considered the father of coordination chemistry.

Acknowledgements I would like to thank the Department of Chemistry and Biochemistry of North Dakota State University (NDSU) for continued support of my historical research, and the NDSU Interlibrary Loan Department for its assistance in obtaining various elusive and obscure sources. A special acknowledgement goes to George B. Kauffman (1930-2020), who has contributed so very much on the topics presented above. I was lucky enough to have communicated with George in the early 2000s and I hope that I have been able to add a bit to his tremendous work on this history.

References

1. Frankland E (1852) XIX. On a new series of organic bodies containing metals. Philos Trans R Soc Lond 142:417–444
2. Couper AS (1858) XII. On a new chemical theory. Phil Mag 16:104–116
3. Kekulé A (1858) Ueber die Constitution und die Metamorphosen der chemischen Verbindungen und über die chemische Natur des Kohlenstoffs. Ann Chem Pharm 106:129–159
4. Kauffman GB (1972) Werner, Kekule, and the demise of the doctrine of constant valency. J Chem Educ 49:813–817
5. Hofmann AW (1865) Introduction to modern chemistry. In: Experimental and theoretic. Embodying twelve lectures delivered in the royal college of chemistry, London. Walton and Maberly, London, p 163
6. Brock WH (2000) The chemical tree. A history of chemistry. W. W. Norton, New York, p 457
7. Van't Hoff JH (1874) Voorstel tot Uitbreiding der Tegenwoordige in de Scheikunde gebruikte Structuur-formules in de Ruimte; benevens een daarmee samenhangende Opmerking omtrent het Verband tusschen Optisch Actief Vermogen en chemische Constitutie van Organische Verbindingen. J. Greven, Utrecht. Reprinted in French: van't Hoff JH (1874) Sur les formules de structure dans l'espace. Arch Neerl Sci Exact Nat 9:445–454
8. Le Bel JA (1874) Sur les relations qui existent entre les formules atomiques des corps organiques et le pouvoir rotatoire de leurs dissolutions. Bull Soc Chim Paris 22:337–347
9. Auwers K, Meyer V (1888) Untersuchungen über die zweite van't Hoff'sche Hypothese. Chem Ber 21:784–817
10. Kauffman GB (1967) Alfred Werner's coordination theory. Educ Chem 4:11–18
11. Kauffman GB (1994) Theories of coordination compounds. Alfred Werner's Triumph. In: Kauffman GB (ed) Coordination chemistry. ACS Symposium Series, American Chemical Society, Washington, DC, pp 3–33
12. Berl E (1942) Some personal recollections of Alfred Werner. J Chem Educ 19:153–154
13. Kauffman GB (1977) Early experimental studies of cobalt-ammines. Isis 68:392–403
14. Ernst KH, Wild FRWP, Blacque O, Berke H (2011) Alfred Werner's coordination chemistry: new insights from old samples. Angew Chem Int Ed 50:10780–10787
15. Kauffman GB (1974) Early theories of metal-ammines. A brief historical review from Graham to Claus (1837–1856). J Chem Educ 51:522–524
16. Kauffman GB (1959) Sophus Mads Jorgensen (1837–1914). A chapter in coordination chemistry history. J Chem Educ 36:521–527
17. Kauffman GB (1960) Sophus Mads Jørgensen and the Werner-Jørgensen controversy. Chymia 6:180–204

18. Werner H (2013) Alfred Werner: a forerunner to modern inorganic chemistry. Angew Chem Int Ed 52:6146–6153
19. Morgan GT (1920) Alfred Werner. J Chem Soc Trans 117:1639–1648
20. Pfeiffer P (1920) Alfred Werner. 1866–1919. Z Angew Chem 33:37–39
21. Pfeiffer P (1928) Alfred Werner. J Chem Educ 5:1090–1098
22. Alfred Werner—biographical. NobelPrize.org. https://www.nobelprize.org/prizes/chemistry/1913/werner/biographical. Accessed 03 March 2024
23. Kauffman GB (1966) Alfred Werner. Founder of coordination chemistry. Springer-Verlag, Berlin, pp 2–19
24. Kauffman GB (1966) Foundation of nitrogen stereochemistry. Alfred Werner's inaugural dissertation. J Chem Educ 43:155–156
25. Moore TS (1936) The Hantzsch memorial lecture. Arthur Rudolf Hantzsch. 1857–1935. J Chem Soc 1936:1051–1066
26. Paal C (1927) Arthur Hantzsch zum 70. Geburtstage. Z Angew Chem 40:301–303
27. Berl E (1939) Georg Lunge (1839–1923). J Chem Educ 16:451–460
28. Quincke F (1909) Georg Lunge zum 70. Geburtstage. Z Angew Chem 22:1825–1826
29. Berl E (1923) Georg Lunge †. 15 September 1839–3 January 1923. Chem Ztg 47:157–158
30. Hantzsch A, Werner A (1890) Ueber räumliche Anordnung der Atome in stickstoffhaltigen Molekülen. Ber 23: 11–30. English translation: Hantzsch A, Werner, A, Kauffman GB (trans) (1966) On the spatial arrangement of atoms in nitrogen-containing molecules. J Chem Educ 43:156–165
31. Hartley GS (1937) The cis-form of azobenzene. Nature 140:281
32. Dennison DM, Hardy JD (1932) The parallel type absorption bands of ammonia. Phys Rev 39:938–947
33. Cleeton CE, Williams NH (1934) Electromagnetic waves of 1.1 cm wave-length and the absorption spectrum of ammonia. Phys Rev 45:234–237
34. Kauffman GB (1966) Alfred Werner. Founder of Coordination Chemistry. Springer-Verlag, Berlin, pp 19–34
35. Kauffman GB (1972) The stereochemistry of trivalent nitrogen compounds: Alfred Werner and the controversy over the structure of oximes. Ambix 19:129–144
36. Werner A, Kauffman GB (trans) (1967) Contributions to the theory of affinity and valence. Chymia 12:189–216
37. Kauffman GB (1967) Alfred Werner's "Habilitationsschrift." Chymia 12:183–187
38. Adloff JP, Kauffman GB (2007) Marcellin Berthelot (1827–1907), chemist, historian, philosopher, and statesman: a retrospective view on the centenary of his death. Chem Educ 12:195–206
39. Kauffman GB (1976) An ingenious impudence: Alfred Werner's coordination theory. J Chem Educ 53:445–446
40. Werner A (1893) Beitrag zur Konstitution anorganischer Verbindungen. Z Anorg Chem 3:267–330
41. Kauffman GB (1975) Christian Wilhelm Blomstrand (1826–1897) Swedish chemist and mineralogist. Ann Sci 32:13–37
42. Frémy ME (1852) Recherches sur le Cobalt. Ann Chim Phys 35:257–311
43. Ramberg PJ (2003) Chemical structure, spatial arrangement. In: The early history of stereochemistry, 1874–1914. Ashgate, Hampshire, p 281
44. Kragh H (1997) S. M. Jørgensen and his controversy with A. Werner: a reconsideration. Brit J Hist Sci 30:203–219
45. Ramberg PJ (2003) Chemical structure, spatial arrangement. In: The early history of stereochemistry, 1874–1914. Ashgate, Hampshire, pp 305–313
46. Werner A, Miolati A (1893) Beiträge zur Konstitution anorganischer Verbindungen. Z Phys Chem 12:35–55
47. Werner A, Miolati A (1894) Beiträge zur Konstitution anorganischer Verbindungen. Z Phys Chem 14:506–521

48. Werner A, Miolati A (1896) Beiträge zur Konstitution anorganischer Verbindungen. III. Z Phys Chem 21:225–238
49. Kauffman GB (1970) Arturo Miolati (1869–1956). Isis 61:241–253
50. Werner A (1907) Über 1.2 - Dichloro - tetrammin - kobaltisalze. (Ammoniak - violeosalze). Ber 40:4817–4825
51. Kauffman GB (1966) Alfred Werner. Founder of coordination chemistry. Springer-Verlag, Berlin, p 48
52. King VL (1942) A rough but brilliant diamond. J Chem Educ 19:345
53. Werner A (1911) Zur Kenntnis des asymmetrischen Kobaltatoms. I. Ber 44:1887–1898
54. Werner A (1911) Zur Kenntnis des asymmetrischen Kobaltatoms. II. Ber 44:2445–2455
55. Werner A (1911) Zur Kenntnis des asymmetrischen Kobaltatoms. III. Ber 44:3272–3278
56. Werner A (1911) Zur Kenntnis des asymmetrischen Kobaltatoms. IV. Ber 44:3279–3284
57. Werner A (1914) Zur Kenntnis des asymmetrischen Kobaltatoms XII. Über optische Aktivität bei kohlenstofffreien Verbindungen. Ber 47:3087–3094
58. Alfred Werner—nominations. NobelPrize.org. https://www.nobel-prize.org/prizes/chemistry/1913/werner/nominations/. Accessed 05 March 2024
59. The Nobel Prize in chemistry 1913. NobelPrize.org. https://www.nobelprize.org/prizes/chemistry/1913/summary/. Accessed 03 March 3 2024
60. Kauffman GB (1966) Alfred Werner. Founder of coordination chemistry. Springer-Verlag, Berlin, pp 96–100

5 Studies of Chiral Centers Other Than Carbon. William Jackson Pope and Kurt Mislow: Pioneers of Stereochemistry

Arthur Greenberg

Abstract

William Jackson Pope and Kurt Mislow were pioneers in stereochemistry contributing to the most fundamental concepts and terminology in the field. Their ground-breaking research activities were separated, for the most part, by over a half century. Contemporary instrumentation and techniques limited Pope's studies to static stereochemistry while Mislow's studies focused significantly on stereochemical dynamics. In common were their research on chiral centers other than carbon.

5.1 Introduction

William Jackson Pope and Kurt Mislow were ground-breaking pioneers in the field of stereochemistry. Pope spent his most productive and final years at Cambridge University and Mislow completed a quarter of a century at Princeton following fifteen years at New York University. He then enjoyed almost three decades as a highly productive emeritus professor. Obviously, a comparison between these two scientists is far from straightforward. While quantitative chemical composition, concepts of valence, purification techniques, and polarimetry were reasonably well developed, most of Pope's most creative research occurred prior to 1916, the year that Walter Kossel established the octet rule, followed one month later by Gilbert N. Lewis's publication of electron dot formulas highlighting covalent and ionic compounds. Pope's major research was accomplished before physical organic chemistry became a tool of mechanistic organic chemistry. It predated the development of commonly employed spectroscopic applications (especially

A. Greenberg (✉)
Department of Chemistry, University of New Hampshire, Durham, NH, USA
e-mail: art.greenberg@unh.edu

A. Greenberg and D. E. Lewis (eds.), *The Birth of the 3rd Dimension in Chemistry*, Perspectives on the History of Chemistry, https://doi.org/10.1007/978-3-031-97743-5_5

nuclear magnetic resonance), and before computational chemistry evolved into a valued component of the chemists' armamentarium. Although X-ray crystallography originated in 1912 and began to flourish in the 1920s, applications to stereochemistry arrived many years later. The first direct determination of the absolute configuration of an enantiomer, sodium rubidium (+)-tartrate tetrahydrate, was made by Johannes Martin Bijvoet, professor of chemistry at the University of Utrecht, and co-workers in 1951 who solved the phase rule. Interestingly, J. H. van 't Hoff was a lecturer at the Veterinary school in Utrecht in 1874. The Veterinary School joined the University in 1918. With the breakthrough research of Herbert Hauptman, Jerome Karle, and Isabella Karle, direct X-ray crystallography gained increased ability to directly determine absolute configuration. These applications began to bear fruit in the 1950s, but the process was so time-consuming that absolute configuration correlations continued to hold sway for many years. Furthermore, World War I interceded during a very active period in Pope's career, when the efforts of Great Britain's chemical industry and universities focused significantly on war-related technology. Mislow's early life was dislocated as the result of the Nazi's persecution of Jews and World War II that eventually brought his family from Germany to New York City. This said, both Pope and Mislow contributed mightily to the very conceptions and lexicon of stereochemistry. Perhaps most striking was the research of each of these distinguished chemists into the chiralities of central atoms other than carbon and this will be the area of greatest coverage in this chapter. Pope and his colleagues made profound breakthroughs in synthesis and static stereochemistry. Mislow and his colleagues studied molecular dynamics, experimentally and computationally. It is quite clear that the research productivities of both Pope and Mislow were so varied and prodigious that the presentation here is perforce an incomplete outline. Any attempt to compare their impacts would be something of "a fool's errand" due to the vastly different eras of their activities.

5.2 Brief Biographical Background for William Jackson Pope

William Jackson Pope was born in London on March 31, 1870. His mother was Alice Pope, and his father, William Pope, was a city merchant [1–3]. His modest education was provided by City Guilds and his technical training at Finsbury Technical College, where he came under the influence of Henry (Fig. 5.1).

Edward Armstrong (1848–1937). Armstrong, one of the outstanding organic chemists of the period, published sixty papers on the structure and reactions of naphthalene as well as studies in crystallography and was a co-founder in 1871 of Finsbury Circus. This institution became Finsbury Technical College (1879), subsequently the Central Technical College, which opened 1884, and ultimately became the City and Guilds College of the Imperial College of Science and Technology. Beginning his science training at Finsbury Technical College, Pope followed Armstrong to the Central Technical College. Pope never formally attained

Fig. 5.1 William Jackson Pope (courtesy National Portrait Gallery, London U.K.)

a university degree. Nevertheless, his brilliance was recognized at an early stage, and he was appointed, in 1897, Head of the Chemistry Department at Goldsmith's Institute at New Cross and simultaneously Lecturer in Crystallography at Central Technical College. In 1901, he was appointed Head of the Department of Pure and Applied Chemistry at the new School of Technology in Manchester. The new building was designed and equipped by James Dewar among other distinguished chemists. During this seven-year period, Pope discovered new compounds of gold and platinum and developed Valency-Volume Theory with W. Barlow, abandoned not long afterward. In 1902, Pope was elected Fellow of the Royal Society and in 1905 when the school became the Faculty of Technology in the University of Manchester, Pope formally received his first university degree along with other department heads. During this period, he worked with his Manchester colleague William H. Perkin as they synthesized 1-methylcyclohexylidene-4-acetic acid (**1**), chiral due to the absence of a plane or center of symmetry but lacking an asymmetric carbon [4]. Although one might be tempted to consider the carbon bearing the methyl group as chiral if one follows the two attached cyclohexyl bonding out to the *cis* or *trans* attachment, what was needed was a newer nomenclature. Specifically, Cahn, Ingold, and Prelog coined the concept of a "chiral axis" present in molecules such as **1–5** (see Scheme 5.1) [5]. Indeed, van't Hoff recognized the potential enantiomers in an allene such as **2** (represented by fused carbon tetrahedra, **2a**). Pope and Perkin recognized resolving dissymmetric allenes as a worthwhile goal, but the reactivities of these compounds precluded this possibility.

Unhappy with the administration in Manchester, Pope was appointed in 1908 Chair of Chemistry at Cambridge, the offer first made to and declined by Perkin. Pope finished his career at Cambridge. Another interesting contribution to stereochemistry was Pope's resolution, in 1914, of enantiomers containing a single asymmetric carbon [6]. Although this was a fundamental concept in 1874, it was a fact that no examples having fewer than three carbons (e.g., lactic acid) were known by the end of the twentieth century. This prompted Arnold Eiloart, author of the 1898 English translation of van 't Hoff's book, to postulate that three carbons were a necessary condition for optical activity [7, 8]. He reasoned that three attached carbons were the minimum condition for a helix. As a believer in the outdated ether theory, Eiloart imagined the mirror-image wakes of clockwise and counterclockwise helical rotations as the source of dextro- and levorotatory optical activities. Pope's publication brought this quaint notion to its final resting place four decades after the initial proposals by van 't Hoff and Le Bel proving that asymmetric carbon atoms need not be bonded only to carbon atoms.

The outbreak of World War I found Pope in Australia at a major meeting. His delayed return to Cambridge included an extended stay in India. During the war, Pope and the university were involved in war-related research including explosives, fuel, and mustard gas as well as organic arsenicals. At the end of the war,

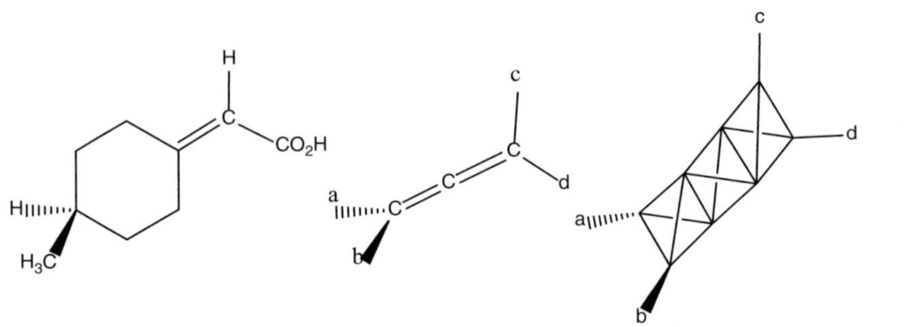

1 **2** **2a**

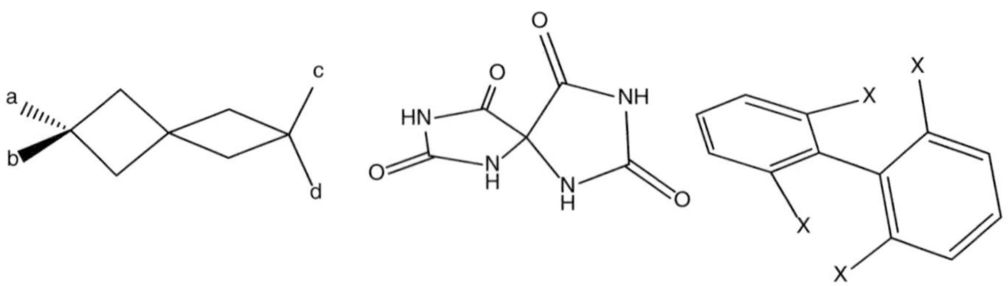

3 **4** **5**

Scheme 5.1 Chiral molecules lacking asymmetric carbons

a grateful petroleum industry donated 210,000 pounds sterling (about 13,400,000 pounds sterling today) to Pope's department. Following the war, Pope was elected President of the Chemical Society and subsequently President of the Society of Chemical Industry. In 1922, he was elected President of IUPAC. In late 1933, Fritz Haber, who had been forced by the Nazis to resign from the position of Director of the Kaiser Wilhelm Institute for Physical Chemistry and Electrochemistry in Berlin, wrote to Pope from Paris requesting positions at Cambridge for himself, two employees as well as family members (9). Pope and Frederick G. Donnan, Department Head, University College London, successfully arranged these positions starting in late 1933. Haber spent two months in Cambridge in very ill health. He died of a heart attack in January 1934 in Basel enroute to Italy to escape the cold, damp English winter. It has been speculated that Pope's own research on chemical warfare agents during World War I helped his understanding of Haber's wartime research on toxic gases for the Germans during this period [9]. During the final twelve years of his life, Pope suffered ill health and pain and died on October 17, 1939, ironically six weeks after the outbreak of World War II.

Gibson [2] noted Pope's very early interest in the biological aspects of chirality and refers specifically to Pope's Cantor Lectures of February and March 1901. John Read also comments on Pope's rich and subtle sense of humor [3]. Gibson relates:

> As usual, sense of humour accompanies his profound scientific insight and originality.

Specifically, considering *Homo sapiens* to be "*Dextro*-man," Pope's "*Laevo*-man" would be immune from infections from bacteria, virus, and protozoa, his clothes would button from the wrong side, and he would die of starvation since he could not metabolize proteins and starches. Gibson quotes Pope:

> The laevo-man has apparently not made his appearance on our planet. He can be produced hypothetically from the dextro-man by removing the latter into four-dimensional space and turning him through half a whole revolution about three-dimensional space; on restoring the individual to three-dimensional space without further rotation he would be found to have undergone conversion into the laevo-form.

It is reasonable to conjecture that Pope was familiar with Edwin Abbot's 1884 book *Flatland* [10] even as Pope's method for interconverting three-dimensional enantiomorphs is more specific. Pope also assembled a famous collection of important alchemical oil paintings, engravings, and prints. After his death, these were acquired by the Fisher Scientific Company, Pittsburgh PA, in October 1939 [11, 12].

5.3 Brief Biographical Background for Kurt Mislow [13–18]

Kurt Martin Mislow was born in Berlin, Germany on June 5, 1923. His father was a businessman who traveled extensively in Europe. The rise of Hitler made life increasingly untenable and dangerous for German Jews. Mislow's family fled as refugees from Düsseldorf, first to Milan, Italy in 1936, where the adolescent Kurt took his schooling in Italian. The family moved to England in 1938, initially interned on the Isle of Man. Just as Mislow was preparing to enter the University of Cambridge, an affidavit secured by his uncle, the famous photographer Alfred Eisenstadt, arrived and the family moved to New York City. Receiving an attractive offer to attend Tulane University, young Mislow relocated to New Orleans where he graduated in 1944. He then moved on to pursue his Ph.D. studies at California Institute of Technology with Linus Pauling as his mentor. Mislow's doctoral dissertation was comprised of three fairly unrelated chapters [19]. The first chapter reported syntheses of four potential antimalarial compounds under the guidance of Dr. Joseph B. Koepli (1904–2004), an organic chemist and research associate at Cal Tech (1932–1971). Koepli has a fascinating history including his appointment as Science Advisor to the State Department during the early 1950s. This was a fraught period characterized by the McCarthy hearings and Koepli's activities did not escape scrutiny during a period often referred to as "The Red Scare" [20]. This collaboration with Koepli produced Mislow's first co-authored paper [21]. The third chapter was an investigation of the mechanism of the Hammick reaction [19]. The second chapter [19], the largest, focused on a major interest of Pauling's, the non-covalent interactions (van der Waals forces, hydrogen bonding) between biological molecules and active sites and was perhaps a stimulus to Mislow's early interests in stereochemistry. For example, the similarity in size and shape between benzoic acid and 2-thiophenecarboxylic acid was consistent with the strong inhibition by the latter of the serological (antibody–antigen) activity of benzoic acid. Mislow made a huge number of extremely careful melting points of various mixtures of pairs of carboxylic acids in order to construct eutectic diagrams to assess solid solutions (and best fit between molecules). Aside from the dissertation, this work appears to be unpublished.

With Pauling's recommendation, Mislow was appointed Lecturer at New York University in 1949. Mislow's deep engagement with stereochemistry was initially stimulated [14, 17] by his reading of George W. Wheland's studies and appears to have begun to blossom about this time [22]. These interests converged in his first sole-authored paper [23], published in 1950. Mislow found the chemistry department of that period to be poorly equipped for competitive research and was disappointed to be denied promotion to Assistant Professor in 1950, despite receiving a Research Corporation grant in 1948, in what he regarded as a petty administrative issue [22]. However, he was promoted to Assistant Professor the following year along with an additional Research Corporation grant. With increasing research success, his outlook clearly brightened [22]. He was promoted to Associate Professor (1956) and Professor (1960) (Fig. 5.2).

Fig. 5.2 Kurt M. Mislow (courtesy Princeton University)

Two early papers authored by Mislow [24, 25] made a significant impact upon the stereochemistry-and wider organic chemistry communities as modifications of van 't Hoff's rules of 1875. A molecule generalized as **6**, in which the X substituents are large enough to prohibit rotation around the central bond in the biphenyl component, forcing essentially perpendicular orientation between the two phenyl rings, lacks asymmetric centers, and can never adopt a symmetric (achiral) conformation (see Scheme 5.2). Nonetheless, the molecule continuously racemizes due to rapid rotation of the two Cabc substituents (in the same direction—one possible mode) about their bonds to biphenyl. Mislow described this molecule as either: (1) "a *meso* diastereomer for which no symmetric conformation is possible," or (2) "a non-resolvable racemic mixture of conformational type" [21]. A closer look is warranted. Two conformations of *meso*-tartaric acid are non-dissymmetric and thus achiral: **7a** having a mirror plane perpendicular to the central C–C bond, thus reflecting the two $CH(OH)CO_2H$ groups, and **7b** having an inversion center. The presence of these conformers causes *meso*-tartaric acid to be optically inactive. If one considers the 1,1′,6,6′-tetrasubstituted biphenyl **8** in which NO_2 is large enough to fix the two biphenyl rings in the perpendicular conformation, the only non-dissymmetric conformer (D_{2d} symmetry) has two symmetry planes intersecting along the central bond axis. In **6** the symmetry elements present in **7a**, **7b**, and **8** are mutually incompatible and no symmetry elements are present in any conformation. A static structure is chiral but can be racemized rapidly by simple, low-barrier rotation as described above. Although the simplest chiral alkyl substituent is *sec*-butyl [24], Mislow and Bolstad exemplified their observation through synthesis of (−)-menthyl (+)-menthyl 2,2′,6,6′-tetranitro-4,4′-diphenate (**9**) (Scheme 5.2) [25].

The comparison of the significance of Pope's molecule **1** and Mislow's molecule **9** is quite a reasonable one since each delved into the epistemology of stereochemistry, modifying accepted concepts dating to 1875. During Pope's era, the stereochemical breakthroughs were static studies of molecules. Thus, the issue

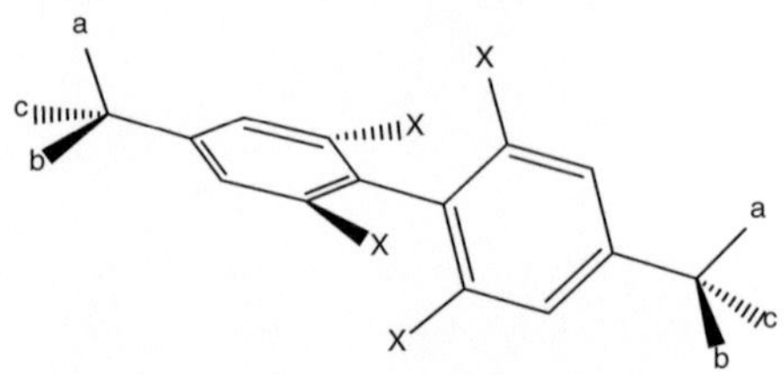

6

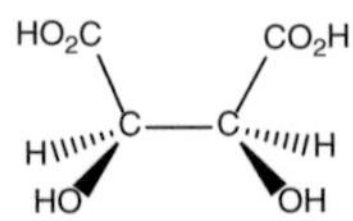

7a

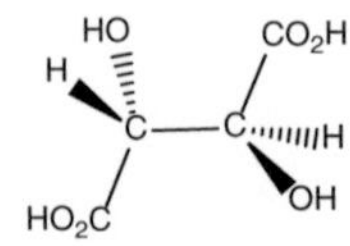

7b

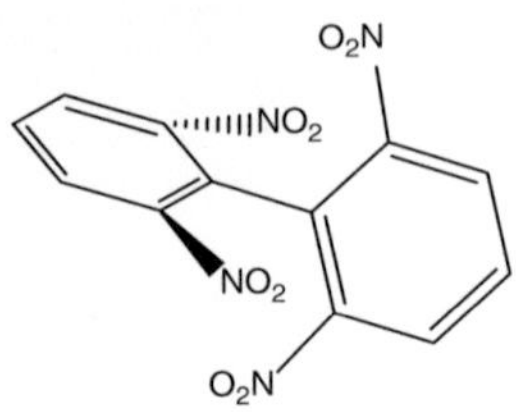

8

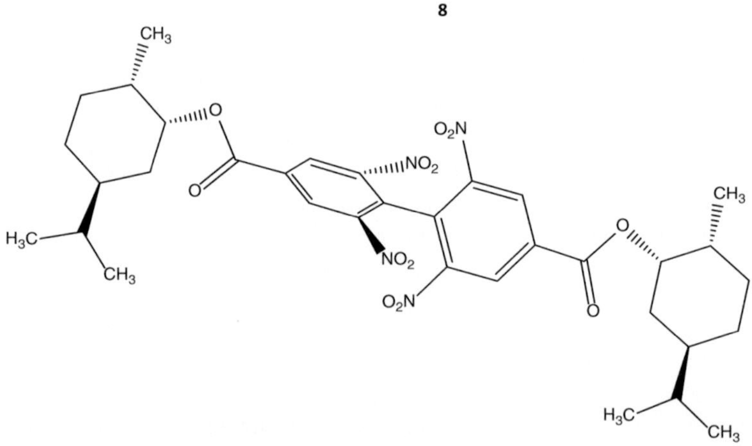

9

Scheme 5.2 Molecule **6**: hypothetical biphenyl derivative—a *meso* diastereomer lacking any symmetric conformation; **7a** and **7b** conformations of *meso*-tartaric acid having a horizontal plane of symmetry and an inversion axis, respectively; 2,2′,6,6′-tetranitrobiphenyl (**8**) is sterically relegated to a non-dissymmetric (D_{2d}) conformation. Mislow's complex molecule **9** substitutes **8** with (−)-menthyl and (+)-menthyl groups at the 4 and 4′ positions to produce an actual example of **6**

of dynamics, including conformations of the six-membered ring in **1**, are not an issue. In contrast, the non-resolvable optically inactive **9** is, in part, a result of its internal motions. This point will be emphasized further in our discussion of chiral centers other than carbon by Pope and over half a century later by Mislow. Further discussion of the significance of **1** and **9** may be found in the encyclopedic *Stereochemistry of Organic Compounds* [26].

In 1964, Kurt Mislow, accompanied by some of his NYU research group, accepted a full professorship at Princeton University. As Hugh Stott Taylor Professor of Chemistry, it is quite clear that Mislow was a very demanding research director {"uncompromising scholar" according to Dennis Dougherty, Cal Tech professor and Mislow Ph.D. student [15]}. Yet, further insight comes from the reflection by his former NYU colleague Professor David Schuster [16]. Schuster notes that when one of Mislow's most promising NYU students, Michael Glass, was suffering from terminal cancer, his Ph.D. advisor wrote his thesis so that Glass could obtain his Ph.D. posthumously. Mislow chaired the Princeton Chemistry Department from 1968 to 1974. He assumed emeritus status in 1988. He was the recipient of numerous awards and recognitions. Among these were election to the National Academy of Sciences (1972), the American Academy of Arts and Sciences (1974), the James Flack Norris Award in Physical Organic Chemistry (1975), AAAS Fellow (1980), first recipient of the Prelog Medal (1986), and the Arthur C. Cope Award (1995). The Prelog medal must have brought particular satisfaction since, according to Pierre Laszlo, Mislow's colleague during the 1960s as well as my own Ph.D. mentor, he was particularly admiring of the pioneering stereochemist and Nobel laureate Vladimir Prelog [27]. Professor Laszlo makes clear his own admiration of Mislow's efforts to place Princeton in the top rank of American chemistry departments [27]. Bart Kahr, Professor at NYU and former Mislow Ph.D. student, described him thusly: "His commitment to truth in science and righteousness in human relations was unwavering whether or not the matter was consequential" [15].

5.4 William Jackson Pope: A Very Brief Outline of Selected Career Research Impacts

Gibson very briefly summarizes William Jackson Pope's career research impacts as follows [2]:

I. Fundamental investigation of the relationships between chemical constitution and crystalline form.
II. Characterization of externally compensated substances in solid and liquid forms.
III. Determination of conditions governing the spontaneous resolution of sodium ammonium racemates.

IV. Experimental investigations leading to proof that elements other than carbon, i.e., nitrogen, sulfur, selenium, and tin having an asymmetric environment, can give rise to optically active compounds.

V. Establishment of conditions for the successful resolution of externally compensated compounds.

VI. Proof that a carbon atom having an asymmetric environment of four different atoms or groups, none of which need contain a carbon atom, is sufficient to determine the possibility of the compound being obtained in optically active forms.

VII. Establishment of the principle of enantiomorphism or mirror image isomerism, and hence of a compound exhibiting optical activity, in that molecular configuration need not be asymmetric, excepting a center of symmetry or a plane of symmetry, it need only be dissymmetric in the sense that it is not identical with its mirror image. In addition to **1**, a later example was the 5,5′-spirohydantoin **4** [28].

It is worthwhile, with respect to Point I, to refer Pope's discussion of enantiomorphism as it pertains to crystals of sodium chlorate or quartz [29]. Both of these substances form crystals that are enantiomorphs. However, in the melt or in solution they are optically inactive. Hence, the molecules (or salts) are themselves not dissymmetric but the arrangement in the crystals is the source of enantiomorphism. Pope found that crystallization of sodium chlorate in dextrose solution consistently favored one enantiomorphic crystal. Pope also addressed the contemporary ambiguity of the term "racemic" [30]:

> It is obvious, however, from a study of the literature of the subject, that this term racemic has no generally accepted definition …

He went on to discuss the contrasts in melting point behavior between what are now known as racemic compounds (crystals containing equal numbers of each enantiomer in the unit cell) and conglomerates (where each enantiomer forms separate crystals and eutectic mixtures occur). Clearly, Pope was an early pioneer exploring the fundamental bases of stereochemistry and contributing to its lexicon. Point VI was briefly mentioned earlier. The synthesis and optical resolution of chloroiodomethanesulfonic acid (**10**) by Pope and Read in 1914 established that a single asymmetric carbon was sufficient for optical activity [6, 8]. Although they had synthesized chlorobromomethanesulfonic acid six years earlier, they were not at that time successful in resolving the racemic mixture. Optical resolution of **10** (Scheme 5.3) demonstrated the point described in VI. Point VII was explicitly demonstrated by the synthesis of molecule **1**.

The research so briefly summarized in Point IV will form the body of the direct comparison between Pope's and Mislow's researches discussed in some detail in the present chapter. As noted earlier, the tools available to Pope in the late nineteenth and early twentieth centuries limited his studies to statical ones. The "proof

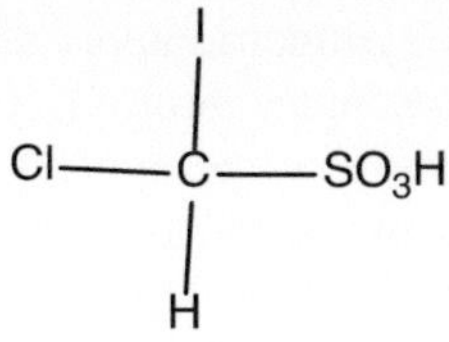

10

Scheme 5.3 Chloroiodomethanesulfonic acid (**10**): the first optically pure molecule containing a single asymmetric carbon

of the pudding," so to speak, relied upon clear resolution and characterization by optical rotation of enantiomers.

5.5 Kurt Mislow: A Very Brief Outline of Selected Career Research Impacts

Hargittai [14], Siegel [13], B. Hargittai, and I. Hargittai [17] have provided cogent summaries of Mislow's research accomplishments and these will be outlined even more briefly here.

I. Siegel provides an overall assessment of Mislow's "application of non-numerical mathematical models to conceptualize form and function."
II. Balancing steric effects and angle strain allowed further understanding of the limitations of the $4n + 2$ aromaticity rule in understanding the stabilities of monocyclic conjugated polyenes. Thus, while benzene ("[6]-annulene") has ideal hexagonal geometry and is strainless, its congeners [10]-annulene and [14]-annulene, while also satisfying the $4n + 2$ rule, are very destabilized by repulsion between non-bonded hydrogen atoms and or ring strain.
III. Restricted rotation in *ortho*-substituted biphenyls with larger substituents (e.g., **8**) directly related dynamics to racemization. Dynamic nuclear magnetic resonance as well as computational chemistry (i.e., molecular mechanics) were tools developed and exploited in the Mislow research group.
IV. Restricted rotation in more complex species such as "propeller molecules" opened the door to a plethora of stereoisomers whose observation via dynamic NMR depended upon the magnitude of the rotational barriers and the temperatures of observation.
V. Development of precise classification of stereoisomers and a lexicon of stereochemical relationships, for example, homotopic, enantiotopic, and diastereotopic groups.
VI. In collaboration with Frank A. L. Anet and their students Steve S. Miura and Jay Siegel, these workers provided a chemical application of the long-time "parlor entertainment" termed "La Coupe de Roi" (Fig. 5.3) [31]. Actually, they demonstrated a "reverse Coupe de Roi" as depicted in Scheme 5.4. At

first blush, cutting a fully symmetric apple selectively into two right halves or two left halves seems counter-intuitive. However, the "chiral agent" is the person who determines how the apple is cut. Thanks to steric effects, the two metacyclophanes in the chemically resolved enantiomeric **11A** or **11B** cannot rotate past each other. Dimerization of optically pure **11A** or optically pure **11B** provides achiral **12** (C_{2v} symmetry). In contrast, condensation of **11A** with **11B** provides achiral **13** (C_{2h} symmetry). Five years later, Cinquini et al. reported the first example of an achiral molecule (**14**, C_{2v} symmetry) cleaved chemically into homochiral halves (3 steps, Scheme 5.5) [32]. The first of two Baeyer–Villiger reactions provides a racemic mixture where enzyme-catalyzed resolution (not a parlor trick by a human being) provides chiral discrimination to generate **15**. The second Baeyer–Villiger reaction provides **16**, readily separated from its achiral diastereomer, and subsequently reduced to optically pure **17**.

VII. Investigation of the dynamics of chiral compounds of nitrogen, phosphorus, sulfur, and arsenic will be the subject of the next section. This included the introduction of graphs mapping static and dynamic relationships, including pseudorotation, between stereoisomers of third-row elements. For example, while pentacoordinate carbon in a nucleophilic substitution reaction is a transition state on its reaction coordinate, nucleophilic substitution on a phosphorane yields a five-coordinate intermediate that can scramble substituents.

Fig. 5.3 "La Coupe de Roi" illustrated by two chiral halves each of two apples cut clockwise and counterclockwise *Photo* Susan and Arthur Greenberg

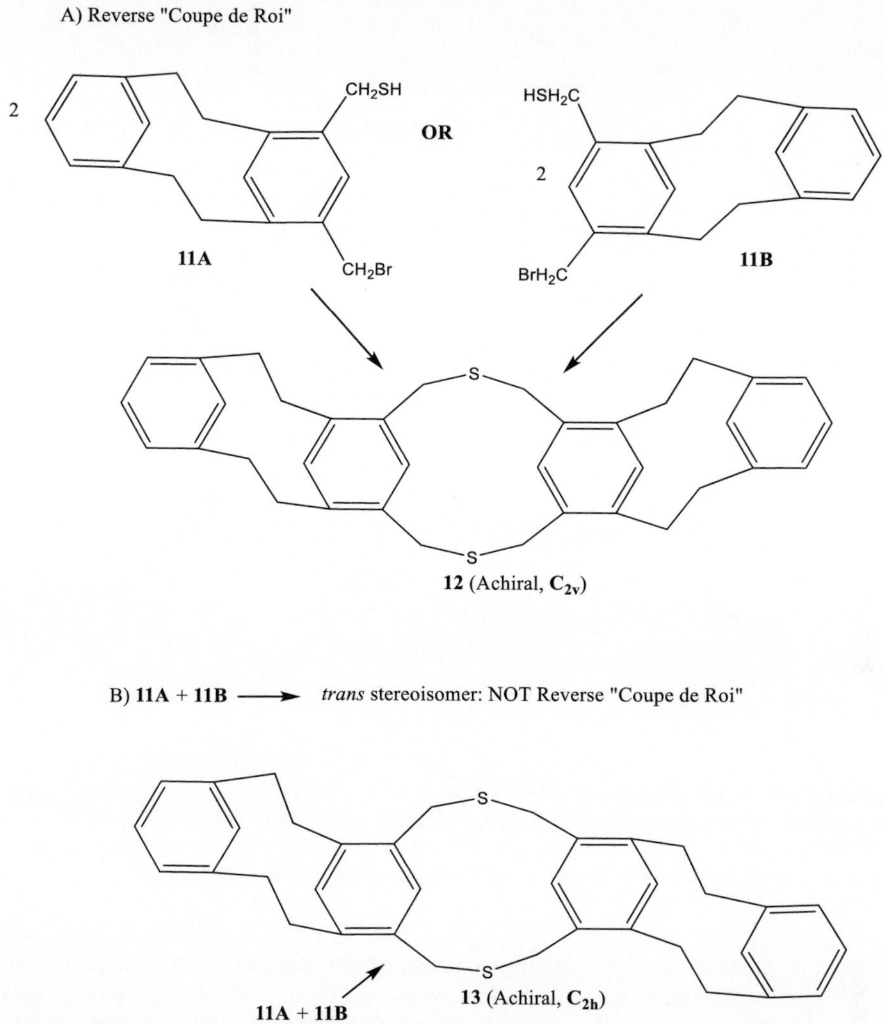

Scheme 5.4 Example of a synthetic reverse "Coupe de Roi," yielding an achiral (C_{2v}) molecule (**12**), employing homochiral, sterically constrained components (two **11A** or two enantiomeric **11B**), in contrast to a non-reverse "Coupe de Roi," combining the two enantiomeric components (**11A** plus **11B**) yielding an achiral (C_{2h}) molecule (**13**)

VIII. Studies of hindered 1,1′-binaphthyl compounds and chiral centers such as phosphorus have provided a fundamental basis for generations of dissymmetric catalysts crucial to the isolation of optically pure substances including catalysts widely applied in organic chemistry, organometallic chemistry, and the pharmaceutical industry.

IX. A lifelong interest in molecular topology in smaller molecules as well as biopolymers including DNA actually started quite early. In 1947, shortly after arriving at NYU, in a letter to Pauling [22], Mislow speculates about

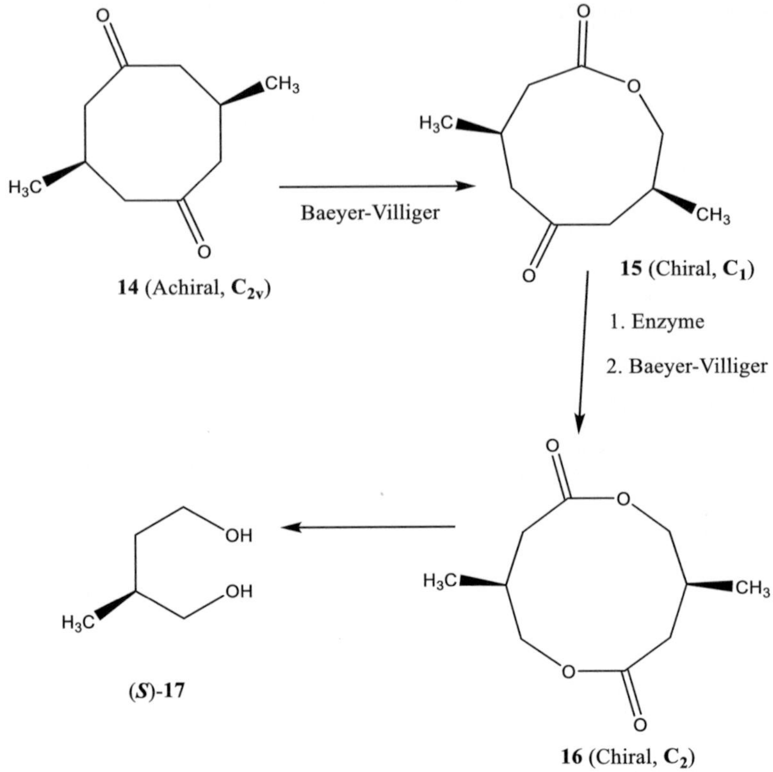

Scheme 5.5 Actual synthetic "Coupe de Roi" starting with *cis*-3,7-dimethylcyclooctane-1,5-dione (**14**, achiral, C_{2v}), performing a Baeyer–Villiger oxidation to generate racemic **15** (C_1), resolved enzymatically. Subsequent Baeyer–Villiger oxidation of a single enantiomer yields optically pure **16** (C_2) which is converted to (*S*)-2-methyl-1,4-butanediol [(*S*)-**17**, C_1]

the possibility of generating a catenane by starting with a large monocyclic molecule such as cycloeicosane [$(CH_2)_{20}$], threading with a long difunctional molecule and ring closure. Some twenty years later, success in directed, as opposed to statistical, syntheses [33] started to play a significant in the rise of nanotechnology.

5.6 Static and Dynamic Stereochemistry at Chiral Centers Other Than Carbon

A. Nitrogen

There is a very comprehensive published presentation of studies of spatial arrangements around nitrogen from 1874 through 1914 [34]. One significant source of confusion was the seeming duality of nitrogen valence: three in ammonia (NH_3) and, apparently, five in ammonium chloride (NH_4Cl). Lewis structures and lone pairs were as-yet unknown. This section will, however, limit itself to some of the contributions made by Pope and Mislow.

Pope and Peachey were the first to unambiguously resolve asymmetric ammonium salts [35, 36], thus proving that carbon was not the only asymmetric center. John Read was acknowledged for his experimental contributions to this breakthrough. This had been a challenge during the previous two decades and Pope and Peachey realized that the problems occurred due to the attempts to isolate ammonium hydroxides ($HONR_1R_2R_3R_4$) as well as working in aqueous solutions. The inability to resolve such compounds led Le Bel, for example, to postulate the possibility of intramolecular motion of substituents on nitrogen in such compounds, thus permitting self-racemization. This reasoning was dismissed by Pope and Peachey who speculated that part of the problem might have been dissociation and reassociation to produce racemic mixtures. They also rejected the Kekulé notion of an "additive compound" [e.g., $NH_3 \cdot HCl$]. The success of Pope and Peachey in resolving salts (bromides, iodides) of α-benzyl phenyl alkyl methyl ammonium (**18**) was achieved by employing (+)-3-bromocamphorsulfonic acid as resolving agent and then exchanging the anion with halides. This proved to be an extremely successful reagent for some of Pope's other optical resolutions. Isolated crystals of the bromide were found to be enantiomorphs. Since these salts were clearly ionic, Pope and Peachey [36] made the case that the bonding about nitrogen could be envisioned schematically as **19** making some provision for the "free" bond. This was still a version of the quinquivalent nitrogen atom. With similar models for sulfur (**20**) and tin (**21**) (see Scheme 5.6), Pope and Harvey commented as follows [36]:

> The abnormal condition of an asymmetric atom of which one bond is free seems to be accompanied by no particularly marked tendency towards optical inversion.

Tin compound **21**, as depicted below, is seemingly an anomaly and will be discussed in the section on Group IV compounds. The "free" bond mentioned refers to that between the ammonium ion and the counterion. Kauffman and Bernal [37] detailed an apparent total absence of direct communication between Pope and Alfred Werner (1866–1919) that could have advanced both of their research agendas. It is ironic since John Read began his research with Werner in Zurich in 1905, completed his Ph.D. in 1907, and spent the next eight years working with Pope [3].

$C_6H_5CH_2N(CH_3)(C_3H_5)(C_6H_5)I$

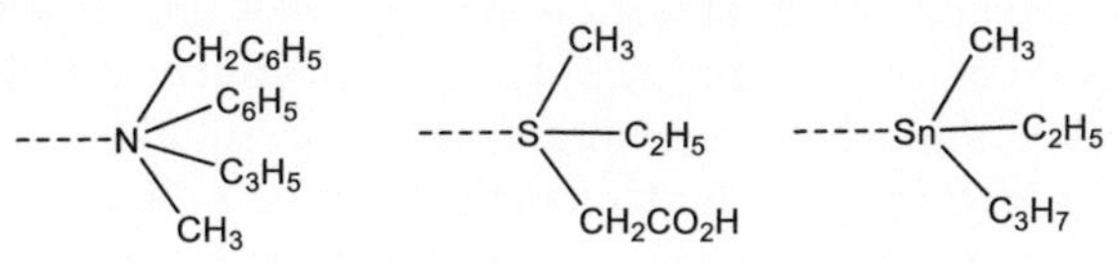

18 **19** **20** **21 (?)**

Scheme 5.6 Early representations of pentavalent nitrogen (**18** and **19**), tetravalent sulfur (**20**), and tetravalent tin (**21**)

In 1890, Werner and his mentor Arthur Hantzsch (1857–1935) published a tetrahedral structure for the nitrogen in ammonium salts [34, 38]. A decade later, Werner appears to have been unaware of Pope's resolution of asymmetric ammonium salts that would have supported the Hantzsch-Werner model [37]. In turn, Pope appears to have been unaware of the Hantzsch-Werner work which would have helped clarify his postulated structure [37]. In addition, had Werner been aware of Pope's (+)-3-bromocamphor-9-sulfonic acid as a highly successful resolving agent, his own optical resolutions might have occurred more rapidly. In 1911, Werner finally employed this reagent to resolve *cis*-bromobis(ethylenediamine)cobalt(III) salt as well as the *cis*-chloro derivative [39]. Of course, Werner's coordination theory led to his 1913 Nobel Prize. During the first decades of the twentieth century and in prior years, clean optical resolution was truly the *sine qua non* of proving chirality. In 1916, Walter Kossel provided graphical proof of the octet rule [40]. One month later, Gilbert N. Lewis depicted the structure for ammonia as three electron pairs (single covalent bonds) shared with hydrogen and one unshared pair which, when shared with a proton, clearly led to four equivalent N–H covalent bonds in ammonium and electrostatic attraction with a counter ion such as chloride [41]. Tetrahedral geometries in tetramethyl- and tetraethylammonium ion salts were finally unequivocally proven in 1928 by the X-ray studies of R. W. G. Wyckoff [42].

All the above-cited works were static structural studies since there were no instrumental methods to observe intramolecular motion. Late nineteenth- and early twentieth-century representations of ammonia and other three-coordinate nitrogen compounds frequently suggested nitrogen at the center of an equilateral triangle [34]. Thus, any NXYZ amine would be assumed to be achiral. This was consistent with the fact that no NXYZ amine had ever been optically resolved during this early period. In 1934, microwave study of ammonia indicated pyramidal geometry [43] and rapid inversion and it started to became clear that chiral amines were racemizing at rates far too fast for resolution under ambient conditions. The occupancy of a non-bonded electron pair at the corner of a tetrahedron in ammonia and other simple amines was one consequence of the intellectual synthesis by Gillespie and Nyholm [44]. Treating the non-bonded pair as a space-occupying substituent rationalized simple NXYZ molecules as chiral. Typically, simple acyclic amines have inversion barriers of ca. 5 kcal/mol and may be observed using microwave or infrared spectroscopies [45]. They invert far too rapidly for the dynamic NMR studies that came of age during the 1960s. Nonetheless, a very clever experiment by Saunders and Yamada [46], monitored the extremely rapid inversion of **22** via NMR through its equilibrium in acidic solution with its conjugate acid **22–H$^+$**. The exchange of diastereotopic benzylic protons in **22–H$^+$** provided the handle for monitoring inversion occurring in the low steady-state concentration of **22** (Scheme 5.7). The first amine to be optically-resolved was Tröger's base (**23**) by Prelog in 1944 (Scheme 5.7) [47]. The bridged system fixed the dissymmetric structure, although dilute acid opened the methylene bridge leading to racemization by inversion of the eight-membered ring, followed by ring closure. Lehn's systematic studies and comprehensive review noted the increase in amine inversion barriers upon substitution by electronegative atoms or groups and very large increases upon incorporating nitrogen in alkyl-substituted three-membered aziridine rings (ca. 19–20 kcal/mol). There is even the ability to kinetically resolve amines with inversion barriers higher than ca. 23 kcal/mol (e.g., oxaziridines) under ambient laboratory conditions [45].

Physical organic chemistry first assumed significance during the 1930s and 1940s and, in a sense, Pope's most productive research period largely predated its emergence. Mislow creatively employed the methodologies of physical organic chemistry as well as the revolutions in instrumentation and computational chemistry. During the 1960s and 1970s, among Mislow's interests were comparison of the similarities and differences between nitrogen inversion dynamics with inversion dynamics in phosphorus, arsenic, and sulfur molecules [48]. Dynamic NMR was employed to investigate a linear free energy relationship (Hammett plot) in a series of 1-aryl-2,2-dimethylaziridines [49]. Aryl groups attached to the nitrogen lower inversion barriers by some 7 kcal/mol. This brought useful monitoring temperatures down to around −60 °C—very accessible for NMR study. Electron-withdrawing substituents on the aryl substituent have a dramatic rate-accelerating effect ($\Delta G^{\ddagger}_{-60}$ ranged from 12.5 to 8.2 kcal/mol; $\rho = 2.8$). Interestingly, one might have expected as hybridization of N goes from sp^3 to sp^2 in the planar transition state, that electron-withdrawing substituents would raise the inversion barrier. However, the authors noted increased delocalization of the lone pair onto the aryl group at the transition state which should be enhanced by electron-withdrawing substituents.

B. Phosphorus and Arsenic

In 1911, Meisenheimer and Lichtenstadt accomplished the first isolation of a chiral phosphorus compound, ethylmethylphenylphosphine oxide (**24**), using (+)-3-bromocamphor-9-sulfonic acid [50] (Scheme 5.8). In 1912, Pope and Gibson [51] were unsuccessful in resolving asymmetric phosphonium salts and arsonium

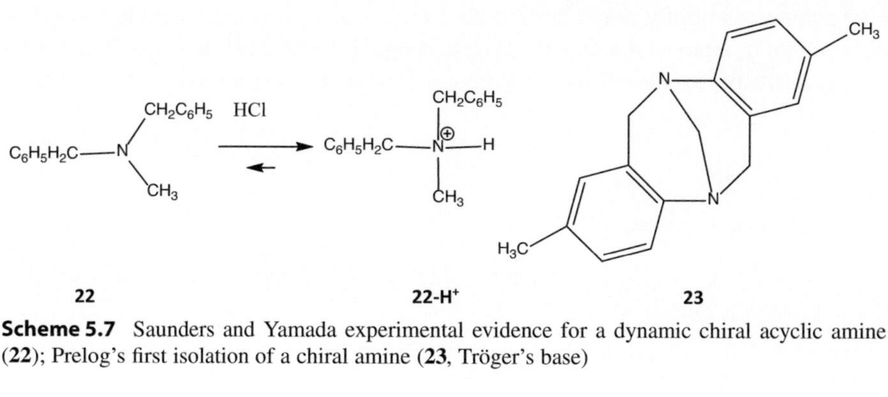

22 **22-H⁺** **23**

Scheme 5.7 Saunders and Yamada experimental evidence for a dynamic chiral acyclic amine (**22**); Prelog's first isolation of a chiral amine (**23**, Tröger's base)

salts [37]. Although they had separated the (diastereomeric) bromocamphorsulfonate salts of **25** (Scheme 5.8), for example, they could not resolve the derived enantiomeric iodides. The first quaternary phosphonium salt was resolved in 1947 [52]. Resolution of chiral phosphines, however, remained an elusive goal for decades [53, 54]. They were sought for their potential as chiral agents in organic chemistry, but especially in organometallic chemistry where the ability to complex with metals provided the opportunity to create chiral organometallic compounds useful as catalysts. Complications arise due to the ease of oxidation to form the corresponding phosphine oxides: the P=O bond is a very strong one. Moreover, as described below, while phosphines have much higher inversion barriers to racemization, they have finite lifetimes under laboratory conditions. For example, methylphenyl-*n*-propylphosphine racemizes in decalin at 130 °C with a half-life of 301 min, corresponding to a barrier of 28–30 kcal/mol [55]. The first reductions of a chiral phosphine oxide to a chiral phosphine were published by Horner and Balzer in 1965 employing trichlorosilane and they discovered the remarkable fact that in the weak base pyridine retention predominated while in the strong base triethylamine inversion predominated [56]. Once again, exploiting (−)-menthol, Mislow's group developed enantiospecific synthesis of optically-active phosphine oxides using the Grignard pathway illustrated in Scheme 5.9 [57, 58], a method pioneered by Anderson in the synthesis of optically active sulfoxides (see below). The absolute configuration of ***(R)*$_P$-26** had been established earlier by X-ray crystallography and configuration correlation with ***(R)*$_P$-27** proved that this reaction proceeded with complete inversion. (Note that according to Cahn-Ingold-Prelog rules, oxygen doubly-bonded to phosphorus is treated as a single P–O bond, hence lower priority than the O–Men substituent.) Reduction of ***(R)*$_P$-27** to phosphine ***(S)*$_P$-28** and its oxidation back to ***(R)*$_P$-27** involved no change in stereochemistry at phosphorus. Mislow's contributions to this rapidly developing field included the introduction of hexachlorodisilane and perchloropolysilanes as reducing agents and clarification of the retention and inversion conditions including in chiral phosphetane oxides [59–62].

Mislow's group continued during this period to research the effects of substitution on the inversion barriers of tri-substituted phosphines including chiral phosphines. As briefly noted earlier, the barriers are considerably higher in phosphines than in amines. A range of $\Delta G^{\ddagger}{}_{130}$ roughly 29–36 kcal/mol was found for a series of substituted phosphines; high enough to permit separation of enantiomeric

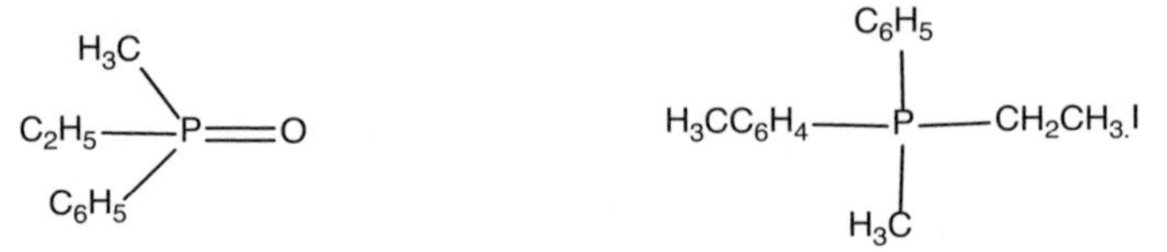

24 **25**

Scheme 5.8 Ethylmethylphenylphosphine oxide (**24**): the first resolved chiral phosphorus compound; a failed attempt at resolving a chiral phosphonium saly (**25**)

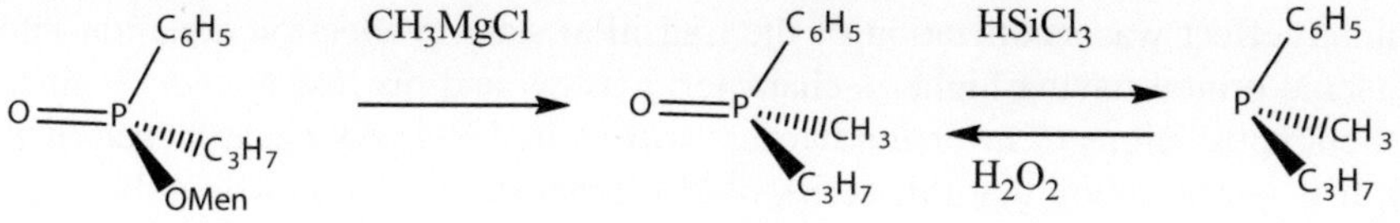

(R)P-26 **(S)P-27** **(R)P-28**

Scheme 5.9 Mislow's exploitation of the Grignard reaction of a diastereomeric phosphine oxide to isolate an optically active phosphine with a high inversion barrier

"invertomers" under ambient conditions [63]. The effects of the substituents attached to phosphorus were significantly weaker than those in amines but in the same direction: aryl groups attached to phosphorus lowered the barrier by ca. 2–4 kcal/mol per attached phenyl group and enhanced electron withdrawal by the aryl substituent (e.g., *p*-CF_3) lowering the barrier by less than 2 kcal/mol. These are similar effects to those observed in the inversion rates of sulfoxides found earlier and described below. A special case was phosphole (**29**) where the inversion barrier was reduced to 16 kcal/mol attributed to aromatic stabilization (2p–3p) π overlap in the planar transition state [64] (Scheme 5.10).

Tetracoordinate carbon-containing compounds undergoing nucleophilic substitution reactions (classical S_N2) pass through a five-coordinate transition state enroute to products. In contrast, third-row elements such as phosphorus accommodate five-coordinate species as reaction intermediates. Five-coordinate phosphorus is known to undergo pseudorotation, wherein equatorial and axial substituents change positions. Mislow employed graph theory to trace the stereochemical consequences for acyclic as well as cyclic phosphorus compounds of different ring sizes [65].

The first inversion barrier for an arsine (**30**) (Scheme 5.10) was determined via polarimeter to provide $\Delta G^{\ddagger}{}_{217.6} = 42.4 \pm 0.5$ kcal/mol, corresponding to $t_{1/2}$ at 200 °C of approximately 740 h [66]. A very striking finding was the discovery that substitution of an alkyl group attached to arsenic by a trimethylsilyl substituent lowered the inversion barrier by some 18 kcal/mol [67]. Indeed, the effects of electropositive substituents on phosphorus and arsenic were parallel [68]. Here the

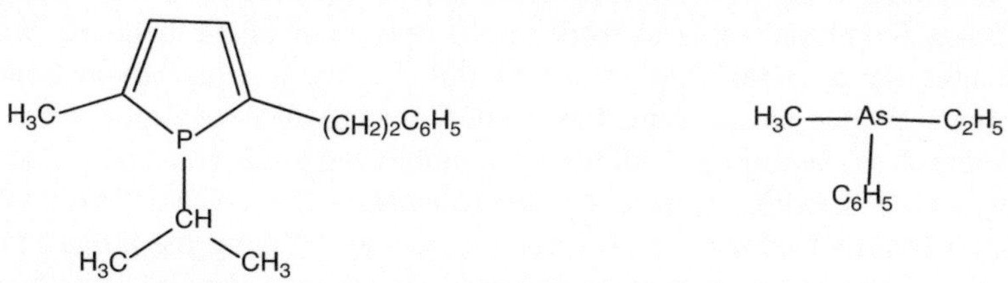

29 **30**

Scheme 5.10 A substituted phosphole (**29**) has a relatively low inversion barrier due to aromatic stabilization of the planar transition state. Arsines, such as ethylmethylphenylarsine (**30**), have considerably higher inversion barriers than the corresponding phosphines

dominant effect was stabilization of the transition state by electron donation into an arsenic orbital having higher s-character. Mislow and his researchers obtained excellent correlations of inversion barriers with Allred-Rochow electronegativities [68]. It is worth noting that this series of electronegativities relates directly to the degree of shielding experienced by valence electrons and would appear to relate most simply to the ability of a substituent to "lose" (donate) electrons. While the Allred-Rochow electronegativities are clearly not orthogonal to other scales, the Pauling scale, for example, is derived from bond energies related to polar covalence. Lambert and Jackson [69] lowered the energy of activation for arsenic inversion to 27 kcal/mol through observation via dynamic NMR of the interconversion of *meso*- and (*R*, *S*)-1,2-dimethyl-1,2-diphenyldiarsine. They attributed the effect as resulting from (p–d) π stabilization.

C. Sulfur and Selenium (also Oxygen)

A very helpful review on chiral sulfur compounds was published by Barbachyn and Johnson [70]. The first resolution of an optically active sulfur compound was reported by Pope and Peachey in 1900 [71]. The compound they isolated in optically pure form was obtained through anion exchange employing platinic chloride ($PtCl_6$) and the structure was depicted as **31** (Scheme 5.11). Much as ammonium salts were rationalized as requiring pentacoordinate nitrogen, the structure depicted by Pope and Peachey involved "truly tetravalent sulfur" with presumably the tetrahedral structure of tetravalent sulfur. Noting the poor solubility in nonpolar solvents and solubility in water, they understood that it was an ionic salt. The modern depiction of this salt is **31A**, having a central tricoordinate sulfur. Two years later, Pope and Neville resolved the first dissymmetric selenium compound, **32** [72], insoluble in water or alcohol but very soluble in acetone. As foul smelling as volatile sulfur compounds are, selenium compounds are even more notorious and Pope and Neville noted the effect graphically [72] and thus a much less volatile compound, methylphenyl selenide, was employed as starting material for the successful synthesis of **32**. However, these researchers remained interested in what effect upon optical rotatory power would "merely" exchanging selenium for the sulfur atom of **31** cause? The story of their work with ethylmethylselenide is related with gusto by John Read [3]: …on account of the "sensational odor, it was found necessary to work on the open roof of the building, with the wind behind the operator." Unfortunately, this was the day of Darwin Centenary celebrations, garden parties ruined as "demoralizing whiffs of vapor swept down upon defenseless Cambridge." All noses eventually led to the University Chemistry Building and the headlines appeared in the *Cambridge Daily News*: "**What Was It? Suspected Drains Exonerated. Science the Sinner**." Taking the project to open fen country, Pope and Neville received permission from a farmer to conduct their experiments on his property, on the bank of the nearby river. Unimpressed by the promise of "a thundering big stink," the farmer offered the researchers the opportunity to smell the farm's muck heap. In the end, a very brief visit to the site of the experiment convinced the farmer that science ruled the day. Another interesting surprise awaited Pope and Neville. Starting with either the *d*- or *l*-salt of **32** with *d*-bromocamphorsulfonate, exchanging anions in a KI solution of mercury(II)iodide, they discovered that the solutions became optically inactive. The analogous anion exchange with the corresponding sulfur compound also produced racemization. Their explanation involved postulation of symmetrical hexavalent octahedral sulfur and selenium compounds **33** and **34** (see Scheme 5.12). Similar treatment of the optically active ammonium salt **18** did not cause racemization. The authors dismissed the possibility of pentavalent nitrogen becoming heptavalent nitrogen with a pentagonal bipyramidal structure. Ultimately, the future explanation of the racemization mechanisms would depend upon formation of free trivalent sulfur or selenium and inversion on the laboratory time scale.

In 1925, Phillips reported the first optical resolution of a sulfinate, **35** [73], and in 1926 the first optical resolution of a sulfoxide, **36** (Scheme 5.13) [74]. Since Lewis structures and the octet rule were only about a decade old, these molecules introduced a significant problem. The conventional Lewis structure of **35**, for example, might include the double bond between sulfur and oxygen depicted in the structure below. However, sulfur would be involved with ten valence shell electrons, hence violating the octet rule. A partial solution to this problem was postulation of a "semipolar double bond," depicted in **35A**, which combines electrovalency with covalency. However, there still remained the problem of enantiomerism. Phillips rationalized this with the tetrahedral structure **37** for the sulfoxide (as well as the sulfinate) in which sulfur occupies one of

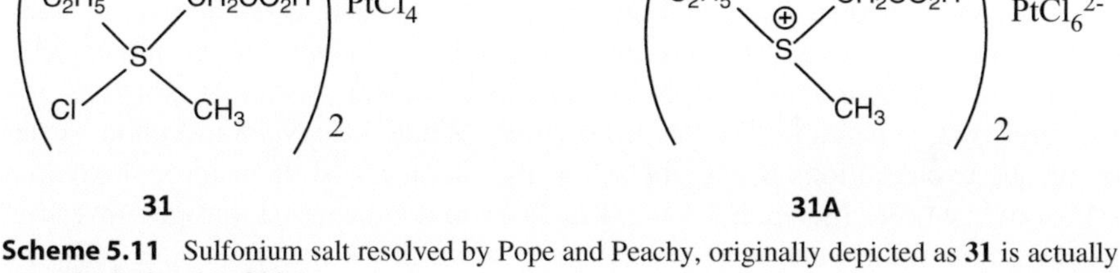

31 **31A**

Scheme 5.11 Sulfonium salt resolved by Pope and Peachy, originally depicted as **31** is actually properly depicted as **31A**

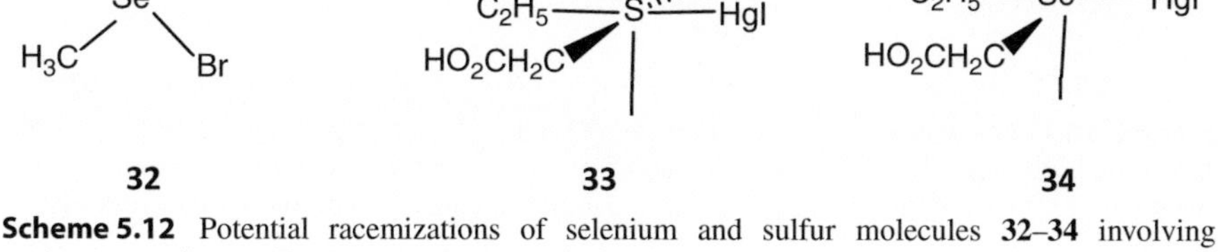

32 **33** **34**

Scheme 5.12 Potential racemizations of selenium and sulfur molecules **32–34** involving exchange reactions with mercury(II)iodide in KI solution

the apices of a tetrahedron (Scheme 5.13) [74]. This question, of course, never seriously arose over dissymmetric amines because none had ever been optically resolved.

Anderson pioneered the Grignard pathway to enantio-pure sulfoxides [75] that was later applied to phosphine oxides. Employing the separable diastereomeric salt (−)-menthyl (−)-*p*-toluenesulfinate, reaction with ethylmagnesium iodide, provided (+)-ethyl *p*-tolylsulfoxide. Mislow and co-workers [76] later rigorously established the mechanistic details first logically postulated by Anderson. Mislow and co-workers began exploring the dynamics of dissymmetric sulfur over six decades ago. While thermally induced racemization had been known for many years, Mislow et al. explored HCl-catalyzed racemization of sulfoxides [77]. Their interesting conclusion involved addition of chlorine to the apical positions of a trigonal bipyramid producing a non-dissymmetric transition state or intermediate. Subsequent work was more focused upon dynamics. Thus, the enthalpy of activation for racemization, demonstrated to be first-order intramolecular inversion, was 26 kcal/mol for 1-adamantylethylmethylsulfonium perchlorate [78]. While this value is slightly lower than the inversion barrier for phosphines, clearly solvent effects would be expected to play a larger role in the ion. In contrast, the racemization (inversion) barriers for the sulfoxides (35–42 kcal/mol) are somewhat higher than those of the phosphines [79] and considerably higher than for the sulfonium salt. Observing the effects of *p*-phenyl substituents upon the rates of inversion, electronic effects were relatively weak (electron-withdrawing *p*-substituents very slightly increasing rates) and steric effects played a larger role [79].

Oxonium ions (e.g., R_3O^+) are reactive intermediates that have been known for many decades and with the advent of super acid solutions and low-temperature NMR studies they have been observed spectroscopically. While VSEPR theory had long predicted trigonal pyramidal geometry about oxygen, crystallographic study of the water network in ice was for a time the only confirmation, albeit complicated by the hydrogen-bonded associations. Trigonal pyramidal structure was validated by Lambert and Jackson in 1968 [80]. While the oxygen inversion barrier of acyclic oxonium ions is extremely low, they synthesized the isopropyloxonium salt of oxirane (**38**) (Scheme 5.14). By analogy to the increased nitrogen-inversion barriers in aziridines due to increased angle strain in the transition state [45], the

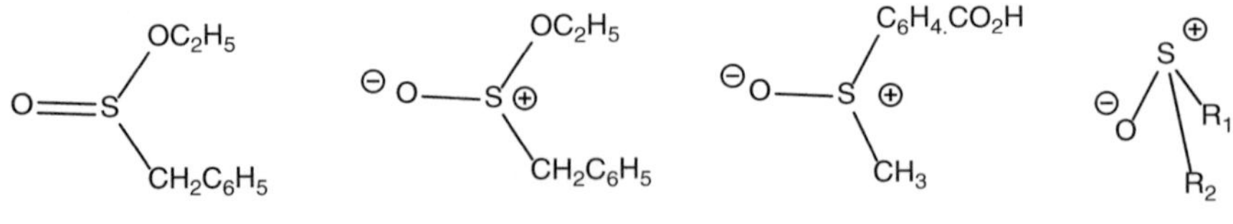

35 **35A** **36** **37**

Scheme 5.13 First optically resolved sulfinate (**35**) and the first optically resolved sulfoxide (**36**). A "semipolar double bond" (**35A**) was a means to satisfy the octet rule in **35**. One way to rationalize enantiomerism in sulfoxides was to postulate a tetrahedral structure with sulfur at an apex (**37**)

same effect was anticipated in this oxonium ion. At −50°, they observed coalescence corresponding to an inversion Arrhenius activation energy of 10 ± 2 kcal/mol. In 2023, computational design coupled with targeted synthesis provided the chiral oxonium ion **39** in which the chirality is due entirely to pyramidal asymmetric oxygen stabilized by π-complexation with a benzene ring as well as steric hindrance [80] (Scheme 5.14). The oxygen inversion barrier is over 26 kcal/mol [81].

A very helpful and accessible review has been published for the carbon congeners silicon, germanium, tin, and lead [82]. Kipping, Pope's former associate, optically resolved the first silicon compound, benzylethylpropylsilicol, in very low yield following a very laborious process [83]. Sommer and Frye successfully synthesized **40** and **41** (Np = α-naphthyl) in 1959 and demonstrated substitutions with pure retention and pure inversion [84]. The first optical resolutions of dissymmetric germanium compounds (**42**) were reported in 1963 [85, 86], using diastereotopic (−)-menthol salts (Scheme 5.15).

In 1900, Peachey and Pope claimed isolation of the first optically active tin compound (**43**) [87, 88] (Scheme 5.16). However, in 1928, Kipping, Pope's earlier associate, was not able to optically resolve a number of asymmetric tin compounds [89]. This brings us back to the curious representation of a tetravalent tin compound depicted by Pope as **21**. The mystery was solved by Peddle and Redd [90]. Employing the diastereotopic methyls in **44** and dynamic NMR, they found that the molecule retained optical activity in non-polar solvents, but in DMSO-d_6, racemized (Scheme 5.16). The racemization was concentration-dependent, hence bimolecular collision, exchange was occurring leading to a net inversion of configuration. However, no such racemization occurred in the tetraalkyltin **45** up to at least 160 °C [90]. This explained the curious representation **21** and the lack of reproducibility of the Peachey and Pope claim for **43** [91]. Indeed, the weak

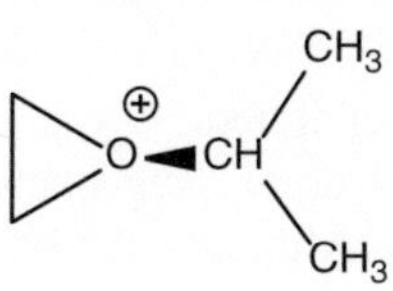

38

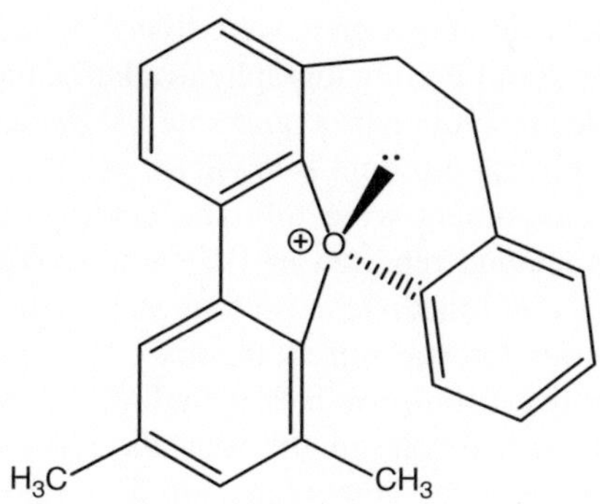

39

Scheme 5.14 Oxonium ions are exceedingly short-lived. Just as aziridines have higher barriers to inversion due to increased strain in the transition state, the oxiranyl ring in **38** allowed NMR observation of racemization at −50 °C. The cleverly designed oxonium ion **39** is observable under ambient conditions

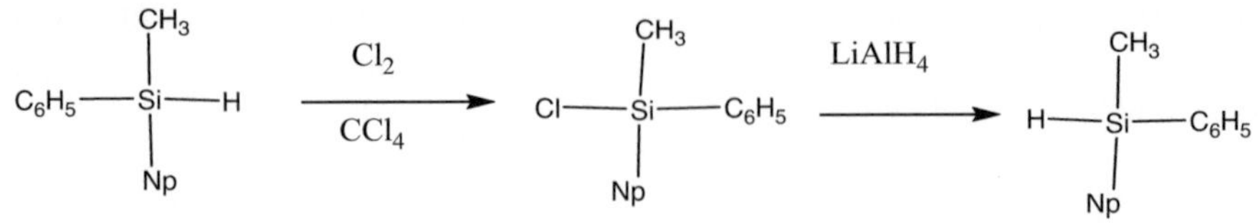

(+)-40 (-)-41 (-)-40

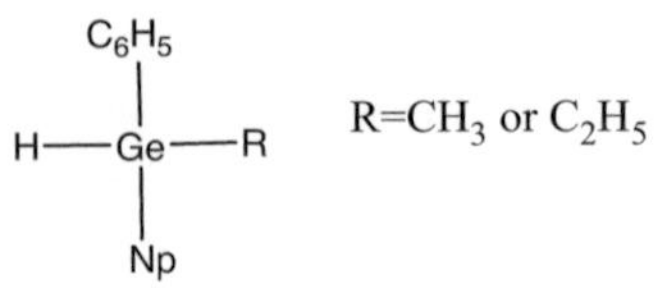

42

Scheme 5.15 Demonstrations of inversion and retention in optically active silicon compounds (**40** and **41**). The first optically active germanium compound (**42**)

Sn-I bond is rapidly exchanged with other anions in aqueous solution. Pope and Peachey had made the unusual observation of crystallizing one diastereomeric salt from the mother liquor but boiling solvent off left only optically inactive substances. Their rationalization was that heating the residue caused its racemization. However, the inability to repeat their work [91] ultimately discredited it. Interestingly, during this period, there were many who thought that dissymmetric carbon compounds with a carbon-halogen bond could not be resolved [92]. Indeed, one could have been tempted to write structures such as **46** (Scheme 5.16), analogous to **21**, for chiral halogen derivatives. The uncertainty remained until over three decades into the future, when evidence accumulated for the S_N1 reaction and concomitant racemization through planar carbocations. Although dissymmetric tetraalkyllead compounds were first reported in 1917 [93], with no reactive "handles" or chiral chromatography available, they were not resolved. In 1933, Austin attempted resolution of a quaternary chiral lead via diastereomeric isomers of compound phenyl-*n*-propyl-*o*-tolyllead *p*-(*sec*-octylphenylether) [94]. Unfortunately, the diastereomers were oils and could not be reliably separated. A quaternary lead compound reported in 1979 was monitored via NMR employing potentially diastereotopic methylene protons and methyl protons [95]. Although the possibility of inversion was raised, the results were ambiguous. An X-ray crystal structure showed five-coordinate lead including complexation with an amino group that was a component of one of the four alkyl substituents. Apparently, compounds with dissymmetric lead atoms have not been optically resolved to this day.

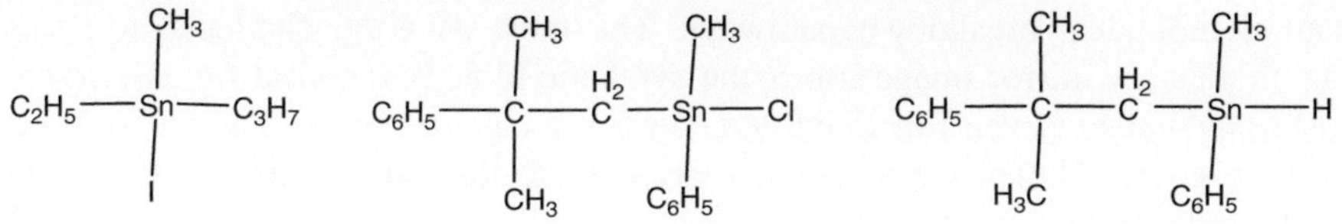

43 44 45

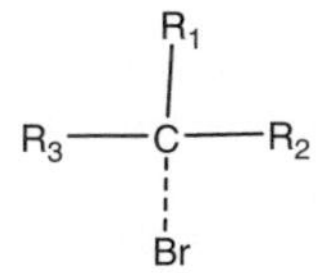

46

Scheme 5.16 Reported (1900) resolution of the tin iodide **43** could not be reproduced (1928). The mystery was solved when it was reported (1970) that the Sn–Cl bond in **44** could ionize in DMSO leading to racemization while no such racemization was reported for **45**. This harkened back to earlier hypotheses of a labile C–halogen bond in a molecule such as **46**

5.7 An Additional Note

William Jackson Pope was officially nominated three times for the Nobel Prize in chemistry [96, 97]. Nominators included Silvanus Thompson (1901—the first Nobel Prize year), J.-H. van't Hoff (1907, by the awardee of the first Nobel Prize), and Thomas Lowry (1927). Three other nominations (1915, 1916, and 1932) were declared invalid by the Nobel Committee [96]. Nominations are sealed for fifty years. Hence nominations for awards announced in 2024 cover years through 1973, roughly a midpoint in Kurt Mislow's career. This was a period when the number of excellent scientists continued to grow exponentially.

5.8 Coda: No "*Laevo* Man". But "*Laevo E coli*"?

Reference was made earlier in this chapter to Pope's amusing "*Laevo* Man" living (probably very briefly) in a "*Dextro* Man" world. This fanciful conjecture was included in "The bearings of geometry on the chemistry of fermentation," the four-part Cantor Lectures of 1901 [98]. Over 120 years later should we begin to worry about the artificial creation of "mirror life" and, specifically, "*Laevo*" bacteria?

A four-page opinion piece in the Policy Forum section of *Science*, titled "Confronting risks of mirror life," co-authored by 38 distinguished scientists representing a wide array of relevant disciplines, comments on the feasibility and risks of creating "mirror bacteria" [99]. It notes that chemical syntheses of complex enantiomeric molecules, including kilo-base nucleic acids and large functional proteins are already feasible and their mirror-image versions are likely to resist natural biological breakdown pathways. The most daunting challenge to creating elementary mirror image life is the synthetic biology required for assembling functional cells or even functional structures such as ribosomes. The authors suggest these developments are at least a decade into the future. Unlike *Laevo* Man, who would require mirror nutrients to survive, *E. coli* for example, is known to thrive in media lacking natural chiral nutrients. The opinion piece speculates about potential impacts of "immunity invasion," "ecosystem invasion," "biosafety," and "biosecurity." It reaches the following conclusion:

> The potential risks of creating mirror bacteria cannot be justified by the relatively limited benefits.

The Recommendation

> Unless compelling evidence emerges that mirror life would not pose extraordinary dangers, we believe that mirror bacteria and other mirror organisms, even those with engineered biocontainment measures, should not be created. We therefore recommend that research with the goal of creating mirror bacteria not be permitted, and that funders make clear that they will not support such work.

Many will argue that this recommendation is too alarmist while others will argue against "opening Pandora's box." At the least, it does appear that Pope's "*Laevo* Man" is centuries if not countless millennia in the future unless we can, Edwin-Abbot-style, move "*Dextro* Man" through the fourth dimension.

Acknowledgements The author wishes to acknowledge the inspirational teaching of Professor Kurt Mislow whose fall 1967 Advanced Organic Chemistry course first excited his interest in stereochemistry and aromaticity. He also wishes to acknowledge his Ph.D. mentor, Professor Pierre Laszlo, who always encouraged creative thinking. The author gratefully acknowledges helpful comments by Professors Gregory Girolami and Jeffrey Seeman and support by the History of Chemistry (HIST) Division of the American Chemical Society which sponsored the symposium "The Birth of the 3rd Dimension in Chemistry," at the ACS National Meeting in New Orleans in March 2024 celebrating the sesquicentennial of van't Hoff's and Le Bel's papers. This chapter is derived from the author's presentation at that symposium.

References

1. Gibson CS (1941) Sir William Jackson Pope. 1870–1939. Obit Not Fellows R Soc 10(9):290–324
2. Gibson CS (1946) The life and work of Sir William Jackson Pope. J R Soc Arts 94:668–674. https://www.jstor.org/stable/41363278. Accessed 28 Dec 2023
3. Read J (1947) Humour and humanism in chemistry. G. Bell and Son Ltd., London, pp 284–301
4. Perkin WH, Pope WJ, Wallach O (1909) Optically active substances containing no asymmetric atom: 1-methylcyclohexylidene-4-acetic acid. J Chem Soc 95:1789–1802
5. Cahn RS, Ingold CK, Prelog V (1956) The specification of asymmetric configuration in organic chemistry. Experientia 12:81–94

6. Pope WJ, Read J (1914) The optical activity of compounds of simple molecular constitution. Ammonium *d*- and *l*-chloroiodomethanesulphonate. J Chem Soc 105:811–821
7. van't Hoff JH (1898) The arrangement of atoms in space, second revised and enlarged edition, with a preface by Johannes Wislicenus professor of chemistry at the University of Leipzig, and an appendix: stereochemistry among inorganic substances by Alfred Werner, professor of chemistry at the University of Zürich, translated and edited by Arnold Eiloart, Longmans, Green and Co., London, New York and Bombay, pp 25–26
8. Greenberg A (1993) Stereochemistry and the "ether" in the evolution of molecular structure theory: the musings of a chemist. In Moller's supplanted "screw theory" beginning with a view of the obnoxious and equally outdated cod-liver oil. J Chem Educ 70:284–286
9. Stolzenberg D (2004) Fritz Haber: chemist, Nobel laureate, German Jew. Chemical Heritage Press, Philadelphia, pp 286–290
10. Abbot EA (1884) Flatland: a romance in many dimensions. Seeley & Co., London
11. Read J (1941) The Pope collection of alchemical paintings and engravings. Nature 3721:243
12. Read J (1947) The alchemist in life, literature, and art. Thomas Nelson and Sons, New York, pp 78, 81
13. Siegel J (2018) Kurt M. Mislow (1923–2017): Pioneer of modern stereochemistry. Angew Chem Int Ed 57:4835
14. Hargittai I (1998) Kurt Mislow. In: The Chemical Intelligencer No. 3, pp 16–25
15. Princeton University (2017) Princeton.edu/news/2017/10/27/kurt-mislow-pioneer-stereochemistry-dies-94. Accessed 31 Dec 2023
16. New York University (2017) Mislow memoriam notice: as.nyu.edu/departments/chemistry/news/fall-2017/in-memorium-kurt-mislow.html. Accessed December 31, 2023
17. Hargittai B, Hargittai I (2023) Kurt Mislow centennial—he changed the way people think about stereochemistry. Struct Chem 34:345–349
18. Wang L (2017) Kurt Mislow. Chem Eng News 95:36
19. Mislow KM (1947) Ph.D. dissertation: I. Synthesis of potential antimalarials. Some 8-(3-diethylaminopropylaminol) quinolines; II. Isomorphism of carboxylic acids in relation to serological specificity; III. A study of the Hammick reaction. California Institute of Technology, Pasadena, CA. https://resolver.caltech.edu/caltechETD:etd-06142004-140844. https://doi.org/10.7907/60-TO-BG46. Accessed 30 May 2024
20. Dr. Joseph B. Koepfli: Oralhistories.library.caltech.edu/286/1/Koepfli%2c%20OHO-final.pdf. Accessed 31 Dec 2023
21. Mislow K, Koepfli JB (1946) The synthesis of potential antimalarials. Some 2-substituted 8-(3-dimethylaminopropylamino) quinolines. J Am Chem Soc 68:1553–1556
22. Special Collections and Archives Research Center, Oregon State University, Corvallis, Oregon. Series 1, Box 248, Folder 4.4. Linus Pauling correspondence with Kurt Mislow. The author wishes to thank Anna Dvorak and her student assistant for their help. Accessed 30 May 2024
23. Mislow K (1950) The concept of "internal compensation." Science 112:26–27
24. Mislow K (1954) Limitation of the symmetry criteria for optical inactivity and resolvability. Science 120:232–233
25. Mislow K, Bolstad R (1955) Molecular dissymmetry and optical inactivity. J Am Chem Soc 77:6712–6713
26. Eliel EL, Wilen SH (1994) Stereochemistry of organic compounds. New York. Molecule **1**, with related examples of chiral compounds lacking chiral centers are discussed in Chapter 14. Specific discussion of molecule **9** is found on p 62
27. Laszlo P (2021) A life and career in chemistry. Autobiography from the 1960s to the 1990s. Springer Nature Switzerland AG, Cham (Switzerland), pp 32, 41, 75, 127, 242–244
28. Pope WJ, Whitworth JB (1931) The optically active spiro-5:5-dihydantoins. Proc Roy Soc A 134:357–371
29. Kipping FS, Pope WJ (1898) Enantiomorphism. J Chem Soc 73:606–617
30. Kipping FS, Pope WJ (1897) Racemism and pseudoracemism. J Chem Soc 71:989–1001

31. Anet FAL, Miura SS, Siegel J, Mislow K (1983) La coupe de roi and its relevance to stereochemistry. Combination of two homochiral molecules to give an achiral product. J Am Chem Soc 105:1419–1426
32. Cinquini M, Cozzi F, Sannicolo F, Sironi A (1988) Bisection of an achiral molecule into homochiral halves: the first chemical analogue of "la coupe de roi." J Am Chem Soc 110:4363–4364
33. Schill G (1971) Catenanes, rotaxanes, and knots. Academic Press, New York
34. Ramberg PJ (2005) Chemical structure, spatial arrangement. The early history of stereochemistry, 1874–1914. Ashgate Publishing Company, Burlington, VT, pp 193–241
35. Pope WJ, Peachey SJ (1899) Asymmetric optically active nitrogen compounds. *Dextro-* and *laevo*-α-benzylphenylallylmethylammonium iodide. J Chem Soc 75:1127–1131
36. Pope WJ, Harvey AW (1901) Optically active nitrogen compounds and their bearing on the valency of nitrogen. *d-* and *l*-α-Benzylphenylallylmethylammonium salts. J Chem Soc 79:828–841
37. Kauffman GB, Bernal I (1988) Overlooked opportunities in stereochemistry: neglected connection between Werner's metalammines and Pope's onium compounds. Pure Appl Chem 60:1379–1384
38. Hantzsch A, Werner A (1890) Ueber raümliche anordnung der atome in stickstoffhaltigen molekülen, Ber Chem Ges 23:11–30 [English translation with commentary by Kauffman GB (1966) J Chem Educ 43: 155–165]
39. Werner A (1911) Zur kenntnis des asymmetrischen kobaltatoms. Ber Chem Ges 44:1887–1898
40. Kossel W (1916) Molekulbildung als frage der atombau. Ann Phys 49:229–362
41. Lewis GN (1916) The atom and the molecule. J Am Chem Soc 38:762–785
42. Wyckoff RWG (1928) The crystal structure of the tetramethyl ammonium halides. Z Krist 67:91–105
43. Cleeton CE, Williams NH (1934) Electromagnetic waves of a 1.1 cm wavelength and the absorption spectrum of ammonia. Phys Rev 45:234–237
44. Gillespie RJ, Nyholm RS (1957) Inorganic stereochemistry. Quart Rev Chem Soc 11:339–380
45. Lehn JM (1970) Nitrogen inversion. In: Dynamic stereochemistry. Topics in current chemistry. Springer, Berlin, pp 311–377
46. Saunders M, Yamada F (1963) Measurement of an amine inversion rate using nuclear magnetic resonance. J Am Chem Soc 85:1882
47. Prelog V, Wieland P (1944) Über die spaltung der Tröger'schen base in optische antipoden, ein beitrag zur stereochemie des dreiwertigen stickstoffe. Helv Chim Acta 27:1127–1134
48. Rauk A, Allen LC, Mislow K (1970) Pyramidal inversion. Angew Chem Int Ed 9:400–414
49. Andose JD, Lehn JM, Mislow K, Wagner J (1970) The effect of substituents on the pyramidal inversion of 1-aryl-2,2-dimethylaziridines. J Am Chem Soc 92:4050–4056
50. Meisenheimer J, Lichtenstadt L (1911) Über optisch-aktive verbindungen des phosphors. Ber Chem 44:356–359
51. Pope WJ, Gibson CS (1912) Some mixed phosphonium derivatives. J Chem Soc 101:735–740
52. Holliman FG, Mann FG (1947) Optically active quaternary phosphonium salt. Nature 159:438
53. Zon G, Mislow K (1971) Studies in phosphorus stereochemistry. In: The chemistry of organophosphorus compounds. Springer, Berlin, pp 61–94
54. Herault D, Nguyen D, Nuel D, Buono G (2015) Reduction of tertiary and secondary phosphine oxides to phosphines. Chem Soc Rev 44:2508–2528
55. Horner L, Winkler H (1964) Phosphorus organic compounds. XLI. Activation energies of racemization of optically active tertiary phosphines. Tetrahedron Lett 461–462
56. Horner L, Balzer WD (1965) Phosphorus organic compounds. LXIX. Steric course of deoxygenation of tertiary phosphine oxides to tertiary phosphines with trichlorosilane. Tetrahedron Lett 1157–1162
57. Korpiun O, Mislow K (1967) New route to the preparation and configurational correlation of optically active phosphine oxides. J Am Chem Soc 89:4784–4786
58. Korpiun O, Lewis RA, Chickos J, Mislow K (1968) Synthesis and absolute configuration of optically active phosphine oxides and phosphinates. J Am Chem Soc 90:4842–4846

59. Naumann K, Zon G, Mislow K (1969) Perchloropolysilanes: novel reducing agents for phosphine oxides and other organic oxides. J Am Chem Soc 91:2788–2789
60. Lewis RA, Mislow K (1969) Direct configurational correlation of trialkyl- and triarylphosphine oxides. J Am Chem Soc 91:7009–7012
61. Naumann K, Zon G, Mislow K (1969) Use of hexachlorodisilane as a reducing agent. Stereospecific deoxygenation of acyclic phosphine oxides. J Am Chem Soc 91:7012–7023
62. DeBruin KE, Zon G, Naumann K, Mislow K (1969) Stereochemistry of nucleophilic displacement at phosphorus in some phosphonium salts. J Am Chem Soc 91:7027–7030
63. Baechler RD, Mislow K (1970) The effect of structure on the rate of pyramidal inversion of acyclic phosphines. J Am Chem Soc 92:3090–3093
64. Egan W, Tang R, Zon G, Mislow K (1971) Barriers to pyramidal inversion at phosphorus in phospholes, phosphindoles, and dibenzophospholes. J Am Chem Soc 93:6205–6216
65. Mislow K (1970) Role of pseudorotation in the stereochemistry of nucleophilic displacement reactions. Acc Chem Res 3:321–331
66. Senkler Jr GH, Mislow K (1972) Barrier to pyramidal inversion in ethylmethylphenylarsine. J Am Chem Soc 94:291
67. Baechler RD, Casey JP, Cook RJ, Senkler Jr GH, Mislow K (1972) Effect of ligand electronegativity on the inversion barriers of arsines. J Am Chem Soc 94:2859–2861
68. Mislow K (1973) Pyramidal inversion barriers of phosphines and arsines. Trans NY Acad Sci 35:227–242
69. Lambert JB, Jackson GFIII (1968) The stereomutation of arsenic. J Am Chem Soc 90:1350–1351
70. Barbachyn MR, Johnson CR (1984) Optical activation and utilization of compounds containing chiral sulfur centers. In: Morrison JD (ed) Asymmetric synthesis, vol 4. Academic Press Education, New York, pp 227–261
71. Pope WJ, Peachey SJ (1900) Asymmetric optically active sulfur compounds. *d*-Methylethylthetine platinichloride. J Chem Soc 77:1072–1075
72. Pope WJ, Neville AW (1902) Asymmetric optically active selenium compounds and the sexavalency of selenium and sulfur. *d*- and *l*-phenylmethylselentine salts. J Chem Soc 81:1552–1563
73. Phillips H (1925) Investigation on the dependence of rotatory power on chemical constitution. Part XXVII. The optical properties of *n*-alkyl *p*-toluenesulphinates. J Chem Soc 127:2552–2587
74. Harrison PWB, Kenyon J, Phillips H (1926) The dependence of rotatory power on chemical constitution. Part XXIX. The resolution of sulphoxides into their optically active forms. J Chem Soc 129:2079–2090
75. Anderson KK (1962) Synthesis of (+)-ethyl *p*-tolylsulfoxide from (−)-menthyl (−)-*p*-toluenesulfinate. Tetrahedron Lett 93–95
76. Bickart P, Axelrod PM, Jacobus J, Mislow K (1967) The absolute configuration of methylallylsulfoxide. Stereochemistry of the Grignard synthesis of optically active sulfoxides. J Am Chem Soc 89:697–698
77. Mislow K, Simmons T, Melillo JT, Ternay A Jr (1962) The hydrogen chloride—catalyzed racemization of sulfoxides. J Am Chem Soc 86:1452–1453
78. Scartazzini R, Mislow K (1967) Pyramidal inversion of sulfonium ions. The racemization of 1-adamantylethylmethylsulfonium perchlorate. Tetrahedron Lett 2719–2722
79. Rayner DR, Gordon AJ, Mislow K (1968) Thermal racemization of diaryl, alkyl aryl, and dialkyl sulfoxides. J Am Chem Soc 90:4854–4860
80. Lambert JB, Jackson GFIII (1968) The stereochemistry and inversion of trivalent oxygen. J Am Chem Soc 90:1349–1350
81. Smith O, Popescu MV, Hindson MJ, Paton RS, Burton JW, Smith MD (2023) Control of stereogenic oxygen in a helically chiral oxonium ion. Nature 615(7952):430–435
82. Belloli R (1969) Resolution and stereochemistry of asymmetric silicon, tin, germanium, and lead compounds. J Chem Educ 46:640–644

83. Kipping FS (1907) Organic derivatives of silicon. Part II. The synthesis of benzylethylpropylsilicol, its sulphonation and the resolution of the *dl*-sulphonic derivatives into optically active compounds. J Chem Soc 91:209–240
84. Sommer LH, Frye CL (1959) Optically active organosilicon compounds having reactive groups bonded to asymmetric silicon. Displacement reactions at silicon with pure retention and pure inversion of configuration. J Am Chem Soc 81:1013
85. Brook AG, Peddle GJD (1963) Resolution of an asymmetric organogermanium compound, methyl-α-naphthylphenylgermanium. J Am Chem Soc 85:1869–1870
86. Eaborn C, Simpson P, Vargas ID (1966) Organogermanium compounds. Part VIII. The preparation of optically active germanium compounds. J Chem Soc A 1133–1136. Reference is made to the initial brief report: Bott RW, Eaborn C, Vargas ID (1963) Chem Ind (London) 614
87. Pope WJ, Peachey SJ (1900) Asymmetric optically active tin compounds. Dextromethylethyl-*n*-propyl tin iodide. Preliminary note. Proc Chem Soc 16:42–44
88. Pope WJ, Peachey SJ (1900) The racemization of optically active tin compounds. Dextromethylethyl-*n*-propyl tin dextrobromocamphorsulphonate. Proc Chem Soc 16:116–118
89. Kipping FB (1928) An attempt to prepare optically active derivatives of quadrivalent tin. J Chem Soc 131:2365–2372
90. Peddle GJD, Redd G (1970) The determination of the stereochemical stability of organotin compounds by nuclear magnetic resonance. J Am Chem Soc 92:365–369
91. Redl G (1970) Reinvestigation of the resolution of methylethyl-*n*-propyltin iodide. J Organomet Chem 22:139–140
92. Roberts C (1896) The development and present aspects of stereo-chemistry. DC Heath & Co., Boston, p 40
93. Grüttner G, Krause E (1917) Die haftfestigkeit der alkylgruppen am bleiatom gegenüber der abspaltenden wirkung von halogen: darstellung gemischter bleialkyle mit 3 und 4 verschiedenen radikalen. Ber Chem Ges 50:202–211
94. Austin PR (1933) Studies of organic lead compounds. V. Asymmetric derivatives. J Am Chem Soc 55:2948–1953
95. van der Kooi O, Wolters J, van der Gen A (1979) Chiral organolead compounds. Part I. [2-(Dimethylaminomethyl)phenyl](4-methoxyphenyl)(4-methylphenyl) lead iodide. Recl Trav Chim Pays-Bas 98:353–355
96. en.wikipedia.org/wiki/List_of_nominees_for_the_Nobel_Prize_Prize_in_Chemistry. Accessed 8 Sept 2024
97. A. Greenberg is grateful to Jeffrey Seeman for directing his attention to the Nobel Prize nominations of William Jackson Pope
98. Pope WJ (1901) The bearings of geometry on the chemistry of fermentation. Lectures 1–4; J Soc Arts (London) 49:677–683, 690–697, 701–708, 713–718
99. Adamala KP et al (2024) Confronting risks of mirror life. Science. https://doi.org/10.1126/science.ads9158. Accessed 20 Dec 2024

Missed It by that Much! How Butlerov Almost Proposed the Tetrahedral Carbon

6

David E. Lewis

Abstract

The Structural Theory of Organic Chemistry, first proposed by Archibald Scott Couper (1831–1892) and (Friedrich) August Kekulé (1829–1896) in 1858, and then clarified and extended by Aleksandr Mikhailovich Butlerov (1828–1886) in 1861, was a necessary prelude to the van't Hoff and Le Bel proposals of the tetrahedral carbon. At the time of Butlerov's proposal of his theory, he took pains to make it clear that his proposals were related to *chemical*, not *physical* atoms, making it more acceptable to those chemists for whom the concept of physical atoms was anathema. On the basis of his theory, Butlerov and his student, Vladimir Vasil'evich Markovnikov (1837–1904), predicted the numbers of isomers of organic compounds. The details about chemical bonding were to await the next century, but like his contemporaries, Butlerov used the concept of "affinities" in his explanations about isomerism in organic chemistry. As part of his theoretical considerations, Butlerov used a model to rationalize the number of equivalent valencies of the carbon atom, using a tetrahedron—an object with four identical faces—as an analogy. However, he did not propose anything more with the analogy. The key development needed to allow the advance by was the realization by van't Hoff and Le Bel that *physical* atoms existed and that their positions could be deduced, which was made easier by their realization that atoms that were *chemically* close to each other in the molecule might also be *physically* close to each other…

It takes encouragement from his peers to persuade an organic chemist to pursue the study of the history of chemistry. This chapter is dedicated to one such individual, Dr. William B. Jensen, with gratitude.

D. E. Lewis (✉)
Department of Chemistry and Biochemistry, University of Wisconsin-Eau Claire, Eau Claire, WI, USA
e-mail: lewisd@uwec.edu

A. Greenberg and D. E. Lewis (eds.), *The Birth of the 3rd Dimension in Chemistry*, Perspectives on the History of Chemistry, https://doi.org/10.1007/978-3-031-97743-5_6

6.1 Introduction

The concept of the tetrahedral carbon could not have arisen as a serious chemical theory unless the Structural Theory of Organic Chemistry, which emerged during the three decades between 1831 and 1861 [1], had achieved general acceptance.

Following the emergence of organic chemistry in the third decade of the nineteenth century, organic chemists began to search for theories that would help to organize the study of the new science. The first foray was the proposal of the benzoyl radical by Justus von Liebig[1] (1803–1873) and Friedrich Wöhler[2] (1800–1882) in 1832 [2], which gave rise to Radical Theory, based on the concept of a radical first proposed by Antoine Laurent Lavoisier[3] (1743–1794) in his *Traité élementaire de chimie* [3].

And yet, despite his involvement in this first move towards the systematization of the study of organic chemistry, in 1835 Wöhler lamented to his friend and mentor, Jöns Jacob Berzelius (1779–1848),

> Organic chemistry nowadays almost drives one mad to me it appears like a primeval tropical forest full of the most remarkable things a dreadful endless jungle into which one dare not enter for there seems no way out [4].

The definition of a radical was further clarified in a short joint paper by Jean-Baptiste André Dumas[4] (1800–1884) and Liebig in 1837 [5]; Dumas became a

---

[1] Biographies of Liebig: (a) Shenstone WA (1896). Justus von Liebig. His Life and Work. MacMillan & Co., New York. (b) von Liebig G (1890). Eigenhändige biographische Aufzeichnungen. Ber dtsch chem Ges.23:817–828 [in German]. (c) Brock WH (1997). Justus von Liebig. The Chemical Gatekeeper. Cambridge University Press, Cambridge, UK. (d) Jackson CM (2023). Molecular World: Making Modern Chemistry, MIT Press, Cambridge, MA. Ch 1–3, pp. 10–106. For an image of Liebig from the Liebig Museum, see: https://www.liebig-museum.de/liebig/.

[2] Biographies of Wöhler: (a) Hofmann AW (1882). Zur Erinnerung an Friedrich Wöhler. Ber dtsch chem Ges.15:3127–3290 [in German]. (b) Keen R (2005), The Life and Work of Friedrich Wöhler (1800–1882). Büttner J (Ed). Verlag Traugott Bautz GmbH, Nordhausen. For an image of Wöhler, see: https://commons.wikimedia.org/wiki/File:Portrait_of_Friedrich_W%C3%B6hler_-_DPLA_-_6c0ff4bba3b2237cecb2af31dc8fd0bc.jpg.

[3] Biographies of Lavoisier: (a) Berthelot M (1890). La Révolution Chimique Lavoisier. Alcan F, Ed. Ancienne Librairie Germer-Baillière et Cie, Paris. (b) McKie D (1935). Antoine Lavoisier: The Father of Modern Chemistry. J. B. Lippincott & Co., Philadelphia. (c) McKie D (1952). Antoine Lavoisier: Scientist, Economist, Social Reformer. Henry Schuman, New York. (d) Thorpe E (1911). Essays in Historical Chemistry. Macmillan and Co., London. Antoine-Laurent Lavoisier. Contemporary Review, December 1890. Ch. VI, pp. 123–148. (e) Thorpe E (1911). Essays in Historical Chemistry. Macmillan and Co., London. Priestley, Cavendish, Lavoisier and *La Révolution Chimique*. Ch. VII, pp. 149–184. (e) Poirier J-P (1996). Lavoisier. Chemist, Biologist, Economist. University of Pennsylvania Press, Philadelphia; English translation from the French of Poirier J-P (1993). Antoine Laurent de Lavoisier, 1743–1794. Editions Pygmalion/Gérard Watelet, Paris. For an image of Lavoisier, see: https://commons.wikimedia.org/wiki/File:Lavoisier.jpg.

[4] For biographies of Dumas, see: (a) E. J. R. (1884). Obituary notices of Fellows deceased. Proc Roy Soc 37:x–xxvii. (b) Hofmann AW (1880). Jean Baptiste André Dumas. Nature (Extra Number, February 6, 1880):i–xl. (c) Tiffeneau Prof (1934). J. B. Dumas 1800–1884. Laboratoires G.

collaborating editor of the *Annalen der Pharmacie* in 1838. Liebig further clarified the concept of a radical in 1838 [6], where he made the observation,

> We call cyanogen a radical (1) because it is a non-varying constituent in a series of compounds, (2) because in these latter it can be replaced by other simple substances, and (3) because in its compounds with a simple substance, the latter can be turned out and replaced by equivalents of other simple substances. [Author's translation]

Within a few years, Radical Theory came under challenge by Substitution Theory, which was championed by Dumas, whose series of papers [7] began in 1834, Auguste Laurent[5] (1807–1853) whose series of papers began in 1836 [8], and Charles Frédéric Gerhardt[6] (1816–1856), whose papers [9] appeared in 1839. By 1840, Dumas had begun publishing his Theory of Types [10], which became dominant as it was expanded by the work of Alexander William Williamson[7] (1824–1904), who added the "Water-Type" [11] and August Wilhelm (von) Hofmann[8] (1818–1892), who added the "Ammonia Type" [12] to the theory. By

---

Beytout, Paris [in French]. For an image of Dumas, see: https://upload.wikimedia.org/wikipedia/commons/d/dd/Jean_Baptiste_Andr%C3%A9_Dumas.jpg.

[5] For biographies of Laurent, see: (a) Yorke PJ (1854). President's Address. Quart J Chem Soc 7:144–160 (esp. 149–157). (b) de Milt C (1951). Auguste Laurent: Guide and inspiration of Gerhardt. J Chem Educ 28(4):198–204. (c) de Milt C (1953). Auguste Laurent, founder of modern organic chemistry. Chymia 4:85–114. For an image of Laurent, see: https://upload.wikimedia.org/wikipedia/commons/5/50/Auguste_Laurent2.JPG.

[6] For biographies of Gerhardt, see: (a) Grimaux E, Gerhardt C [Jr] (1900). Charles Gerhardt, sa vie, son oeuvre, sa correspondence 1816–1856. Masson et C[ie], Paris [in French]. (b) Partngton JR (1964). A History of Chemistry. MacMillan & Co, London. pp. 405–424. For an image of Gerhardt, see: https://commons.wikimedia.org/wiki/File:Gerhardt_Charles.jpg.

[7] For biographies of Williamson, see: (a) Divers E (1907). Obituary notices of Fellows deceased. Proc Roy Soc 78A:xxiv–xliv. (b) Foster GC (1905). Alexander William Williamson. Born May 1st 1824; Died May 6th 1904. J Chem Soc Trans 87:605–818. Foster GC (1911). Alexander William Williamson. Ber dtsch chem Ges 44:2253–2269 [in German]. (c) Harris J, Brock WH (1978). From Giessen to Gower Street: Towards a biography of Alexander William Williamson (1824–1904). Ann Sci 31(2):95–130. For an image of Williamson, see: https://upload.wikimedia.org/wikipedia/commons/1/16/Williamson_Alexander.jpg.

[8] There are numerous biographies of Hofmann: (a) Volhard J, Fischer E (1902). August Wilhelm von Hofmann: ein Lebensbild, Deutsche Chemische Gesellschaft: Berlin. (b) Volhard J (1902). August Wilhelm von Hofmann. I. Ber dtsch chem Ges 35(2):1–196. (c) Fischer E (1902). August Wilhelm von Hofmann. II. Ber dtsch chem Ges 35(2):197–284. (d) Meinel C (1992). August Wilhelm Hofmannm—"Reigning Chemist-in-Chief." Angew Chem Int Ed Engl 31:1265–1282. (e) Travis AS (1992). August Wilhelm Hofmann (1818–1892). Endeavour 16:59–65. (f) Ronge G (1972). Hofmann, August Wilhelm von. Neue Deutsche Biographie 9:446–450. (g) Playfair L (1896). Hofmann Memorial Lecture. Personal reminiscences of Hofmann and of the conditions which led to the establishment of the Royal College of Chemistry and his appointment as its professor. J Chem Soc Trans 69:575–579. (h) Abel FA (1902). The history of the Royal College of Chemistry and reminiscences of Hofmann's professorship. J Chem Soc Trans 69:580–596. (i) Perkin WH (1902). The origin of the coal-tar colour industry, and the contributions of Hofmann

1853, the field had advanced to the point that Gerhardt proposed the "Newer Type Theory" [13].

6.2 1858–1861: Structural Theory Emerges

The first two major proponents of the Theory of Chemical Structure were (Friedrich) August Kekulé[9] (1829–1896) [14], and a young Scotsman, Archibald Scott Couper[10] (1831–1892) [15], who were the first to propose the theory. Their papers describing Structural Theory were published six weeks apart in 1858 by Kekulé and Couper. Kekulé's paper was received by the *Annalen* on March 16, 1858, and appeared in May [16]; Couper's paper [17] was presented before the Académie des Sciences on June 14 the same year, and appeared in the *Comptes rendus*. The timing of the two disclosures led to Kekulé being credited with priority for the discovery.

At the time he submitted his paper, Kekulé was Privat-Docent at Heidelberg; in October the same year he accepted the call to the University of Ghent, in Belgium, as Professor of Chemistry. As a Privat-Docent, Kekulé was independent, and free to publish his work when and where he saw fit. His publication detailed his new theory, but it was conservative in its approach and did not break completely with Type Theory. This is evident from Kekulé's representation of three organic reactions: (1) the hydrolysis of glycolide (whose structure as drawn by Kekulé corresponds to the monomeric α-lactone rather than the dimeric dilactone, 1,4-dioxane-2,5-dione) to glycolic acid, (2) the hydrolysis of succinimide to succinamic acid, and (3) the reaction between isocyanic acid and ammonia to give urea. Kekulé's versions of these reactions are compared with the modern versions in Fig. 6.1. Kekulé may well have concluded that his theory stood a better chance of being widely accepted if he showed how it fit in with the dominant theory of the time.

When he was ready to submit his theory for publication, Couper was a student in the Paris laboratory of (Charles) Adolphe Wurtz[11] [18] (1817–1884), and so he required Wurtz' approval to publish his work. Wurtz considered that the *Comptes rendus* was the most appropriate place to publish the work, but he was not a member of the Académie at the time, so under the rules of the Académie, he could not present the paper himself as a non-member. He needed to find an Académie member to present the paper. His search for a member to read the paper before

and his pupils. J Chem Soc Trans 69:596–637. For an image of Hofmann, see: https://digital.sciencehistory.org/works/m6cis8h.

[9] For an image of Kekulé, see: https://upload.wikimedia.org/wikipedia/commons/f/fa/Frkekul%C3%A9.jpg.

[10] For an image of Couper from the Edgar Fahs Smith collection, see: https://www.sciencehistory.org/education/scientific-biographies/august-kekule-and-archibald-scott-couper/.

[11] For an image of Wurtz, see https://upload.wikimedia.org/wikipedia/commons/2/22/Charles-Adolphe_Wurtz.jpg.

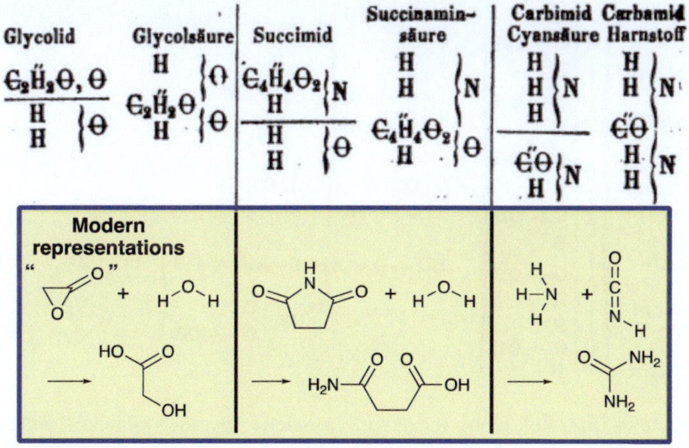

Fig. 6.1 Three reactions from Kekulé's paper and their modern forms (Ref. [15], p. 139)

the Académie took a little time, although, as Albert Ladenburg[12] noted in a letter to Richard Anschütz,[13] Wurtz describes himself as having "bungled this a little" [19].[14]

The resultant delay between Couper presenting his manuscript to Wurtz for publication and its presentation to the Académie by Dumas was long enough for Kekulé's paper to be published first. Couper blamed Wurtz for the delay and the loss of priority (although not altogether unjustifiably, as Ladenburg's letter to Anschütz suggests). However, Couper's reproof was delivered in such an insolent and disrespectful way that it prompted the usually genial Wurtz to dismiss him summarily from the laboratory. Couper returned to Britain, and in December 1858 he accepted an appointment as second assistant in the laboratory of (later Lord) Lyon Playfair.[15] Five months later, however, he suffered a nervous breakdown, and he was institutionalized as a private patient for two months (he had independent means). Immediately upon his release, he went on a fishing trip, where he suffered a relapse (attributed at the time to sunstroke) that led to him being institutionalized for another three years. His health was now broken, and he spent the rest of his life being cared for by his mother. It took Anschütz' work to rehabilitate his scientific reputation [15].

[12] For an image of Ladenburg, see: https://en.wikipedia.org/wiki/Albert_Ladenburg#/media/File:Albert_Ladenburg.jpg.

[13] For an image of Anschütz, see: https://en.wikipedia.org/wiki/Richard_Ansch%C3%BCtz#/media/File:Richard_Ansch%C3%BCtz_1898.jpg.

[14] "Neglected" or "mishandled," give a better sense of what Ladenburg wanted to convey.

[15] For an image of Playfair, see: https://en.wikipedia.org/wiki/Lyon_Playfair,_1st_Baron_Playfair#/media/File:Lyon_Playfair.jpg.

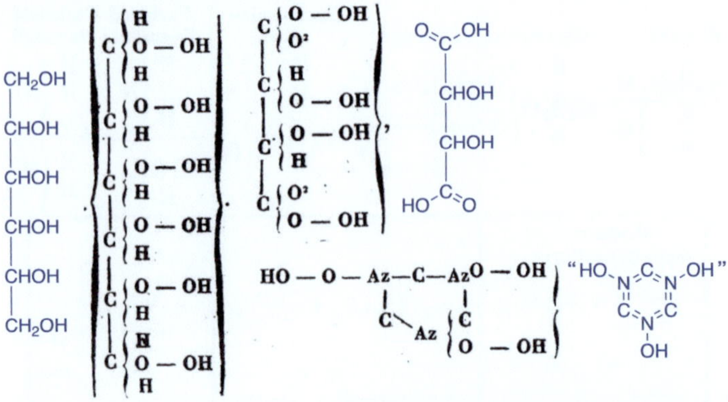

Fig. 6.2 Structures from Ref. [16b]: (clockwise from left) "glycose" (glucitol), tartaric acid, cyanuric acid. Couper's structure of cyanuric acid assumes pentavalent nitrogen atoms in the ring

Unlike Kekulé, Couper broke completely with Type Theory in his paper. In it, he provided structures that are remarkably like modern line drawing structures; the three Couper structures in Fig. 6.2 are based on the equivalent weight for oxygen (O = 8), necessitating the use of inseparable pairs of oxygen atoms; these formulas are translated into their modern versions as the blue structures in the same figure.

6.3 Aleksandr Mikhailovich Butlerov (Бутлеров Александр Михайлович, 1828–1886)

The next advance in Structural Theory came in 1861, when the young Russian Professor, Aleksandr Mikhailovich Butlerov[16] [20] (Бутлеров Александр Михайлович, 1828–1886), from the relatively provincial Kazan Imperial University about 800 km east of Moscow on the Volga River, presented a paper, "Einiges über die chemische Structur der Körper," at the 36. *Versammlung deutscher Naturforscher und Ärzte zu Speyer* on September 19, 1861. At the urging of Emil Erlenmeyer, the newly-installed Editor of the *Zeitschrift für Chemie und Pharmacie*, Butlerov permitted his theory to be published in that journal [21].

6.3.1 Butlerov's Education

Butlerov is one of the most admired organic chemists in Russia; he is celebrated in the form of the Butlerov Prize and Gold Medal, presented by the Russian Academy of Sciences. Butlerov was born into the minor nobility and educated at Kazan

[16] For an image of Butlerov, see: https://en.wikipedia.org/wiki/Alexander_Butlerov#/media/File:Butlerov,_A._M._1828-1886.jpg.

University. As a child and a young man, he displayed an aptitude for the sciences, with a special interest in entomology. He entered Kazan University in 1844 at a time when seminal discoveries were being made there. At that time, the teaching of chemistry was in the hands of two individuals. with very different interests in chemistry: Karl Karlovich Klaus (Клаус Карл Карлович, Carl Ernst Claus, 1796–1864),[17] the Professor of Chemistry, and Nikolai Nikolaevich Zinin (Зинин Николай Николаевич, 1812–1880),[18] the Professor of Chemical Technology.

Klaus [22] had been trained as a pharmacist on Dorpat (now Tartu, in Estonia) and had moved to Kazan, where he established a pharmacy and became highly respected for his encyclopedic knowledge of botany and herbal medicines. He was also a talented artist, and it was in his capacity as an artist and botanist that he traveled on scientific expeditions into the Volga-Ural region of Russia. It was during one of these expeditions that his interest in the platinum metals was stimulated. After spending time working as a pharmacist in Saratov and Kazan, then six years in Dorpat where he took the *kandidat filosophii* and *Magistr Filosofii* degrees, he returned to Kazan in 1837, where he became Professor of Chemistry. Klaus' interests were, not surprisingly, predominantly in inorganic chemistry, where Berzelius' electrochemical (dualistic) theory held sway.

Zinin [19, 23] was among the first Russian students who were sent on *komandirovka* (paid study leave) to western Europe after the ascension of Nicholas I to the throne. These *komandirovky* were meant to provide an opportunity for students to attend lectures, but Zinin was one of the first to use the opportunity to engage in research.

Zinin had graduated in physics under the mathematician Nikolai Ivanovich Lobachevskii (Лобачевский Николай Иванович, 1792–1856) [24],[19] the inventor of non-Euclidean geometry, who was Rector of the university at the time. After graduating, Zinin was appointed to teach hydrostatics. However, chemistry at the time was very poorly taught by the incompetent (but very politically astute) Ivan Ivanovich Dunaev (Дунаев Иван Иванович, 1788–1843). Thanks to the University Statute of 1935, the university was able to dismiss Dunaev in 1837. In 1835, Zinin, the physicist, was appointed to teach chemistry "in support of Duvaev." At that time, one did not have to have a degree in a subject to be appointed to teach it! The year after Dunaev's dismissal, Zinin was sent on a three-year *komandirovka*. By the time he returned to Russia, he had become a convert to the unitary theory of Gerhardt, which was replacing the older Dualistic Theory of Berzelius. On his return to Russia Zinin was appointed to the Chair of Technology because Klaus had been appointed to the Chair of Chemistry in his absence. Zinin's *komandirovka* research had been in organic chemistry, mainly in Liebig's laboratory in Giessen, so it was natural that the teaching of organic chemistry should devolve upon him.

[17] For an image of Klaus, see: https://commons.wikimedia.org/wiki/File:ClausKE.png.
[18] For an image of Zinin, see: https://en.wikipedia.org/wiki/Nikolay_Zinin#/media/File:Nikolai_Nikolajewitsch_Sinin.JPG.
[19] For an image of Lobachevskii, see: https://commons.wikimedia.org/wiki/File:Lobachevsky_03_crop.jpg.

Butlerov's entry to Kazan in 1844 came at a time when seminal discoveries were occurring there. That was the year that Klaus discovered the element ruthenium [25], and just two years after Zinin had discovered the reduction of nitroarenes with hydrogen sulfide to give aminoarenes [26]. Butlerov was more attracted to Zinin's modern view of chemistry than to Klaus' dualistic view, and was also more attracted to the relatively new discipline of organic chemistry. He was happy to begin working with Zinin. But, when Zinin departed for the Medical-Surgical Academy of St. Petersburg in 1848, this left the teaching of chemistry in the hands of Klaus, and Butlerov's zeal for chemistry waned. He returned to his first love, entomology, to write his dissertation for the degree of *kandidat* [27] on the butterflies of the Volga-Ural region of Russia (the region of Russia bounding the Volga and Ural Rivers, where Europe meets Asia), which he submitted in October the same year.

However, Zinin's departure left the university without someone to teach organic chemistry—the university needed a replacement. Just as they had done when the physicist, Zinin, was appointed to teach chemistry, the University Council decided that the needs of the university trumped the training of the student, and *kandidat* Butlerov, the entomologist, became *kandidat* Butlerov, the chemist. He began teaching chemistry part-time in 1849. However, in order to advance in academic rank, he had to acquire the degree of *Magistr Khimii*, the minimum requirement for appointment as Extraordinary Professor, and the degree of *Doktor Khimii*, which would permit his appointment as to a Professorial Chair.

In February 1851, Butlerov submitted his dissertation [28] for the degree of *Magistr Khimii*. The same year, he became an assistant to Klaus, with full responsibility for teaching chemistry. The next year, Klaus left Kazan to take up the Chair of Pharmacy at Dorpat, leaving Butlerov with the responsibilities of a department Head. On March 13, 1853, Butlerov submitted a dissertation, "*Ob efirnykh maslakh [On essential oils]*," to Kazan University for the degree of *Doktor Khimii*. Because the Chair of Chemistry was vacant at the time, the Professor of Technology, Modest Yakovlevich Kittary [29][20] (Киттары Модест Яковлевич, 1825–1880), and the Professor of Physics, Aleksandr Stepanovich Savel'ev (Савельев, Александр Степанович, 1820–1860) [30][21] were appointed as examiners. Kittary gave a positive assessment of the dissertation, but Savel'ev did not. He took the position that the dissertation lacked the novel, in-depth results that were needed to meet the required standard for the *Dr. Khim.* It is noteworthy that Butlerov published neither of his two chemistry dissertations, perhaps indicating that he understood and even agreed with Savel'ev's position.

After their split decision was referred to the full Physics-Mathematics Faculty, a debate ensued that ended with Butlerov being denied the degree, but being permitted to submit the dissertation elsewhere. On January 3, 1854, he asked permission from the Rector of Moscow University to submit his dissertation, along with his

[20] For an image of Kittary, see: http://www.ksu.museum.ru/chmku/eng/s6.htm.
[21] For an image of Savel'ev, see: https://www.eduspb.com/node/1151.

Magistr Khimii Diploma, and to take the examinations for the *Doktor Khimii*. The request was granted. The dissertation was officially received by Moscow University on January 16, 1854, and Butlerov subsequently passed the examinations and successfully defended the dissertation; his degree was awarded on June 2, 1854.

6.3.2 Butlerov's First Independent Academic Position: The "Bemerkungen" Paper and His Attitude Towards Structural Theory

On his return to Kazan, the University Council appointed Butlerov as Extraordinary Professor of Chemistry by a 23-1 vote on September 25, 1854. Thirty months later, on March 20, 1857, the Council voted 20-1 to appoint him Professor of Chemistry. Butlerov was awarded a *komandirovka* for the 1857–1858 academic year, and this became a truly life-altering event. This had not been possible when he was a newly minted *kandidat* in 1849 because of the university's staffing needs.

During this *komandirovka*, Butlerov met Kekulé and Erlenmeyer, and he spent the period between December 1857 and May 1858 in the Paris laboratory of Adolphe Wurtz. This was a pivotal time to be in Paris. The Chemical Society of Paris was in its infancy, and discussions of the most recent and important developments in organic chemistry—especially theoretical developments, thanks to the work of Dumas, Gerhardt and Laurent, in particular—were the routine subjects of discussion. Among the "hot topics" of the time were questions of the structure of compounds and the valency ("atomicity") of elements.

In August 1857, Kekulé had submitted a paper to the *Annalen* that was published in the second issue of the year, and in which he proposed the tetravalency of carbon [31]. Butlerov, along with at least some of the Paris organic chemists, adopted the Kekulé position. When Kekulé and Couper published their theories, Butlerov chose the Kekulé version. Couper had taken the position that Structural Theory should completely replace Type Theory, and he completely abandoned the trappings of Type Theory in his paper. Kekulé, on the other hand, chose to disclose his theory as an extension of Type Theory, some of whose trappings he retained. This difference in emphasis may explain why the (inherently conservative) organic chemistry community preferred the Kekulé theory.

In December 1858, Butlerov submitted a "Remarks (Bemerkungen)" paper critical of Couper's theory to the *Annalen*; it appeared in the first issue of 1859 [32]. Couper's paper had begun by asserting that Theory is the goal of scientific research, but Butlerov saw identifying new chemical laws as the goal of science—for Butlerov, a theory was a series of substances from whose behavior which laws could be derived. More importantly, at the time he wrote this paper, Butlerov was inherently mistrustful of inductive reasoning, and he made the point that Couper had extrapolated far beyond what the experimental data would support. He was also reluctant to accept that Gerhatdt's types needed to be replaced. He recognized that double decompositions (the basis of Gerhardt's theory) gave no useful information about the internal structure of substances, he did not see Couper's theory as the answer to the problem.

In Paris, Butlerov had been an active participant in discussions of the hot topics of organic chemistry of which atomism [33] and questions of structure were foremost. During this time, he also became a member of the Société chimique de Paris. Although he was not yet convinced of the validity of the new Structural Theory when he returned to Russia, he was well acquainted with the arguments on both sides of the question. He returned to Russia still unconvinced of the validity of Structural Theory, at least in the forms proposed by Kekulé and Couper, but fully engaged in the question of the structures of organic compounds. In 1859, he singled out Couper's version of the theory for criticism in his *Bemerkungen* paper in the *Annalen* [32] because he disagreed with Couper's opening premise:

> The end of chemistry *is its theory*. The guide in chemical research *is a theory*. [Author's translation]

Butlerov's stand is clear from his opening sentences from the *Bemerkungen* paper:

> The establishment of a theory, says Couper, is the purpose of scientific investigation. In my opinion the development of a theory is the necessary consequence of previous investigations, but the purpose is rather the knowledge of the laws by which chemical metamorphoses take place. By knowing the laws, we become acquainted with the general causes which condition them—the original properties of matter. [Author's translation]

Butlerov also believed that in his theory, Couper had extrapolated further than was warranted from the known facts, as the next sentences show:

> *Experimental* investigations will give us the basis of a true chemical theory, which will be the mathematical theory of what we call chemical affinity, the molecular force. But since affinity is not only the cause of metamorphosis, but also that of the definite grouping of the elementary atoms in a chemically composed molecule, this affinity must be studied not only during the molecular movement it produces, but also in the state of equilibrium of matter. So far it shows a great want of research of this kind, and it is especially the investigations of the physical properties of bodies, possible without the destruction of the molecule, which can give us the means of discovering this grouping. [Author's translation]

And yet, less than three years after publishing this paper, Butlerov had become a committed Structuralist.

6.3.3 Epiphany: Becoming a Proponent of Structural Theory

We still have the hand-written notes of Butlerov's course for the academic year 1859–1860 [34], taken by his closest acolyte and (it is fair to assert) favorite student, Vladimir Vasil'evich Markovnikov (Марковников Владимир Васильевич,

1837–1904)]. They provide us an insight into Butlerov's position on the question at the end of 1860 and represent the last time that Butlerov taught the course based on Type Theory—in less than a year, Butlerov had delivered his paper to the *36 Versammlung deutscher Naturforscher und Ärzte zu Speyer* in September 1861 and published it in Erlenmeyer's *Zeitschrift für Chemie*. In this paper, he used the term, "chemical structure," for the first time, and he proposed that a substance was characterized by one and only one chemical structure.

In Russia, this paper has come to be known as the "Speyer paper." Butlerov included it in his report to the university on his *komandirovka*, so it was also published in the *Uchenye zapiski Kazanskogo universiteta [Scientific Notes of Kazan University]* [35]. This move was unprecedented, required the approval of the Physical–Mathematical Faculty By the 1862–1863 academic year, Butlerov was teaching organic chemistry based entirely on Structural Theory [36]. Following the publication of the Speyer paper, Butlerov focused his research program on developing his theory and finding support for it.

It is important to remember that Butlerov speaks of the *chemical structure* rather than the *physical structure* of a substance, a point he made explicitly in the Speyer paper [37]. The following translation from the Russian version [35], published after Butlerov's return from abroad arguments are as below, is subtly different from the translation from the German by Kluge and Larder [38].

> True, we do not know the connection that exists between the mutual chemical action of the atoms of its *components* and their mechanical arrangement inside a complex particle. We do not know, for example, whether atoms that directly act chemically on each other are directly adjacent to one another. Nevertheless, even when we leave aside the concept of physical atoms, it cannot be denied that the chemical properties of a complex body are determined primarily by the chemical relationship of the elements of its components. Based on the idea that each chemical atom included in the composition of the body takes part in the formation of this latter and acts here with a certain amount of chemical force (affinity), I call the distribution of the action of this force, as a result of which chemical atoms, indirectly or directly influencing each other, combine into a chemical particle, the *chemical structure*.
>
> I would for now consider it possible to change the well-known rule, which states that the nature of a complex particle is determined by the nature, quantity and arrangement of its elementary constituent parts, as follows: *the chemical nature of a complex particle is determined by the nature of its elementary components, their quantity and chemical structure*.[Author's translation; emphasis mine]

Here, Butlerov asserts that one can determine how the affinities of the individual atoms are connected to each other, but not how the atoms are located physically with respect to each other. Consequently, understanding differences in the chemical affinities of polyvalent atoms such as carbon, assumed a position of central importance.

To test his ideas, Butlerov fell back on a question that became central to the Kazan Chemical School under his leadership—how many isomers are there for a given molecular formula, and how well does the theory predict the number of isomers corresponding to a particular molecular formula?

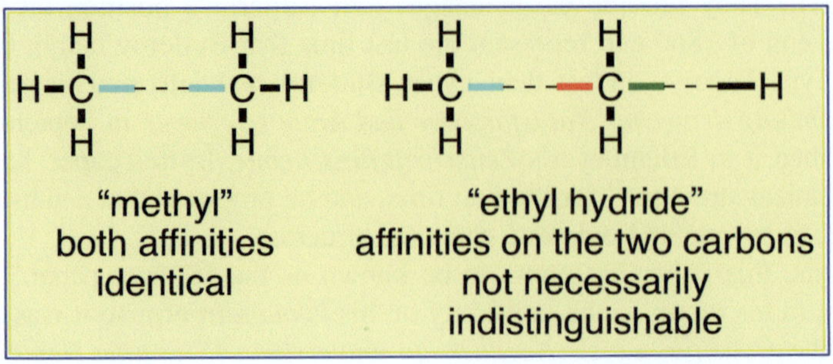

Fig. 6.3 Are the affinities used by carbon to form the carbon–carbon bond of "methyl" and "ethyl hydride" distinguishable?

This means, of course, that the information that they were relying on had to be correct. From the theoretical standpoint, the first question to answer was about the four affinities of a tetravalent atom, with two possible answers: either (1) the four affinities are distinguishable, or (2) the four affinities are indistinguishable. The choice of proposition (1) was based, in part, on the fact that there were two molecules with the molecular formula, C_2H_6: "methyl," with the structure CH_3–CH_3, and "ethyl hydride," with the structure C_2H_5–H. On the assumption these two compounds were isomers, Butlerov concluded that the two carbon atoms of "methyl" are identical, so the affinities used to bind them of together must also be identical In "ethyl hydride," the carbon atoms (one methyl and one methylene) are different, so the affinities binding them together are not necessarily indistinguishable (Fig. 6.3).

6.4 1862–1864: The Question of Variable Affinities

Butlerov extended his arguments to polyvalent atoms in two papers in 1862, one in the *Uchenye Zapiski* [39], which was then reprinted as a German translation in the *Zeitschrift* [40]. In these papers, Butlerov addressed the question: are the affinities used by carbon to form the carbon–carbon bond in ethyl hydride distinguishable? This question was presaged in 1860 by Louis Pasteur (1822–1895) [41],[22] who wrote in his *Leçons de chimie professées en 1860* [42],

[22] For an image of Pasteur as a young man, see: https://commons.wikimedia.org/wiki/File:Louis_Pasteur_en_1857.jpg.

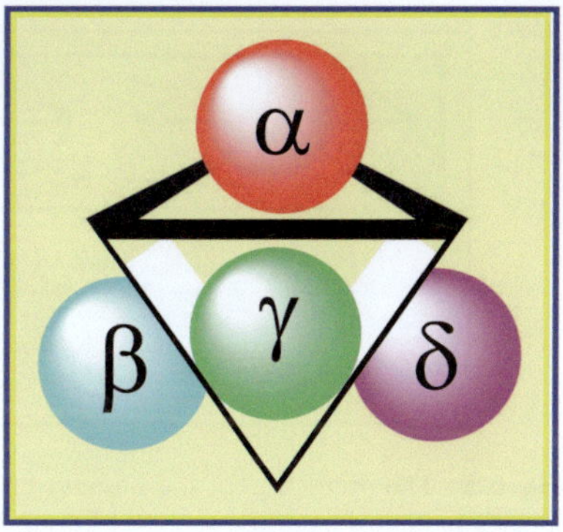

Fig. 6.4 Butlerov's model of a carbon atom with four different affinities

> Les atomes de l'acide droit sont-ils groupés suivant les spires d'une hélice dextrorsum ou placés aux sommets d'un tétraèdre irregulier, ou disposés suivant tel ou tel assemblage dissymétrique déterminé? Nous ne saurions répondre à ces questions.[23]

In the 1862 papers, Butlerov begins with the problem of identifying a suitable model that he could use to illustrate his thoughts. It is probably a step too far, but it is tempting to speculate that Pasteur's thoughts may have influenced Butlerov's thinking as he was wrestling with the question of affinities of carbon. For whatever reason, he chose to use a tetrahedron as the model for the carbon atom, and its bonds to monovalent elements as the base for his logical arguments. He begins with the assumption that the four units of affinity are different:

> Thus, if we take a crude example, and assume that a tetravalent atom of carbon has all 4 affinity units different, we imagine it in the form of a tetrahedron, in which each of the 4 faces is capable of binding 1 atom of hydrogen, without being able to designate each face according to the method of attraction it exerts and determine its position, we, nevertheless, can assert that this application is different for each face, and we can characterize this difference, although we do not know which of the faces this or that method of action belongs to. [Author's translation]

We can represent this model graphically as shown in Fig. 6.4.

Using hydrogen and chlorine as representative monovalent elements, Butlerov asserts that, if the affinities of carbon are distinguishable, the methyl chloride derived by replacing one of the hydrogen atoms of methane, at positions α, β, γ or δ, with chlorine should exist as four possible isomers: CH_3Cl_α, CH_3Cl_β, CH_3Cl_γ and CH_3Cl_δ. Likewise, the chloroform derived by replacing all the hydrogen atoms of methane, except those at positions α, β, γ or δ, with chlorine should also exist

[23] Are the atoms of the right-handed [dextrorotatory] acid grouped according to the turns of a right-handed helix, or placed at the vertices of an irregular tetrahedron, or arranged according to this or that specified dissymmetric assembly? We could not answer these questions.

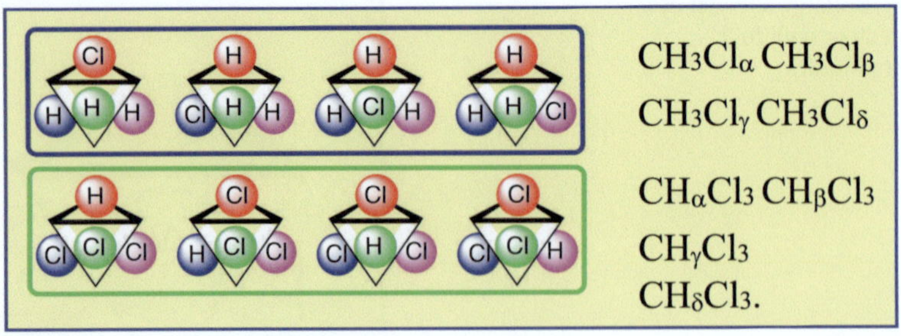

Fig. 6.5 The prediction of the number of isomers of methyl chloride and chloroform using Butlerov's model of a carbon atom with four different affinities

as four possible isomers: $CH_{\alpha}Cl_3$, $CH_{\beta}Cl_3$, $CH_{\gamma}Cl_3$ and $CH_{\delta}Cl_3$. These predicted isomers are shown graphically in Fig. 6.5.

When two of the hydrogen atoms of methane are replaced by chlorine to get dichloromethane, the number of possible isomers changes, depending on how many of the affinities are identical to each other. When all four affinities are different (i.e., there are 1 + 1 + 1 + 1 distinguishable affinities), Butlerov predicted the existence of six possible isomers: $CH_{\alpha}H_{\beta}Cl_{\gamma}Cl_{\delta}$, $CH_{\alpha}Cl_{\beta}H_{\gamma}Cl_{\delta}$, $CH_{\alpha}Cl_{\beta}Cl_{\gamma}H_{\delta}$, $CCl_{\alpha}H_{\beta}H_{\gamma}Cl_{\delta}$, $CCl_{\alpha}H_{\beta}Cl_{\gamma}H_{\delta}$, and $CCl_{\alpha}Cl_{\beta}H_{\gamma}H_{\delta}$. These six possibilities are shown as the top row of Fig. 6.6.

If there are two indistinguishable affinities and two other distinguishable affinities (i.e., 2 + 1 + 1 distinguishable affinities), Butlerov predicted that there should be four isomers of dichloromethane (the second row in Fig. 6.6). With two different kinds of affinities on carbon, Butlerov distinguished two possible arrangements: (1) as two pairs of indistinguishable affinities (i.e., 2 + 2 distinguishable affinities), which predicted to three possible isomers of dichloromethane, as shown in the third row of Fig. 6.6, or (2) as one unique and a set of three indistinguishable affinities (i.e., 1 + 3 distinguishable affinities), which predicted two possible isomers, as shown in the bottom row of Fig. 6.6.

Extrapolating from these predictions, Butlerov concluded that carbon has two different types of affinity (Fig. 6.7).

Butlerov proposed that affinities of carbon used for bonding between the cyano groups of cyanogen are *primary affinities.* The affinities of carbon used for bonding between the methyl groups of "methyl" *[ethane]* and for bonding of carbon to iodine in methyl iodide, for example, he described as *secondary affinities.* The carbon–carbon bond in acetonitrile uses one primary and one secondary affinity.

Acetonitrile did, however, reveal an inconsistency in Butlerov's primary/secondary affinity model. The cause of the inconsistency came from Dumas' observation that acetonitrile (with one primary and one secondary carbon affinity used) is converted into ethane (with two secondary carbon affinities used) and potassium cyanide by treatment with potassium metal [43]. If we assume that the secondary affinity used by carbon to bind to the primary affinity of the cyano carbon, $C^{[2°(1°)]}$,

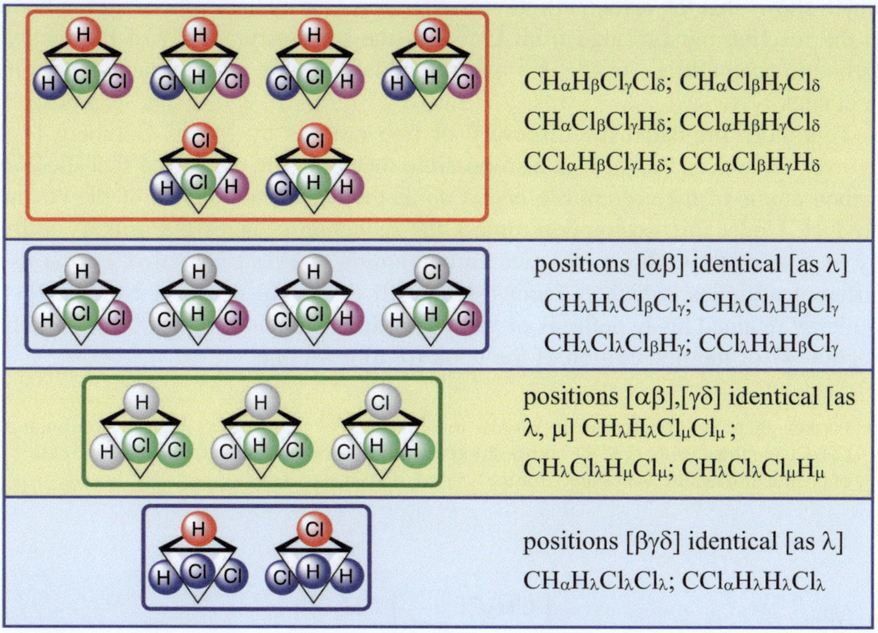

Fig. 6.6 Prediction of the number of isomers of dichloromethane using Butlerov's model of a carbon atom with four, three or two (as two identical pairs or an identical triplet and one unique affinity) different affinities

Fig. 6.7 Butlerov's definition of primary and secondary affinities

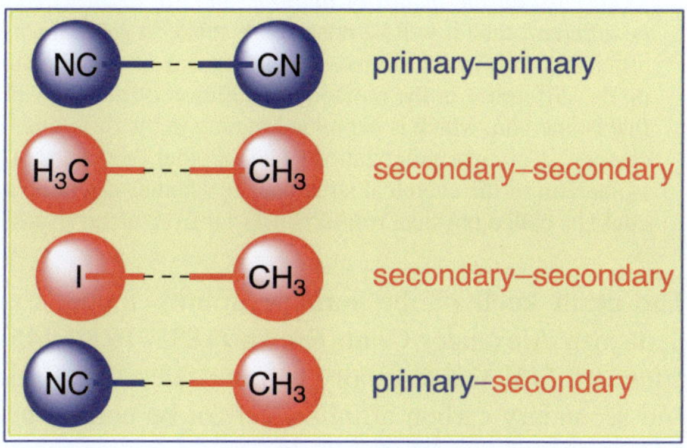

is not identical to the secondary affinity of the methyl carbon bonded to the secondary affinity of carbon or iodine, $C^{[2°(1°)]}$, we can write the equation for the reduction in this form:

$$2H_3C^{[2°(1°)]}C^{[1°]}N + 2K \rightarrow H_3C^{[2°(2°)]}C^{[2°(2°)]}H_3 + 2KC^{[1°]}N$$

which shows that the carbon affinities used to form the ethane carbon–carbon bond in the reaction must change from $C^{[2^\circ(1^\circ)]}$—the secondary affinity of the methyl carbon in acetonitrile—to $C^{[2^\circ(2^\circ)]}$—the secondary affinity of both methyl carbons in ethane.

This work pre-dated the discovery of free radicals by Moses Gomberg [44] by over three decades, so it is also possible that Butlerov conceived that the two carbon atoms of the acetonitrile ended up as the two carbon atoms of the ethane product. Under this assumption, unless the reduction changes the primary affinity to a secondary affinity, the two carbon atoms of ethane ("methyl") must use different affinities, which is inconsistent with the initial premise. Nevertheless, Butlerov retained his hypothesis of variable affinities in his 1864 textbook [37]. In Section 47 of the first edition of the book (p. 61), we find:

> Further, there is some reason to suspect the existence of two isomers with the formula CH_3Cl and three isomers of the composition $C_2H_3Cl_3$—For the first formula, only one case of chemical structure is possible, for the second, only two cases:
>
> $$\left\{\begin{array}{l} CH_2Cl \\ CHCl_2 \end{array}\right. \text{ and } \left\{\begin{array}{l} CH_3 \\ CCl_3 \end{array}\right.$$
>
> *To explain such cases of isomerism, if their reliability is confirmed, it will be necessary to accept that the action of the four units of affinity belonging to the carbon atom may be different*; then it will be possible to speak in general about the difference in the units of affinity of polyvalent atoms. The isomerism of the substance CH_3Cl, would thus depend on the difference in the methods of addition of the chlorine atom. Finally, the question of that isomerism, which is expressed mainly in the difference in the optical properties of substances (in circular polarization), and whether this series of phenomena is explained by the isomerism of the chemical structure, or whether it will be a special kind of isomerism that could be called physical isomerism, is far from clear. [Author's translation; emphasis mine]

The death knell of the variable affinity model came from the M.D. thesis of Scotsman Alexander Crum Brown (1838–1922) [45].[24] In this thesis, there is an argument [46] that logically demonstrates that Butlerov's hypothesis of primary and secondary carbon affinities cannot be correct because it makes predictions at odds with experiment and itself.

In 1885, the final disposition of this problem was settled by Belgian chemist, Louis Henry (1838–1913) [47, 48].[25] In his paper, Henry described the preparation

[24] For an image of Crum Brown, see: https://en.wikipedia.org/wiki/Alexander_Crum_Brown#/media/File:Brown_Alexander_Crum.jpg.

[25] For an Image of Henry, see: https://commons.wikimedia.org/wiki/File:Louis_Henry_(1834-1913).jpg.

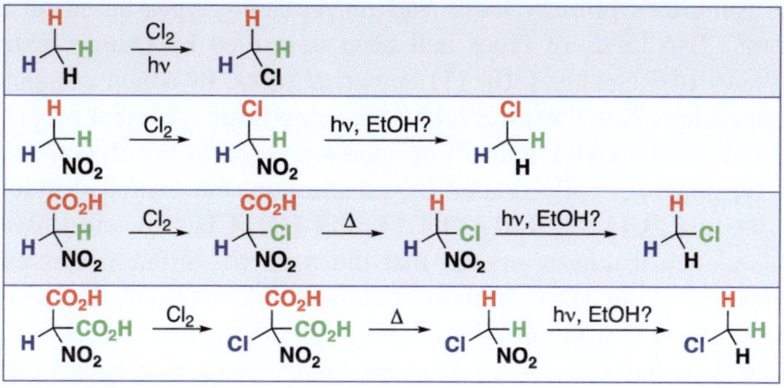

Fig. 6.8 Henry's definitive preparation of chloromethanes by specific replacement of the hydrogen atoms of methane by chlorine

of the chloromethanes obtained by replacement of each of the hydrogen atoms of methane with chlorine. The sequences used are shown in Fig. 6.8, where the hydrogen atoms are color-coded to facilitate following the details of Henry's proof.

In Henry's proof, the first experiment was the chlorination of methane to give chloromethane, whose physical properties were carefully measured to provide a standard. In the second experiment, Henry chlorinated nitromethane, which could be prepared from chloromethane, so the hydrogen replaced by chlorine in this case was clearly not the same as that replaced by chlorine in the direct chlorination. The reduction of the nitro group, presumably by irradiation in the liquid phase [49], gave a chloromethane whose properties were identical in all respects to the standard from the direct chlorination. When Henry started with nitroacetic acid, the radical chlorination gave the chloronitroacetic acid, in which the third hydrogen of methane is replaced by chlorine. Nitroacetic acids are easily decarboxylated to the nitroalkanes on heating, so heating the chloronitroacetic acid gives the chloronitromethane from the previous sequence, and this was then photochemically reduced to chloromethane that had properties identical with the previous samples. The final sequence begins with nitromalonic acid, and provides a chloromethane by a similar approach. This chloromethane, too, was identical in all respects with the previously synthesized samples. Thus, Henry demonstrated unequivocally that all four hydrogens of methane were equivalent, and that the theory of variable affinities was incorrect.

6.5 What Happened?

So why did Butlerov abandon his tetrahedral model? We do not know exactly, but it seems reasonable to assume that his commitment to Structural Theory may have played a major role. In the Speyer paper, Butlerov's focus was on the affinities of polyvalent atoms, and he took pains to emphasize that he was referring to *chemical*, rather than *physical* atoms (Sect. 6.3.3).

Clearly, Butlerov's primary focus was on replacing types based on the affinities of atoms. Two kinds of types had been described by Dumas, extrapolating from the views of Regnault [50]: (1) *chemical types*, in which compounds having different compositions, exhibit retain the same basic chemical properties (e.g., the chlorinated acetic acids), and (2) *mechanical types*, in which compounds have analogous formulas formally related by substitution, but exhibit different chemical properties (e.g., $\mathbf{CH_3}$–H, $\mathbf{CH_3}$–OH, O=**CH**–OH, **CH**$(Cl)_3$, etc.). By using the concept of valence, Butlerov argued that the need for different classes of types became unnecessary. In 1864, Butlerov published his textbook [Error! Bookmark not defined.], which was the first based entirely on Structural Theory,

Two conditions had to be met before the theory of the tetrahedral carbon atom could succeed: First, the existence of *physical* atoms needed to become accepted, and second, the concept that atoms chemically close to each other in the molecule are also physically close to each other needed to be accepted. Once these two concepts were accepted, the logical basis for the tetrahedral carbon atom was in place. Butlerov's focus on the affinities of the carbon atom likely prevented him from accepting the first of these postulates, the *sine qua non* for the development of the van't Hoff–Le Bel tetrahedral carbon atom.

Acknowledgements I am grateful to the Interlibrary Loan Department of the McIntyre Library of the University of Wisconsin-Eau Claire, and John M. McCrackin, **Interlibrary Loan and UW Resource Sharing Coordinator**, for assistance in obtaining various elusive and obscure sources.

References

1. Monographs: (a) Benfey OT (1964) From vital force to structural formulas. In: Hart H (ed) Classical researches in organic chemistry. Houghton-Mifflin, New York. (b) Lewis DE (2010) 150 Years of organic structures. In: Giunta CJ (ed) Atoms in chemistry: from Dalton's predecessors to complex atoms and beyond, vol 1044. ACS Symposium Series, American Chemical Society, Washington, DC, pp 35–57
2. Liebig J, Wöhler F (1832) Untersuchung über das Radikal der Benzoesäure. Ann Pharm 3:249–282. This is followed by the following: Schreiben von Berzelius an Liebig und Wöhler über Benzoyl und Benzoesäure. Ann Pharm 3:282–287
3. Lavoisier AL (1789). Traité élémentaire de chimie. Cuchet, Paris, vol 1, p 293; English Translation by Kerr R (1790) Elements of chemistry, Creech, Edinburgh, reprinted by Dover Press, New York, p 66 (1965)
4. Letter to Berzelius JJ (1835, January 28) In: Bulletin of the atomic scientists (Nov 1949), p 310. Date of letter as given in Mackay AL (1991) A dictionary of scientific quotations 267. This translation of the quotation is from Partington JR (1964) A history of chemistry, vol 4. MacMillan & Co., London, p 327
5. Dumas J, Liebig J (1837) Note sur l'état actuel de la chimie organique. Compt rend hebd séances Acad Sci 5:567–572
6. Liebig J (1838) Ueber Laurent's Theorie der organischen Verbindungen. Ann Pharm 5:1–31
7. (a) Dumas J (1834) Recherches de chimie organique. Ann Chim Phys [2] 56:113–154; (b) Dumas J (1839) Mémoire sur la constitution de quelques corps organiques et sur la théorie des substitutions. Compt rend hebd séances Acad Sci 8:609–622; (c) Dumas J (1839) Ueber

die Chloressigsäure, die Constitution einiger organischer Körper, und über die Substitutionstheorie. Ann Pharm 32:101–119; (d) Dumas J (1840) Ueber die Constitution der Essigsäure und Chloressigsäure. Ann Chem Pharm 33:179–182

8. (a) Laurent A (1836) Théorie des combinaisons organiques. Ann Chim Phys [2] 61:125–146; (b) Laurent A (1836) Essai sur l'action de chlore sur la liqueur des Hollandais et sur quelques éthers. Ann Chim Phys [2] 63:377–389; (c) Laurent A (1837) Suite des recherches diverses de chimie organique. Ann Chim Phys [2] 66:314–335
9. (a) Gerhardt C (1839) Sur la constitution des sels organiques à acides complexes, et leurs rapports avec les sels ammoniacaux. Ann Chim Phys [2] 72:184–214; (b) Gerhardt C (1845) Compt rend hebd séances Acad Sci 20:1031–1038
10. (a) Dumas J (1840) Premier mémoire sur les types chimiques. Ann Chim Phys [2], 73, 73–100; Dumas J (1840); (b) Dumas J (1840) Mémoire sur la loi des substitutions et la théorie des types. Compt rend hebd séances Acad Sci 10:149–178; (c) Dumas J (1840) Ueber das Gesetz der Substitutionen und die Theorie der Typen. Ann Chem Pharm 33:259−300; (d) Dumas J, Stass JS (1840) Ueber die chemischen Typen. Ann Chem Pharm 35:129–173; (e) Dumas J, Peligot E (1840) Ueber die chemischen Typen. Ann Chem Pharm 35:281–292; (f) Dumas J, Piria R (1842) Ueber die chemischen Typen. Ann Chem Pharm 44:66–100
11. (a) Williamson AW (1850) Theory of Ætherification. Philos Mag [iii] 37:350–356; (b) Williamson AW (1851) Suggestions for the dynamics of chemistry derived from the theory of Ætherification. Chem Gaz 9:294–298; (c) Williamson AW (1851) On the constitution of salts. Chem Gaz 9:334–339; (d) Williamson AW (1852) On etherification. Quart J Chem Soc 4:229–239
12. (a) Hofmann AW (1849) Researches on the volatile bases. Quart J Chem Soc 1:159–173; (b) Hofmann AW (1849) Researches on the volatile organic bases. Quart J Chem Soc 1:269–281; (c) Hofmann AW (1849) Researches on the volatile organic bases. Quart J Chem Soc 1:285–317; (d) Hofmann AW (1850) Researches on the volatile organic bases. Quart J Chem Soc 2:300–335; (e) Hofmann AW (1851) Researches on the volatile organic bases Quart J Chem Soc 3:231−240; (f) Hofmann AW (1850) Researches regarding the constitution of the volatile organic bases. Philos Trans 140:93–131. Most of these papers were translated into German: (g) Hofmann AW (1848) Beiträge zur Kenntniss der flüchtigen Basen. Ann Chem Pharm 66:129–149; (h) Hofmann AW (1848) Beiträge zur Kenntniss der flüchtigen organischen Basen. Ann Chem Pharm 67:61–78; (i) Hofmann AW (1848) Beiträge zur Kenntniss der flüchtigen organischen Basen. Ann Chem Pharm 67:129–173; (j) Hofmann AW (1849) Beiträge zur Kenntniss der flüchtigen organischen Basen. Ann Chem Pharm 70:129–149; (k) Hofmann AW (1850) Ann Chem Pharm 73:91–92; (l) Hofmann AW (1850) Beiträge zur Kenntniss der flüchtigen organischen Basen. Ann Chem Pharm 73:180–189; (m) Hofmann AW (1850) Beiträge zur Kenntniss der flüchtigen organischen Basen. Ann Chem Pharm 74:1–33; (n) Hofmann AW (1850) Beiträge zur Kenntniss der flüchtigen organischen Basen. Ann Chem Pharm 74:33–40; (o) Hofmann AW (1850) Beiträge zur Kenntniss der flüchtigen organischen Basen. Ann Chem Pharm 74:117–177; (p) Hofmann AW (1850) Beiträge zur Kenntniss der flüchtigen organischen Basen. Ann Chem Pharm 75:356−368
13. Gerhardt C (1853) Recherches sur les acides organiques anhydres. Ann Chim Phys 37(3):285–342
14. For biographies of Kekulé, see: (a) [M. H.] (1896) August Kekulé. Ber Dtsch Chem Ges 29:1971–1976; (b) Gillis J (1966) Auguste Kekulé et son oeuvre, réalisée à Gand de 1858 à 1867. Mém Acad Roy Belg Classe Sci 37:1–40; (c) Japp FR (1898) Kekulé memorial lecture. J Chem Soc 73:97–138; (d) Anschütz R (1929) August Kekulé I. Leben und Wirken. II. Abhandlungen, Berichte, kritiken, Artikel, Reden. Verlag Chemie GMBH, Berlin; (e) Rimbach E (1903) Das Kekulé-Denkmal in Bonn. Ber Dtsch Chem Ges 36(4):4616–4640; (f) Anschütz R (1961) August Kekulé 1829–1896. In: Farber E (ed) Great chemists. Inter-Science Publishers, New York, pp 697–702
15. For biographies of Couper, see: (a) Anschütz R (1909) Archibald Scott Couper. Arch Geschichte Naturwiss Tech 1:219–261; (b) Anschütz R (1909) Life and chemical work of Archibald Scott Couper. Translated by A. Crum Brown. Proc Roy Soc Edinburgh 29:193–273.

This article was published as a separate pamphlet under the same title in Edinburgh by Robert Grant & Sons, and in London by Williams & Norgate; (c) Dobbin L (1934) The Couper Quest. J Chem Educ 11:331–338

16. Kekulé A (1858) Ueber die Constitution und die Metamorphosen der chemischen Verbindungen, und über die chemische Natur des Kohlenstoffs. Ann Chem Pharm 106:129–159
17. (a) Couper AS (1858) Sur une nouvelle théorie chimique. Comptes rend 46:1157–1160; (b) Couper A-S (1858) Sur une nouvelle théorie chimique. Ann Chim [3] 53:469–489. These papers appeared in English and German: (c) Couper AS (1858) On a new chemical theory. Philos Mag 16:104–116; (d) Couper AS (1859) Ueber eine neuer chemische Theorie. Ann Chem Pharm 110:46–51
18. For biographies of Wurtz, see: (a) Friedel [C] (1885) Notice sur la vie et les travaux de Charles-Adolphe Wurtz. Bull Soc chim Paris 43:i–lxxx; (b) Williamson AW (1885) Obituary notices of Fellows deceased. Charles Adolphe Wurtz. Proc Roy Soc 38:xxiii–xxxiv; (c) Hofmann A (1920) Adolph Wurtz. Ber Dtsch Chem Ges 20:815–996; (d) Rocke AJ (2001) Nationalizing science: Adolphe Wurtz and the Battle for French Chemistry. MIT Press, Cambridge MA
19. Ladenburg to Anschütz, letter dated May 12, 1906. The English translation of part of this letter is cited in Ref. [14b]: "Couper worked with Wurtz in Paris and asked him to pass on to the Academy his paper on the quadrivalence of carbon. Wurtz, who at the time was not a member of the Academy, was obliged to give the paper to some one else who was a member (usually Balard). He bungled this a little and so Kekulé's communication appeared before Couper's was laid before the Academy. On account of this, great wrath of Couper, who took Wurtz to task and became insolent. This displeased Wurtz and he expelled him from the laboratory. Couper seems to have taken this very much to heart and it was believed in Paris that the beginning of his illness dated from this episode. The story itself is authentic: I have it from Wurtz."
20. For biographies of Butlerov in English, see: (a) Leicester HM (1940) Alexander Mikhaĭlovich Butlerov, J Chem Educ 17:203–209; (b) Bykov GV (1970) Butlerov, Aleksandr Mikhailovich. In: Gillespie CC (ed) Dictionary of scientific biography, vol 2. Charles Scribners and Sons, New York, p 620; (c) Brooks NM (2008) Butlerov, Aleksandr Mikhailovich. Complete dictionary of scientific biography, vol 19. Scribner's Sons, Detroit, pp 465–468; (d) Lewis DE (1994) The University of Kazan—provincial cradle of Russian organic chemistry. Part I. Nikolai Zinin and the Butlerov School. J Chem Educ 71(1):39–42; (e) Lewis DE (2012) Early Russian organic chemists and their legacy. Springer Verlag, Heidelberg, pp 47–50
21. Butlerow A (1861) Einiges über die chemische Structur der Körper. Z Chem 4:549–560
22. For biographies of Klaus in English, see: (a) McDonald D (1964) Karl Karlovitch Klaus. Thirty years of research on the platinum metals. Platinum Metals Rev 8:67–69; (b) Pitchkov VN (1996) The discovery of ruthenium. "I named the new body in honour of my Motherland." Platinum Metals Rev 40:181–188; (c) Lewis DE (2012) Early Russian chemists and their legacy. Springer, Heidelberg, pp 42–43; (d) Lewis DE (2016) Klaus at Kazan: the discovery of ruthenium. Bull Hist Chem 41:3–11; (e) Lewis DE (2019) The minor impurity in spent ores of the "Siberian metal": ruthenium turns 1785. Chem Eur J 25:11394–11401
23. For biographies of Zinin, see: (a) Butlerow AM, Borodin AP (1881) Nikolaus Nikolajewitsch Zinin. Bder Dtsch Chem Ges 14:2887–2908; (b) Leicester HM (1940) N. N. Zinin, an early Russian chemist. J Chem Educ 17:303–306; (c) Brooks NM (1995) Nikolai Zinin at Kazan University. Ambix 42:129–142; (d) Brooks NM (2002) Nikolai Zinin and synthetic dyes: the road not taken. Bull Hist Chem 27(1):26–36
24. English biographies of Lobachevskii are relatively few. For an early biography in English, see: Vucinich A (1962) Nikolai Ivanovich Lobachevskii: the man behind the first non-Euclidean geometry. Isis 53:465–481
25. (a) Klaus K (1844) Khimicheskoe issledovanie ostatkov ural'skoe platinovoi rudy i metalla ruteniya [Chemical investigations of the residues of Ural platinum ore and of the metal ruthenium. Uch zap Kazan Un-ta 3:7–136; (b) Claus C (1844/1845) Découverte d'un nouveau metal. Bull Cl Phys-Math Acad Imp Sci St Petersbourg 3:311–316. This paper was reprinted as Claus C (1845) Entdeckung eines neuen Metalles (Ruthenium). Pogg Ann Phys Chem 64:192–197; (c) Claus C (1844/1845) Fortsetzung der Untersuchung des Platinrückstandes,

nebst vorläufiger Ankündigung eines neuen Metalles. Bull Cl Phys-Math Acad Imp Sci St Petersbourg 3:353–371. This paper was reprinted twice under the same title: Claus C (1845) Pogg Ann Phys Chem 65:200–221; J Prakt Chem 34:420–438; (d) Klaus KK (1845) O rutenie [On ruthenium]. Gorn Zhurnal [Mining Journal] Part III 7:159–163

26. (a) Sinin N (1842) Beschreibung einiger neuer organischer Basen, dargestellt durch die Einwirkung des Schwefelwasserstoffes auf Verbindungen der Kohlenwasserstoffe mit Untersaltpetersäure. Bull Sci Acad Imp Sci St Petersbourg 10:273–285; (b) Zinin N (1842) Organische Salzbasen aus Nitronaphthalose und Nitrobenzid mittelst Schwefelwasserstoff entstehend. Ann Chem Pharm 44:283–287; (c) Zinin N (1842) Beschreibung einiger neuer organischer Basen, dargestellt durch die Einwirkung des Schwefelwasserstoffes auf Verbindungen der Kohlenwasserstoffe mit Untersaltpetersäure. J Prakt Chem 27:140–153; (d) Zinin N (1844) Einwirkung von Schwefelammonium auf Nitronaphthalese und Binitrobenzid. Ann Chem Pharm 52:361–362
27. Butlerov A (1849) Dnevnye babochki volgo-ural'skoi fauni [Diurnal butterflies of the Volga-Ural fauna]. Kand. Diss., Kazan
28. (a) Butlerov A (1851) Ob okislenii organicheskikh soedinenii [On the oxidation of organic compounds]. M Khim Diss., Kazan; (b) Butlerov A (1852) Ob okislyayushchiem deistvii osmievoi kisloty na organicheskie veshchestva [On the oxidizing action of osmic acid on organic matter]. In: Butlerov AM Sochichenie, vol 1, pp 13–20; (c) Buttlerow A (1852) Ueber die oxydierende Wirkung der Osmiumsäure auf organische Körper. Bull Classe Phys-Math Acad Sci St-Pétersbourg 10:177–187. Abstracted in 1852 as Buttlerow A (1852) Ueber die oxydierende Wirkung der Osmiumsäure auf organische Körper. J Prakt Chem 56:271–283 and by the editors of Ann Chem Pharm (1852) as Ueber die Einwirkungder Osmiumsäure auf organische Substanzen. Ann Chem Pharm 84:278–280
29. For a brief English-language biography, see: Lewis DE (2012) Early Russian organic chemists and their legacy. Springer Verlag, Heidelberg, footnote p 65
30. There are no English-language biographies of this physicist, who worked on problems associated with the polarization of electric currents. A Russian-language biography is available at https://www.alto-lab.ru/chemists/savelev/. The English translation of the Russian Wikipedia page states: Savel'ev was born on May 27 (June 8), 1820 in St. Petersburg into a merchant family; brother of the archaeologist and orientalist P(avel) S(tepanovich) Savel'ev. In 1840 he graduated as a candidate from the 2nd department (physics and mathematics) of the Faculty of Philosophy of the Imperial St. Petersburg University and in 1841 participated in an expedition organized by the Academy of Sciences to the White Sea and the Arctic Ocean. After receiving a master's degree from St. Petersburg University on July 14, 1845, he was assigned to Kazan University on January 31, 1846, as an adjunct in the department of physics and physical geography. In 1850, he was nominated for the position of extraordinary professor, but due to the lack of a doctorate, this proposal from the Faculty of Physics and Mathematics was rejected. In the same year, he was entrusted with the work of inspecting meteorological stations and physical classrooms of educational institutions in the Kazan educational district. After March 16, 1852, A. S. Savelyev was a Doctor of Physics and Chemistry at Kazan University, becoming the first Kazan professor of physics to receive an academic degree in Russia. However, although he had been been elected as an ordinary professor on May 21, 1854, he was again not confirmed. The following year he moved to Moscow, where from March 13, 1855 he was on the staff of the Konstantinovsky Land Survey Institute. From 1857, he taught in St. Petersburg, in military educational institutions, the commander-in-chief of which was Ya. I. Rostovtsev. He died May 6 (18), 1860. He was buried at the Smolensk Orthodox Cemetery. https://ru.wikipedia.org/wiki/Савельев,_Александр_Степанович. Accessed 18 June 2024
31. Kekulé A (1857) Ueber die s. g. gepaarten Verbindungen und die Theorie der mehratomigen Radicale. Ann Chem Pharm 104:129–150
32. Butlerow A (1859) Bemerkungen über A. S. Couper's neue chemische Theorie. Ann Chem Pharm 110:51–66
33. Rocke AJ (1984) Chemical atomism in the nineteenth century. Ohio State University Press, Columbus, OH

34. Morkovnikov V (1860) Organicheskaya Khimiya, sostavlennaya po lektsiyam ord. prof. A. M. Butlerova stud. Vl. Morkovnikovym [Organic chemistry compiled from the lectures of Ord. Prof. A. M. Butlerov by student Vl. Morkovnikov]; Kazan lit., 1859–1860 (166 p)
35. Butlerov A (1862) O khimicheskom stroenii veshchestv [On the chemical structure of substances]. Uch Zap Kazan Un-ta otd 1:1–11. This paper contains the following footnote on page 1: "Stat'ya eta byla uzhe napechetana v Zeitschrift f. Chemie und Pharmacie 1861, p 549, (Einiges über die chemische Structur der Körper), no tak, kak ona vmeste s sleduyushchei statey, sostavlyayet prilozheniye k otchetu o zagranichnoy pozdke prof. Butlerova, pomeshchennomu vo 2-m otdele etogo vynuska i pri tom obe eti stat'i imeyut neosporimuyu solidarnost', to redaktsiya i sochla neobkhodimym s odobreniya fakul'teta pomestit' v Uchenykh Zapiskakh stat'yu uzhe napechatennuyu v zagranichnom izdanii. [This article was already published in the Zeitschrift f. Chemie und Pharmacie 1861, p 549, (Einiges über die chemische Structur der Körper), but as it, together with the following articles, forms an appendix to the report by Prof. Butlerov, it is placed in the 2nd section of this issue. Both these articles have undeniable solidarity, so the editors have considered it necessary, with the approval of the faculty, to place in Scientific Notes an article already published in a foreign journal.]" Translation by the author
36. Bukhvostov I (1863) Organicheskaya Khimiya, sostavlennaya po lektsiyam ord. prof. A. M. Butlerova stud. Iv. Bukhvostovym [Organic chemistry compiled from the lectures of Ord. Prof. A. M. Butlerov by student Iv. Bukhvostov]; Kazan, 1859–1860 (299 p)
37. Butlerov A (1864) Vvedenie k polnomu izucheniyu organicheskoi khimii [An Introduction to the Complete Study of Organic Chemistry]. PI Fan-der-Flit Publishing House, Kazan
38. For an English translation of the German, see: Kluge FF, Larder DF (1971) A. M. Butlerov: on the structure of substances. J Chem Educ 48(5):287–291
39. Butlerov A (1862) O srodstve mnogoatomnykh pave [On the affinities of polyvalent atoms], Uchen Zap Kazan Un-ta 1 (otd. 1):12–17
40. Butlerow A (1862) Ueber die Verwandtschaft der mehraffinen Atome. Z Chem Pharm 5:297–304
41. Biographies of Pasteur abound; some representative biographies include: (a) Debré P, Forster E (1998) Louis Pasteur. Johns Hopkins University Press, Baltimore; (b) Benz FE (1938) Pasteur, knight of the laboratory. Dodd, Mead & Company, New York; (c) Duclaux É (1920) Pasteur. The history of a mind. Smith EF, Hedges F (translators) W. B. Saunders Company, Philadelphia and London; (d) Vallery-Radot R (1911) The life of Pasteur. Devonshire RL, Osler W (translators). Constable & Co., Ltd., London
42. Pasteur L (1861). Recherches sur la dissymmétrie moléculaire des produits organiques naturels, Leçons de chimie professées en 1860. Ch. Lacure et Cie., Paris, pp 1–48
43. Dumas [JBA] (1847) Action de l'acide phosphorique anhydre sur les sels ammoniacaux. Compt rend hebd Séances Acad Sci 25:383–384
44. (a) Gomberg M (1900) Triphenylmethyl, ein Fall der dreiwertighem Kohlenstoff. Ber Dtsch Chem Ges 33:3150–3163; (b) Gomberg M (1900) An instance of trivalent carbon: triphenylmethyl. J Am Chem Soc 22:757–771
45. For biographies of Crum Brown, see: (a) Kendall J (1932–1935) Obituary notices fell. Roy Soc 1:537–549; (b) Kendall JC (1927) The University of Edinburgh. J Chem Educ 4:565–569; (c) Walker J (1923) Alexander Crum Brown. J Chem Soc Trans 123:3422–3431; (d) Walker J (1923) Alexander Crum Brown 1838–1922. Proc Roy Soc A 105(734):i–v
46. (a) Crum Brown A (1864) On the theory of isomeric compounds. Trans Roy Soc Edinburgh 23:707–719; (b) Crum Brown A (1865) On the theory of isomeric compounds. J Chem Soc 18:230–245
47. For biographies of Henry, see: (a) Delacre M (1914) Notice sur Louis Henry. Ann Acad Sci Lett Beaux-Arts Belg 80:311–409; (b) Bruylants A (1979) Henry (Louis). Biogr Nationale Acad Roy Sci Belles Lett Beaux-Arts Belg Supp 13:414–427
48. Henry L (1886) Sur la valence de l'atome de carbone. Méthode de détermination de la valeur relative de ses quatre unités d'action chimique. Bull Acad Roy Sc Bruxelles 12:644–661

49. The photolysis of nitromethane in condensed phase is initiated by homolysis to methyl radical and nitrogen dioxide: Rebbert RE, Slagg N (1962) Primary processes in the photochemical decomposition of nitromethane. Bull Soc Chim Belges 71:709–721
50. Regnault V (1839) De l'action du chlore sur les éthers hydrochloriques de l'alcool et de l'esprit de bois, et de plusieurs points de la théorie des éthers. Ann Chim Phys Ser 2(71):353–430

René Just Haüy and the Contribution of Crystallography to Three-Dimensional Chemistry

7

Gary D. Patterson

Abstract

Even before the advent of chemical atoms, Natural Philosophers intuited that matter is inherently three-dimensional. Theophrastus (4th BCE) was already a master of Mineralogy and sorted "*Stones*" based on shape, color, hardness, and many other observable characteristics. Pliny the Elder (23–79 CE) compiled a massive *Natural History*, included details about hundreds of minerals, and organized the stones into similarity groups. The foundational document for modern Mineralogy is *De Natura Fossilium* written by Georgius Agricola (1494–1555) in 1546. René Descartes (1596–1650) was a master geometer who applied his novel insights to the structure of matter. Descartes was fully aware of the unique shapes associated with macroscopic crystals and of the mathematics needed to describe them. Robert Hooke (1635–1703) observed the unique geometry of the faces of such crystals in his *Micrographia* (1665). The greatest mineralogist of the eighteenth century in France was Louis Jean-Marie Daubenton (1716–1800) at the Museum of Natural History in the Jardin du Roi in Paris. He was a mentor to René Just Haüy (1743–1822). In 1784, Haüy took these nascent ideas and constructed a full three-dimensional theory of the crystalline state. Many classes of crystals exhibit optical rotation. Such symmetries are called "chiral." They lack a "center of symmetry," and can be either dextrorotatory or levorotatory. A selection of chiral crystals taken from Haüy's classic *Traité de Minéralogie* (1801) will be presented and discussed in terms of their symmetry. Haüy proposed that, in addition to the macroscopic symmetry of the crystal, there was a microscopic "unit cell" with the same symmetry. During the nineteenth century crystallographers developed the full theory of macroscopic crystals. Haüy's successor at the Museum was Gabriel Delafosse (1796–1878). He was the thesis

G. D. Patterson (✉)
Carnegie Mellon University, Pittsburgh, PA, USA
e-mail: gp9a@andrew.cmu.edu

A. Greenberg and D. E. Lewis (eds.), *The Birth of the 3rd Dimension in Chemistry*, Perspectives on the History of Chemistry, https://doi.org/10.1007/978-3-031-97743-5_7

advisor of Louis Pasteur (1822–1895). Pasteur used the techniques of crystallography to understand the existence of both dextro- and levorotatory crystals of tartaric acid salt in 1860. The complete groundwork for the elucidation of the optical activity of organic molecules was available to Le Bel and van't Hoff more than a decade before their "discovery."

7.1 Introduction

In September 1974, the HIST Division hosted a symposium, *van't Hoff–Le Bel Centennial,* at the 168th Meeting of the American Chemical Society in Atlantic City, New Jersey [1]. It is now 150 years since the classic papers by van't Hoff and by Le Bel [2, 3]. The independent realizations that tetrahedral carbon atoms would be optically active if they had four different substituents remains true today! While this advance is rightly celebrated for its influence within the practice of both organic and inorganic chemistry, there is a long history of such ideas within the mineralogical community. The present paper will trace these ideas from Greek books, such as "*On Stones*" (fourth century BCE), through the final demonstration of the optical activity of tartaric acid crystals by Louis Pasteur (1822–1895).

Natural historians have been observing the material world for millennia. One of the outcomes of such activity is the organization of observations into similarity groups. Crystalline minerals have been "found in the field" for a long time. They are characterized by unique shapes and well-defined surfaces. They also differ in "hardness," color, taste, refractive index, and birefringence. Theophrastus (fourth century BCE) was already a serious practitioner of the Natural History of minerals. He carefully observed the many properties of "solids found in the field" and organized this knowledge in a "systematic" way. One of the observable properties was the rhombohedral structure of crystalline minerals. A typical mineral cited by Theophrastus was σμαραγδοσ (emerald) [4] (Fig. 7.1).

The solid geometry needed to describe such crystals was already developed by Euclid. For example, the "perfect solids" included (Fig. 7.2).

One of the key concepts in Natural History is "Transdiction": the combination of excellent data with quantitative characterization and exact mathematics that is conformal with the observations. While this "Royal induction" is not always possible, we will demonstrate that it works for minerals and molecules!

The foundational document for modern Mineralogy is *De Natura Fossilium* written by Georgius Agricola (1494–1555) in 1546 [5]. (My copy is an English translation by Mark and Jean Bandy published by the Geological Society of America in 1955.) It presents a "systematic mineral classification" based entirely on observable properties. His discussion of "stones" provided the extensive data for an attempt at a valid Natural History of Stones.

In order for René Just Haüy to create a consistent and useful theory of crystals, he needed a community of expertise that could both teach him what he needed to

Fig. 7.1 A typical emerald crystal known in antiquity as σμαραγδοσ. It can exist as a hexagonal prism and was highly prized. (Large, di-hexagonal prismatic crystal of 1,390 carats uncut with a deep green color. It is transparent and features few inclusions in the upper 2/3, and is translucent in the lower part. Housed at the Mim Museum, Beirut, Lebanon)

Tetrahedron	Cube	Octahedron	Dodecahedron	Icosahedron
Four faces	Six faces	Eight faces	Twelve faces	Twenty faces
 (Animation, 3D model)	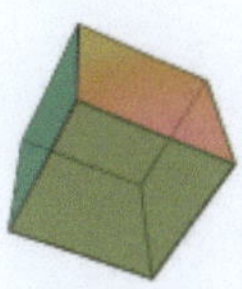 (Animation, 3D model)	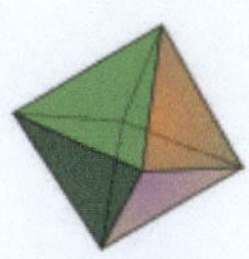 (Animation, 3D model)	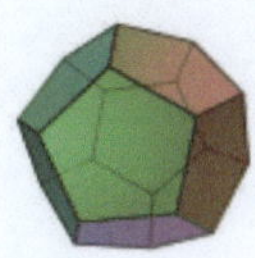 (Animation, 3D model)	 (Animation, 3D model)

Fig. 7.2 Euclid's five perfect solids https://en.wikipedia.org/wiki/Platonic_solid

know and judge the results. The mathematics for such an achievement belonged to René Descartes (1596–1650). *La Géométrie* (1637) provided this base. All the three-dimensional crystalline shapes could be calculated using the analytic solid geometry of Descartes.

One of the most important experimental techniques needed to understand inorganic crystals was three-dimensional optics. Descartes created the first systematic presentation of the properties of light and its interactions with matter. The *Traité de la Lumière* was written in 1632. The system of optics developed by Descartes is now called "ray optics." Light is envisioned to travel in "straight lines." It can be either "reflected" or "refracted" at a surface. Ray optics is still a useful tool. The observation called Snell's Law ($n_1 \sin \theta_1 = n_2 \sin \theta_2$), the refractive index of the incident ray times the sine of the incident angle is equal to the refractive index of the refracted ray times the sine of the refracted angle, was also discovered independently by Descartes [6].

While the macroscopic shape of crystalline minerals could be easily observed, the question still remained about the mesoscopic symmetry of the crystal. Robert Hooke (1635–1703) employed his "microscope" to study the surface of a broken flint crystal (Fig. 7.3).

A systematic study of the facial angles was carried out by Nicolaus Steno (1638–1686). He examined both quartz and hematite (Fig. 7.4).

Steno concluded that the interior angles of the crystal faces remained constant as the overall shape of the macroscopic crystal varied and this induction is still taught as "Steno's Law of Constant Angles." The precise angles were measured with a goniometer (Fig. 7.5).

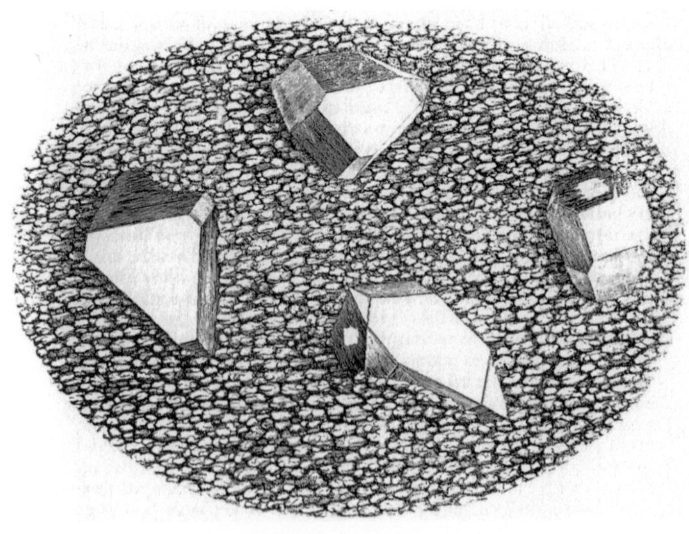

Fig. 7.3 Hooke's drawing of the face of a broken flint, including four larger drawings of individual crystals showing clearly the angles on the faces [7]. Flint is crypto-crystalline quartz (SiO_2)

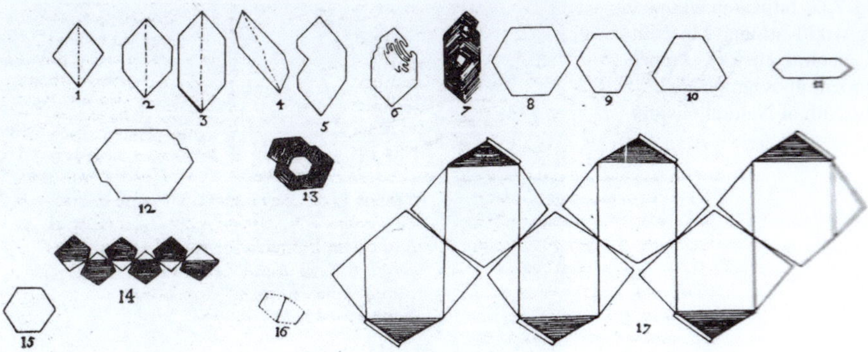

Fig. 7.4 Crystal faces of quartz and hematite (Fe_2O_3) (1669)

Fig. 7.5 Brass contact goniometer of the Carangeot type. [https://en.wikipedia.org/wiki/Goniometer]

7.2 Mineralogy in the Eighteenth Century in France

All the elements for Transdiction involving mineral crystals were available by the eighteenth century. In order to create a collection both large enough and curated enough required money and people. France had both! Louis XV was committed to making France the envy of the world. The Jardin du Roi provided both space and beauty. The human factor was grounded in George-Louis Leclerc, Comte de Buffon (1707–1788). He published the monumental 36 volume *Histoire Naturelle* (1749–1774). Buffon was the mover behind the magnificent Musée d'Histoire Naturelle, located in the Jardin (Fig. 7.6).

One of his most famous disciples was Louis Jean-Marie Daubenton (1716–1800). He was called to Paris in 1742 to join Buffon on the Natural History project. His genius was soon recognized and he was elected to the Académie des Sciences in 1744. Buffon appointed him as Keeper of the King's Cabinet of Natural History.

Fig. 7.6 https://en.wikipedia.org/wiki/National_Museum_of_Natural_History,_France. Modern photograph of the Museum of Natural History

He was also the Professor of Mineralogy at the Jardin du Roi. This community was the best in the world in the eighteenth century for the advancement of Mineralogy [8]. Daubenton became the greatest museum mineralogist of the eighteenth century. He often gave lectures on Mineralogy at the Jardin du Roi [9].

7.3 René Just Haüy (1743–1822)

René Just Haüy was a provincial genius that eventually became an Abbé and was posted to Paris. He loved to walk in the Jardin du Roi. One day he attended lectures by Daubenton on Mineralogy. He immediately grasped his affinity for physical Mineralogy. He pursued the collection and analysis of crystals as an amateur, but he corresponded with a wide community of mineralogists. In 1779, he inadvertently dropped a specimen of "rock crystal" (silicon dioxide). He was struck by the many different faces exposed by this "accident." He read everything he could find, including the *Essai de Crystallographie* (1772) by Jean-Baptiste Louis Romé de l'Isle (1736–1790) [10]. He became the greatest "crystalloclaste" in Europe. He was a master illustrator and draftsman. He measured the angles on the faces and the interfacial angles of his broken crystals. He then created rhombohedrons that included all the known facial forms for each type of crystal (Fig. 7.7).

Eventually he carved such rhombohedrons for all 32 crystal classes. Such sets are still employed in modern Geology departments! This synthesis of the results of laboratory experiments on real crystals is Natural History at its best. But even more remarkable work was yet to come.

Haüy applied his Cartesian geometry to the calculation of all the possible faces and angles for his crystals. He also discovered that there existed preferred planes for cleaving the crystals. He employed his laboratory results to envision the process of cleaving "ideal crystal forms along these axes." There are six ideal crystal forms. The first form is called "isometric" and includes the cube, the regular octahedron, and the rhombic dodecahedron. The second form is called "tetragonal" and includes two types of right square prisms and a square octahedron. The third form

Fig. 7.7 https://www.lindahall.org/about/news/scientist-of-the-day/rene-just-hauy/. Ideal rhombohedron of crystalline quartz carved in pearwood by Haüy. This object lacks a center of symmetry as shown by the two rhombic faces

is called "hexagonal" and includes the hexagonal prism and a pure rhombohedron with equal sides. The fourth form is called "orthorhombic" with all right angles but three different sides. The fifth form is called "monoclinic" and contains three different sides and only two right angles. The sixth form is called "triclinic" and includes three different sides and three different angles. A graphic presentation of this system is (Fig. 7.8).

One of Haüy's greatest inductions was the notion that the macroscopic symmetry was due to the real presence of "unit cells" that retained these symmetries at mesoscopic dimensions. He then envisioned ideal rhombohedra that displayed all the possible faces (Fig. 7.9).

Crystals grow one unit at a time. Once the process has proceeded past a so-called "critical nucleus" stage, crystals grow "on the faces." Haüy envisioned this growth in terms of "integrant molecules" that already displayed the required symmetry. The growth then proceeded one "molecule" at a time. He did not know the size of these units, but he knew they were smaller than could be seen by a visible light microscope. The transdiction that represented the whole as a coherent collection of identical "units" was not a proven proposition. (He could have been wrong.) But history and eventual scientific verification has been "very kind" to this brilliant suggestion.

After he had both created his initial system and carried out the calculations of the structures possible under this assumption, he showed his results to Daubenton. His "inspiration" had borne fruit and Daubenton immediately recognized what Haüy had done. He showed the results to Pierre-Simon Laplace (1749–1827) and both of them encouraged Haüy to prepare a formal paper to be submitted to the Acadèmie. It was a resounding success and he was elected to the Acadèmie in 1783. The paper was published as a book in 1784: *Essai d'une théorie sur la structure des crystaux* (Fig. 7.10).

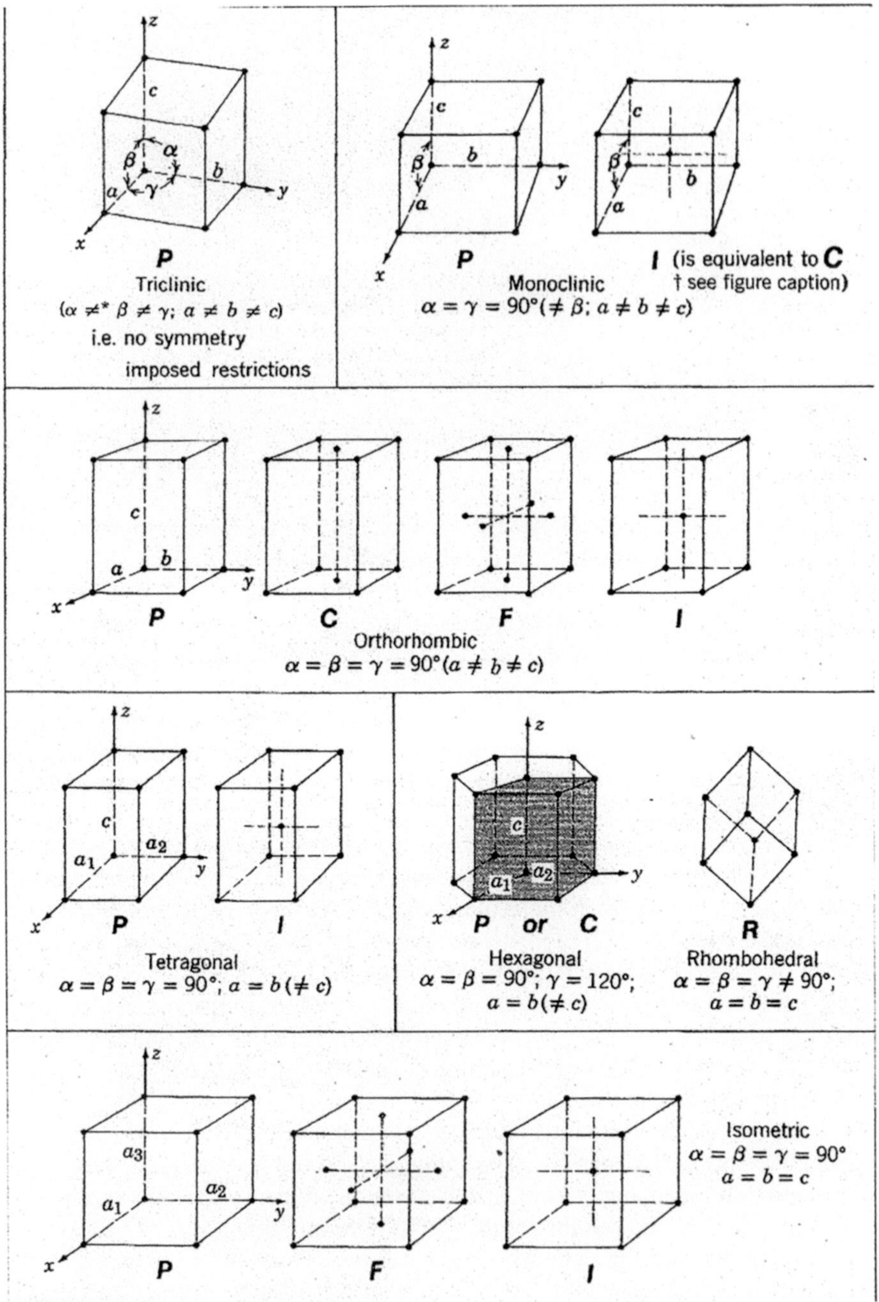

Fig. 7.8 Six different crystal systems [11]

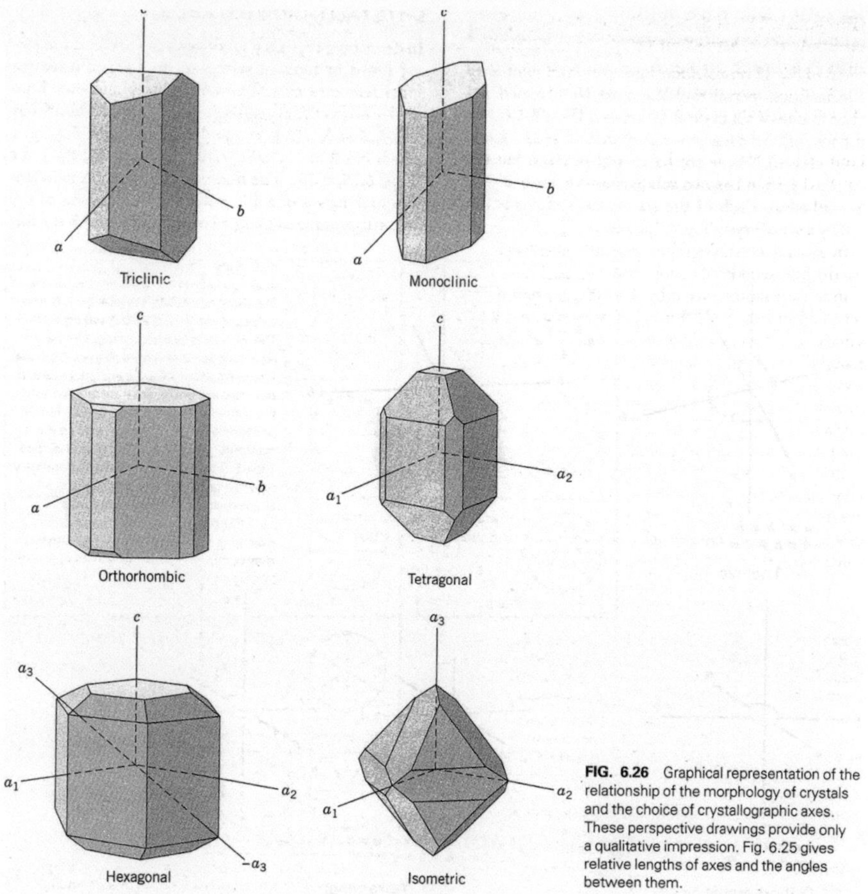

Fig. 7.9 Ideal rhombohedra displaying some of the possible faces consistent with the symmetry [11]

One of the most famous "plates" from this book demonstrated the conceptual progression from a pure cube to other isometric forms (Fig. 7.11).

Haüy's mineral inspiration for Plate 1 was fluorite. A beautiful example of a real fluorspar crystal sample is (Fig. 7.12).

Haüy spent the rest of his life collecting and analyzing mineral specimens. He also traveled throughout Europe explaining his new system of crystallography and being admitted to more than 20 scientific societies. He was lionized in France and was one of the 72 "immortals" inscribed on the Eiffel tower (Fig. 7.13).

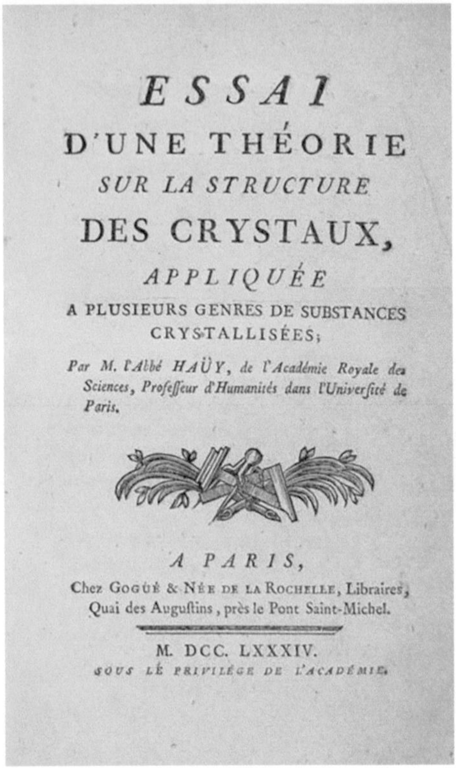
ESSAI
D'UNE THÉORIE
SUR LA STRUCTURE
DES CRYSTAUX,
APPLIQUÉE
A PLUSIEURS GENRES DE SUBSTANCES CRYSTALLISÉES;
Par M. l'Abbé HAÜY, de l'Académie Royale des Sciences, Profeſſeur d'Humanités dans l'Univerſité de Paris.
A PARIS,
Chez GOGUÉ & NÉE DE LA ROCHELLE, Libraires, Quai des Auguſtins, près le Pont Saint-Michel.
M. DCC. LXXXIV.
SOUS LE PRIVILÉGE DE L'ACADÉMIE.

Fig. 7.10 Title page of the Essai (by permission of the Othmer Library of the Science History Institute)

7.4 Optical Crystallography in the Nineteenth Century

The groundbreaking system of Haüy allowed crystallographers to discern which of the six classes a particular crystal was in. But there was substantially more to be learned and France remained one of the centers of progress. One of the areas of major advance was the understanding of the optical properties of solids. Isometric crystals have the same refractive index, *n*, along all three principal axes. All the other classes exhibit more than one refractive index. Tetragonal and Hexagonal crystals are "uniaxial." The refractive index along the "vertical (c)" axis differs from those along the a and b-axes. The final three classes are "biaxial." There are three unique refractive indices. In order to observe the "optical tensor" character of crystals, mineralogists employ a "polarizing microscope [12] (Fig. 7.14)."

The "faces" of optically clear crystals reflect light as well as refract it. The precision of measurement for the crystal facial angles and interfacial angles was greatly improved by William Hyde Wollaston (1766–1828) and his "optical reflection goniometer" [13] (Fig. 7.15).

The precision of the reflecting goniometer was 10 times better than the contact form. This almost immediately led to a reconsideration of the reported angles of Haüy. A detailed discussion of this issue has been presented by John G. Burke

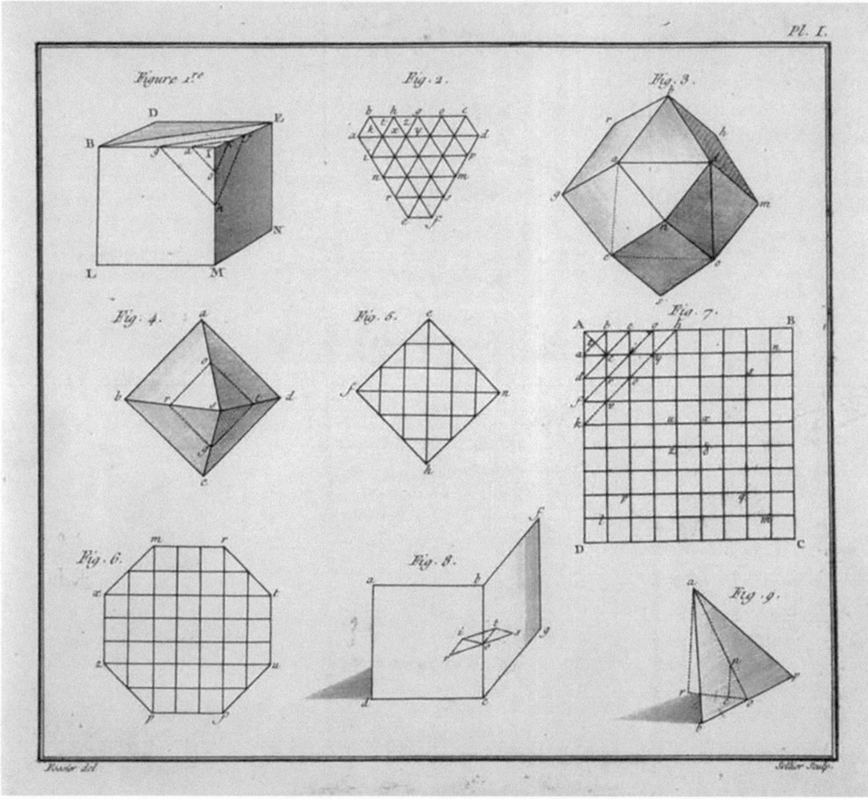

Fig. 7.11 Plate 1 from the Essai (with permission from the Othmer Library of the Science History Institute)

Fig. 7.12 Calcium fluorite mineral specimen https://www.mindat.org/min-1576.html

Fig. 7.13 Montage celebrating Haüy https://www.iucr.org/news/newsletter/volume-30/number-4/rene-just-hauy-and-the-birth-of-crystallography

Fig. 7.14 Brass polarizing microscope from the early nineteenth century https://microscopeinternational.com/ross-london-english-brass-binocular-microscope-antique/

in his *Origins of the Science of Crystals* (1966) [14]. "Perfect" geometric rhombohedrons display "rational" angles for the faces and interfacial planes. Wollaston measured slightly different angles for real crystals "found in the ground" [13]. Physical reality is often slightly different than "ideal theories."

Fig. 7.15 Wollaston reflecting crystal goniometer [13]

7.5 Atomicity and Crystal Structure

While progress in the understanding of the chemistry of pure materials was made in the eighteenth century, most chemists still thought of matter as continuous with either no characteristic lengths or at least a very small one. While there was not enough evidence to convince many chemists, and almost no physicists, that pure substances were composed of real chemical atoms, John Dalton (1766–1844) envisioned such entities and proposed that simple gases were composed of heteronuclear diatomic molecules. The size of these "molecules" was a controversial subject throughout the nineteenth century, but for those scientists that adopted this vision, great progress occurred.

The fact that "integrant molecules" in crystals had specific shapes implied that they were composed of smaller objects that could "bond together" to create specific geometric patterns. Hooke had already proposed such a perspective in the seventeenth century. Modern crystallography includes a chemical atomic unit cell. The modern version for Fluorite (CaF_2) is (Fig. 7.16).

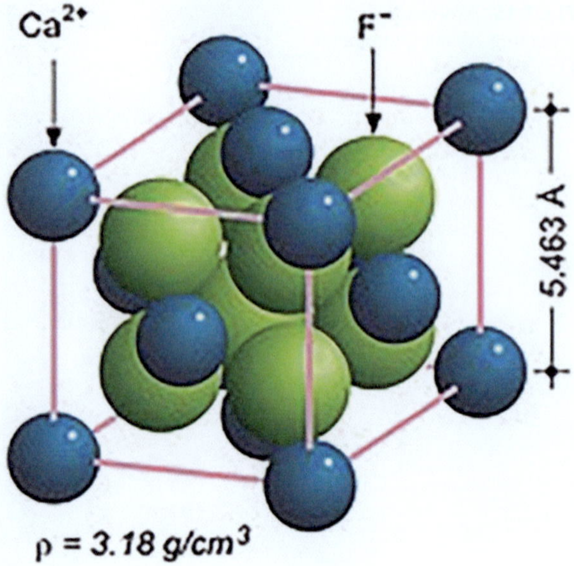

Fig. 7.16 Atomic unit cell for fluorite (https://en.wikipedia.org/wiki/Fluorite#:~:text=Fluorite%20(also%20called%20fluorspar)%20is,isometric%20forms%20are%20not%20uncommon)

7.6 Formal Theory of Crystal Lattices

The final requirement for a full theory of atomic crystals was a more complete analysis of the geometry of "lattices" (see Fig. 7.17). The unit cells can be analyzed in terms of the possible symmetry displayed by the structure. The unit cell of fluorite has a four-fold "rotational symmetry" about the c-axis (A_4). Not all crystal forms have rotational symmetry. For those that do display this symmetry, only 2, 3, 4 and sixfold symmetries are possible for a unit cell that can be replicated indefinitely. There are 3 mirror planes associated with the atoms on the faces of the unit cell and 4 associated with atoms on the "corners" (*m*). There is a center of inversion symmetry (*i*). This symmetry is denoted by the "bar" over the 3. The modern notation for this crystal is $4/m\bar{3}2/m$.

This level of description was constructed by J. F. C. Hessel (1796–1872) in 1830 and published as *Krystallometrie oder Krystallonomie und Krystallographie* (1830). He discovered that there were exactly 32 combinations of symmetry elements that could lead to viable unit cells. A modern summary of these 32 forms and their associated symmetry is given in Figure 7.17.

7.7 The Crystallography of Quartz

Quartz is described by the Hexagonal crystal system and is class 32. It lacks a center of symmetry. As shown in the following figure, there are two non-superimposable forms (Fig. 7.18).

Table 6.3 The 32 Point Groups (Crystal Classes) and Their Symmetry Contents (Those listed in bold are presented in detail in Chapter 9)

Crystal System	Crystal Class	Symmetry Content	Crystal System	Crystal Class	Symmetry Content
Triclinic	1	none	Hexagonal*	3	$1A_3$
	$\bar{1}$	i		$\bar{3}$	$1\bar{A}_3 (= i + 1A_3)$
Monoclinic	2	$1A_2$		**32**	$1A_3, 3A_2$
	m	$1m$		**3m**	$1A_3, 3m$
	2/m	$i, 1A_2, 1m$		**$\bar{3}$2/m**	$1\bar{A}_3, 3A_2, 3m$
Orthorhombic	222	$3A_2$			$(1\bar{A}_3 = i + 1A_3)$
	mm2	$1A_2, 2m$		6	$1A_6$
	2/m2/m2/m	$i, 3A_2, 3m$		$\bar{6}$	$1\bar{A}_6 (= 1A_3 + m)$
Tetragonal	4	$1A_4$		**6/m**	$i, 1A_6, 1m$
	$\bar{4}$	$1\bar{A}_4$		**622**	$1A_6, 6A_2$
	4/m	$i, 1A_4, m$		**6mm**	$1A_6, 6m$
	422	$1A_4, 4A_2$		$\bar{6}m2$	$1\bar{A}_6, 3A_2, 3m$
	$4mm$	$1A_4, 4m$			$(1\bar{A}_6 = 1A_3 + m)$
	$\bar{4}$2m	$1\bar{A}_4, 2A_2, 2m$		**6/m2/m2/m**	$i, 1A_6, 6A_2, 7m$
	4/m2/m2/m	$i, 1A_4, 4A_2, 5m$	Isometric	**23**	$3A_2, 4A_3$
				2/m$\bar{3}$	$3A_2, 3m, 4\bar{A}_3$
					$(1\bar{A}_3 = 1A_3 + i)$
				432	$3A_4, 4A_3, 6A_2$
				$\bar{4}$3m	$3\bar{A}_4\ 4A_3, 6m$
				4/m$\bar{3}$2/m	$3A_4, 4\bar{A}_3, 6A_2, 9m$
					$(1\bar{A}_3 = 1A_3 + i)$

Fig. 7.17 Explanation of the symmetries of the 32 crystal point groups [11]

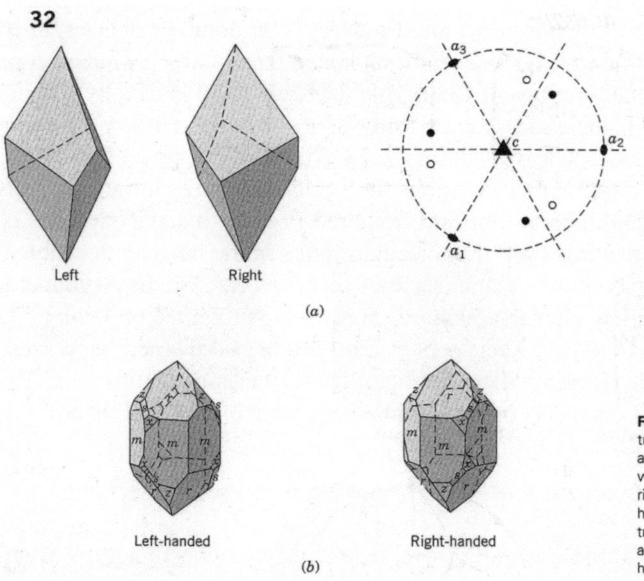

FIG. 9.28 (a) Trigonal trapezohedron {hkil}, positive left- and positive right-handed forms, with a stereogram of the positive right form. (b) Left-handed and right-handed quartz crystals. The trigonal trapezohedral faces are marked by x and their placement determines the handedness of the crystal.

Fig. 7.18 Figure from the Manual of Mineral Science [11] showing the ideal forms for quartz

The possible faces for quartz crystals are denoted by the letters: m, r, s, z, and x. The two x faces mark the non-superimposable faces. The "stereogram" reveals the inherent symmetries of quartz and clearly shows the non-superimposability.

François Arago (1786–1853) was a towering figure in French history, but his role in our story is as a contributor to optics. He was elected to the Académie des Sciences at the age of 23 in 1809. He discovered the rotation of polarized light by quartz crystals in 1811. He invented a polariscope and used it to study many crystals. In 1820, Sir John F. W. Herschel (1792–1871) used his polarizing microscope to separate small crystals of right and left-handed quartz [15]. Like Haüy, Herschel was a good draftsman and included drawings of his crystals.

7.8 Pasteur and Tartaric Acid

What is it about these materials that leads to the phenomenon of optical activity? Symmetry analysis reveals that all crystals that rotate the plane of polarization lack a center of inversion symmetry! Louis Pasteur (1822–1895) separated tartaric acid crystals based on their shape in 1850 [16]. For a molecular crystal, such as tartaric acid, the underlying symmetry already resides in the molecule itself. The fully correct atomic level unit cell for tartaric acid was only determined in 1950. (*Acta Cryst.* (1950) **3**, 341–346.) But W. T. Astbury (1898–1961) did determine the symmetry and dimensions of the unit cell in 1923. It is monoclinic. There are only two molecules per unit cell!

Pasteur did not just "stumble on to his discovery." Sir John Herschel studied the different crystals of quartz and physically separated them under his microscope into dextro- and levorotatory types in 1820. This work was well known by Gabriel Delafosse (1796–1878), the successor to Haüy at the Natural History Museum in Paris. Pasteur was a student of Delafosse and Auguste Laurent (1807–1853). Laurent suggested to Pasteur that he study the optically active tartrates for his thesis. Great teachers and a great topic still required Pasteur to carry out some of the most careful and exacting work in molecular optics in the nineteenth century.

In 1860 Pasteur delivered two famous lectures on his work: "On the Asymmetry of Naturally Occurring Organic Compounds." These were presented on January 20 and February 3 at the Chemical Society of Paris. Pasteur had learned his crystallographic lessons well. He recognized the hemi-hedral faces of his crystals. The crystal class of the (+) crystal is orthorhombic. The modern unit cell determined by X-ray crystallography is [17] (Fig. 7.19).

Pasteur summarized his conclusions at the beginning of his second lecture:

1. The molecule of (+)tartaric acid, however else it may be constituted, is asymmetric, and has that kind of asymmetry which is not superposable with its mirrored image.
2. The molecule of the (−)laevo-tartaric acid is formed by exactly the opposite (mirror image) grouping of the atoms.

Fig. 7.19 Modern crystallographic unit cell of tartaric acid [17]

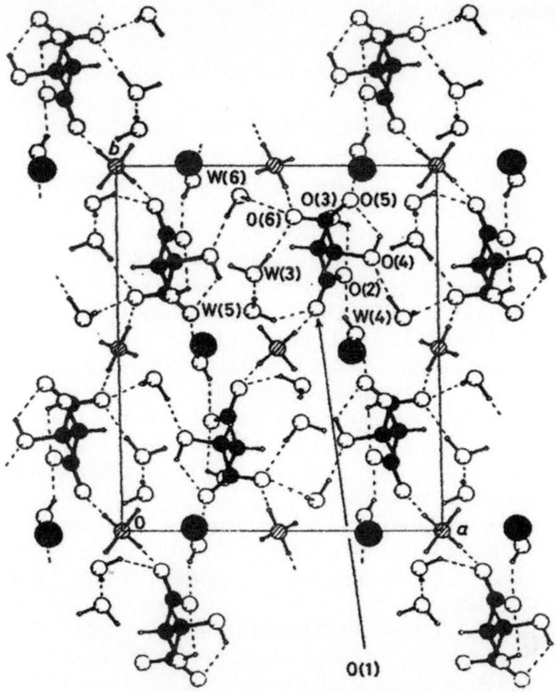

FIGURE 2 Projection of the (+)-(1a) crystal structure on the *ab* plane. Hydrogen bonds are indicated by broken lines

7.9 Concluding Reflections

Chemistry is the study of matter in all its forms. The chemical atomic view of ordinary terrestrial matter has proven to be successful for understanding both the structures and reactions of matter. The description of the structure of matter requires a geometry sufficiently sophisticated to encompass all the tensor properties of matter.

Natural philosophy in the time of van't Hoff and Le Bel had achieved a very high level of mathematical sophistication. No new concepts were needed to make the final "leap."

Phenomenological mineralogy had also become very sophisticated by the end of the nineteenth century. There were many real examples of the crystallographic point groups and many optically active minerals that all lacked a center of symmetry. Was it too much to expect that isolated molecules, either as neat liquids or in solution, would also need to lack a center of symmetry in order to be optically active? Only the courage to envision actual chemical molecules as stable three-dimensional assemblies of atoms was needed.

References

1. Ramsay OB (1975) van't Hoff-Le Bel centennial. ACS Symposium Series 12, Washington, D.C., USA
2. Van't Hoff JH (1874) Sur les formules de structure dans l'espace. Archives Nederlandiases des Sciences Exactes et Naturelles 9:445–454
3. Le Bel JAQ (1874) Sur les relations qui existent entre les formules atomiques des corps organiques, et le pouvoir rotatoire de leurs dissolutions. Bull Soc Chim Fr 22:337–347
4. Caley ER, Richards JF (1956) Theophrastus (4th BCE) on stones. In: Introduction, Greek text, English translation, and commentary. The Ohio State University Press, Columbus, Ohio, USA
5. Agricola G (1546) De Natura Fossilium. Dover Publications, Mineola, NY, USA
6. Descartes R (1965) Discourse on method, optics, geometry, and meteorology. Tr. Paul J. Olscamp, Hackett, Indianapolis, USA
7. Hooke R (1665) Micrographia. Dover Publications, Mineola, NY, USA
8. Roule L (1924) Buffon et la description de la nature. Flammarion, Paris
9. Roule L (1925) Daubenton et l'exploitation de la nature. Flammarion, Paris
10. Romé de l'Isle (1772) Essai de Crystallographie. Didot Jeune, Paris
11. Klein C, Dutrow B (2007) The 23rd edition of the manual of mineral science. John Wiley & Sons, New York, USA
12. Nesse WD (1991) Introduction to optical mineralogy. Oxford University Press, Oxford
13. Wollaston WH (1809) Philos Trans R Soc London 99:253–258
14. Burke JG (1966) Origins of the science of crystals. University of California Press, Berkeley, CA, USA
15. Herschel JFW (1820) Trans Camb Philos Soc 1:43–51
16. Pasteur L (1850) Annales de Chimie et de Physique 28:56–99
17. Kuroda R, Mason SF (1981) J Chem Soc Dalton 6:1268–1273

Part III

Using Stereochemistry to Study Reactions and Structure

L'inversione di Walden: The Undergraduate Thesis of Primo Levi, Chemist and Writer

8

Bart Kahr and Miriam Rossi

Abstract

The Complete Works of Primo Levi in English (3 vol., 2910 p) published in 2015 did not include Levi's undergraduate thesis, *L'inversione di Walden* (1941). Levi, a student at the University of Turin, reviewed the literature of Walden inversion, the surprise that some molecules will turn themselves "inside-out" as if they had passed through a looking glass. The Italian race laws restricted Levi from the laboratory and consigned him to the library. Here, is an English translation of the thesis with some commentary on its relation to Levi's *oeuvre*. Levi's review stopped at 1933, just before mechanisms of nucleophilic substitution were sorted by Ingold and Hughes in the UK. Until then, the primary literature on the mechanism of Walden inversion was unsure, rendering it hard to summarize with the clarity that later would become a hallmark of Levi's genius. An up-to-date thesis was also thwarted by the disadvantaging race laws, and by the bombardment of Turin that made daily life difficult for all. Nevertheless, Levi searched in his sources for a quantitative mechanism, internal clocks, for predicting stereochemistry. Levi's two earliest poems, one before Levi's enslavement in the *Arbeitslager* and one immediately thereafter, show that, the thesis notwithstanding, Levi's gifts as a writer and sensibilities were innate, and they predated his most formidable experiences. Molecular chirality remained a lifelong curiosity as did mirrors and their manufacture. Some have extended Levi's study of enantiomers to the "opposite" world he encountered in the *Lager*. For Levi, mirrors and mirror image molecules were real, parts of

B. Kahr (✉)
Department of Chemistry and Molecular Design Institute, New York University, New York, NY, USA
e-mail: bk66@nyu.edu

M. Rossi
Department of Chemistry, Vassar College, Poughkeepsie, NY, USA
e-mail: rossi@vassar.edu

A. Greenberg and D. E. Lewis (eds.), *The Birth of the 3rd Dimension in Chemistry*, Perspectives on the History of Chemistry, https://doi.org/10.1007/978-3-031-97743-5_8

nature, unlike Auschwitz for which there were few analogies and inadequate language. Precision forbade Levi from comparing enantiomers and mirrors to the *Arbeitslager* which did not invert the normal world but rather turned it upside down (*capovolto*), something mirrors never do.

8.1 Walden Inversion

The mystery of the Walden inversion was captured by Frankland in 1913, 20 years after its observation: [1] "Walden's discovery was certainly a puzzle and a surprise, for it did not fit into any pre-existing theory of optical activity, and had not been anticipated as a corollary to van't Hoff's theorem of the asymmetric carbon atom. It is highly probable that it may mark an epoch in our view with regard to the mechanism of the process of substitution in general" [2]. Frankland is referring to the discovery of Paul Walden (1863–1957) that certain cycles of substitution reactions will turn an enantiomer into its mirror image without racemization. A half a century later, Campbell still called this process "astonishing" [3]. At the turn of the nineteenth century, chemists believed that substitution reactions occurred with the least overall motion favoring direct substitutions with stereochemical retention of configuration. Walden turned stereochemistry inside out.

Walden was born in the Russian Empire in territory that is now Latvia. He was orphaned at an early age but an older brother supported his education [4]. The Baltic region was always at the mercy of the Prussian/German and Russian/Soviet empires to the West and East, respectively. His first schools emphasized German. The Riga Polytechnic, where Walden studied with Wilhelm Ostwald (1853–1932), a pioneering physical chemist with a philosophical spirit, also emphasized German instruction, despite the Russification of the institute by Nicholas II. Ostwald ultimately moved to Leipzig and his professorship was passed to Carl Adam Bischoff (1855–1908) from whom Walden took an interest stereochemistry [5]. This association led to the famous Walden inversion study. Walden nevertheless worked in Leipzig to finish his PhD with Ostwald but elected to return to Riga afterward and was appointed professor in 1894. Then he moved to Saint Petersburg for the credentials that ordinarily would have been required for a professorship. Walden ascended to the director of the Riga Polytechnic but the Russo-Japanese war (1905) caused massive dislocations. As German forces pushed into Latvia in 1915, the university was dismantled and moved within Russia until Walden as rector oversaw the return of the Polytechnic to Riga in 1918. But Russia's civil war concurrent with Latvia's declaration of independence was not a time of restoration.

Walden, having developed the sensibility of a historian by this time, [6] had written his own summary of the Walden inversion in a book, *Optische Umkehrerscheinungen (Walden Umkehrung)* (Optical Inversion Phenomena (Walden Inversion)) published in 1919 [7] and had a chance to express how global dislocations affected his communications about the eponymous mechanism.

According to Walden's preface, his book had its own struggles among warring nations. "This little book," Walden wrote, "also has its story of war and suffering." On a sunny ocean voyage to New York to attend the World Congress of Chemistry in 1912, Walden received an invitation, presumably by telegram, to write a book about his historic discovery for a 1914 volume in series of books on *Wissenschaft* (science). He received the proofs in Riga in July 1914 but then the "war torch flared and set the whole world on fire," severing all contact between Riga and Germany. Students and teachers in Riga were evacuated to Moscow. The project was only completed after the war. The mechanism, however, was not sorted within.

Today, topologists know that a trefoil knot, a closed chiral curve embedded in three dimensions, can turn itself into its mirror image without edge scission by taking an excursion to the fourth dimension and then returning to the familiar world [8]. However, when Walden discovered his chemical inversion mechanism at stereogenic, tetrahedral carbon atoms, no chemist was prepared to suggest the intervention of a fourth geometric dimension. It was clear that enantiomer interconversion had to be worked out wholly in our three dimensions, and without a mirror.

While the S_N2 reaction with inversion of configuration is now a standard part of the university organic chemistry curriculum, and often the first reaction mechanism studied in detail, we argue that understanding how a mirror works is a prerequisite to this lesson that is often ignored. This is unfortunate because mirrors are magical. Every year, when introducing molecular chirality to organic chemistry students, we ask how a mirror works. Specifically, *Why does a mirror exchange left and right, but not up and down?* A class of bright students is typically flummoxed. "How is it possible," asked first by Don Quixote, and later by prodding professors, "that things so trivial...can have the power to perplex and absorb an intelligence as mature as yours?" [9] Mirrors, essentially 2D surfaces, do not exchange left and right, nor up and down, but rather back and front. However, we never experience people exchanging their fronts and backs, turning inside out, faces pushed through skulls and emerging out the other side. Because this never happens for constructions as complex as humans, it *appears* that mirrors exchange left and right. This is a psychological trick. We imagine that if we were to walk behind a mirror to become our image, we would have to switch what our right and left hands were doing. But we don't walk behind mirrors to become our images. The plane of the mirror exchanges our backs with our fronts, a transformation so improbable that it has never happened (to the best of our knowledge) (Fig. 8.1).

Primo Levi surely knew how mirrors worked when, in 1941, as an undergraduate chemistry student at the University of Turin, he selected for his thesis the topic of *L'inversione di Walden* [10]. Molecules are simple enough that they can exchange back and front, by real, yet to be falsified, chemical mechanisms. They can turn into their mirror images and back again.

Levi's thesis now can be found on the website of the *Centro Internazionale di Primo Levi* [11]. It is also available in Levi's two volume *Opere complete* [12] (OC hereafter. First published 1997, *Edizione* 2017 cited herein), as a scanned

Fig. 8.1 **a**. Mirrors exchange fronts with backs. **b** and **c** are unfamiliar isometries (size and shape preserving transformations) that also exchange ups with downs

reproduction. *The Complete Works of Primo Levi* [13] (CW hereafter), the English translation of 2015, does not reproduce the Italian thesis nor translate it.

8.2 Primo Levi

Primo Levi is considered an essential writer of the twentieth century which is unusual because he lacked an international reputation until the English translations of his memoire, *Se questo è un uomo* (*If This is a Man)* [OC I pp141–308; CW pp 2–207], and his incomparable, *Il sistema periodico (The Periodic Table)* [OC I pp 861–1036; CW pp 752–947], where each chapter, named for a chemical element, is loosely connected to an episode of his life or is a work of fiction.[1] For those unfamiliar with Levi, a brief biography may be welcome. These are facts that can be taken from any number of sources, but Barbarulo [14] gives a skeletal biography of the right size for our purposes:

> 1919: Levi was born in Turin in a house where he almost always lived (with involuntary interruptions).
>
> 1937: He enrolled in a course of chemistry at the University of Turin
>
> 1941: He worked briefly as an industrial chemist after earning his degree.
>
> 1943–44: He was arrested and deported to Auschwitz after brief misadventures as a partisan in the mountains. He survived at *Buna,* the synthetic rubber works adjacent to the death camp because of his training as a chemist with working knowledge of German.
>
> 1945: He returned to Turin, after an 8-month travail through devastated Central and Eastern Europe.

[1] For concision, reference to Levi's writings will be given as page numbers bracketed with OC or CW, in place of endnotes.

1946: He worked at a varnish and paint factory and wrote *If This Is a Man* [CW p 2–207] about the *Lager* which had a quiet reception.

1955–1974: He published varied writings while working as general manager of the factory.

1975: He retired from the chemical industry and published *The Periodic Table* [CW p 752–949].

1976–1987: He dedicated himself to writing and his international reputation soared [15] when his books were translated to English.

1987: He died in his house in Turin, found at the bottom of a stairwell.

How can one summarize Levi's achievements in a paragraph or two? *The New Yorker* book critic James Wood did in his review of CW.

> Though his experiences in Auschwitz compelled him to write, and became his central subject, his writing is varied and worldly and often comic in spirit, even when he is dealing with terrible hardship. In addition to his two wartime memoirs, "If This Is a Man" [CW p 2–207] and "The Truce" [CW p 208–397] …and a final, searing inquiry into the life and afterlife of the concentration camp, "The Drowned and the Saved" (1986) [CW p 2407–2574], he wrote realist fiction—a novel about a band of Jewish Second World War partisans, titled "If Not Now, When?" (1982) [CW 1547–1863]—and speculative fiction; also, poems, essays, newspaper articles, and a beautifully unclassifiable book, "The Periodic Table" (1975). [CW p 752–949] [16]

Levi wrote about the weightiest themes with what the celebrated American novelist Philip Roth (1933–2018) called "the moral stamina and intellectual poise of a twentieth century Titan," [17] a balancing act between the emotional response to the worst experiences, and an intellectual remove. Robert Gordon of Cambridge University introduced Levi as surpassing all in chronicling "unbearable events with such accessible economy, elegant wit and humane power, for such a sustained period of time" [18]. The celebrated Italian novelist Italo Calvino (1923–1985) might have been thinking of his friend Levi when he described "literature as an existential function, the search for lightness as a reaction to the weight of living" [19].

Ever gracious, Levi did not charge his reader with understanding and internalizing the incomprehensible. The Nazi's "words and deeds that are not human but, rather, counterhuman, without historical precedent, barely comparable to the cruelest events of the biological struggle for existence…Auschwitz has nothing to do with war; it is not an episode of war, it is not an extreme form of war. War [on the other hand] is an everlasting terrible fact: It is deplorable, it is in us, it has a rationality, we 'comprehend' it (CW p 190)."

Meanwhile, any brief overview of Levi runs the risk of oversimplification, something Levi struggled to avoid in service of the truth.

8.3 L'Inversione di Walden

8.3.1 Backstory

Here we aspire to the following: (1) Explain how we came to take an interest in Levi's undergraduate thesis, (2) translate the thesis to English, (3) evaluate the influence of *L'inversione di Walden* on both Levi's fiction and non-fiction, and (4) evaluate the influenced of Levi's thesis on secondary scholarship.

Some years ago, BK was studying the chiroptical properties of chiral crystals of sodium chlorate grown from a solution containing an equilibrium racemic mixture of a blue triarylmethane cation dye with propeller conformations, following a paper from 1919 by Eligio Perucca (1890–1965) of the *Politecnico di Torino* [20] (Fig. 8.2A). The "handshake" between a chiral salt crystal and a dynamic racemate in solution raised all manner of worthwhile stereochemical and crystal optical questions that we pursued systematically [21, 22].

BK also became interested in Perucca, a scientist otherwise unknown to him. In 2006, the internet was still in its adolescence, and all BK was able to learn about Perucca—who now has a copius Wikipedia page [23, 24]—was from a history of Turinese scientists in which Perucca was described only as *un personaggio bizzarro* [25]. From that instant, we had a dual mission, to understand the nature of Perucca's bizarre crystal *and* the nature of Perucca's bizarre character. But the second trail went cold. To gain a scent, we read from scientists who lived in Turin between the world wars. Foremost was Primo Levi, whose work was already populating our bookshelves.

In his chapter "Potassium," from *The Periodic Table*, Levi describes his training and makes the conditions of the laboratories vivid. He found the preparation of dyes—triarylmethane cation dyes included—in the chemistry lab "entertaining." [CW p 795] But he preferred physics, measurements of "optical rotatory power, and so forth." Dyes and chiroptics. Might he have known Perucca? Levi continues, "I had made desperate attempts, with this and that professor, to enter as a degree student. Some, with twisted mouth, or even rudely, had responded that the racial laws prohibited it; others had had recourse to vague and flimsy pretexts." [CW p 787] In fact, Levi's first major biographer, Myriam Anissimov, identified Perucca

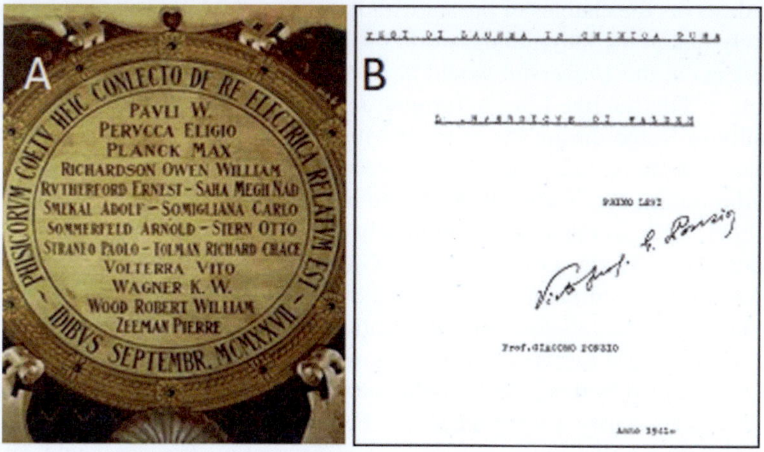

Fig. 8.2 Perucca and Ponzio. A. In 1927, when the leading physicists were gathering to sort out the meaning the quantum theory, there was a prestigious Volta conference at *L'Instituto Carducci* on Lake Como, Italy. The distinguished participants were immortalized on the ceiling of the conference room by the painter Zambelli in 1928. Perucca's name is between Wolfgang Pauli and Max Planck, in alphabetical order: Good neighbors for perpetuity. Photograph courtesy of D. Viterbo (d.). B. The title page of Levi's thesis [OC II p. 1721]

as one of the professors [26–28][2] who refused Levi "with contempt, citing racial laws." [29].

BK asked the biographer Anissimov how she came to know that Perucca was one of these prejudicial professors. She replied by email on 27 January 2007: "By happy circumstance, a friend of Primo Levi from Torino came to see me today in Paris…He said that Eligio Perucca was a perverse and extremely spiteful man. He terrorized students, and was a cruel and inhuman being. This is what the people of Torino say of him today" [30].

In 2007, at a meeting of American Chemical Society in Salt Lake City, BK described experiments on Perucca's crystal, and seemingly confirmed Perucca's eccentric, prejudicial personality with Anissimov's correspondence. In the middle of BK's presentation, a gentleman stood up and asked the speaker to yield the floor. The microphone was passed and the recipient said, paraphrasing from memory, *My name is Davide Viterbo. I am from Torino. I knew Eligio Perucca. He was terrible man* [31–33]. It sounded like a confirmation but Viterbo was unsettled. It is possible to terrorize students non-selectively. Even if Perucca was unkind, that doesn't mean

[2] Levi ultimately found an experimental mentor in the astrophysicist Nicolò Dallaporta (1910–1989), who reported that Levi was an extraordinary student. Dallaporta was delighted to share his laboratory with Levi despite the risks for his career. (see refs. [26]) Together, they studied the dielectric susceptibility of liquids. This work became one of the two *sottotesi* or sub-theses required for graduation. See ref. [27]. The other on X-rays was supervised by Mario Milone and analyzed by Viterbo (see ref. [28]).

that he deserves the charges of racism without evidence. Moreover, Viterbo thought it unlikely that in the labyrinthine world of Italian academia in the 1940s a student such as Levi at the *Università* would have a reason to approach a professor at the *Politecnico*. Viterbo, like Levi, a Turinese Jew and chemist, recounted hiding with his family in Rome during the reign of Mussolini and learning to respond to the safer name of *Paolo*. Our story touched decades-old memories.

Viterbo returned to Turin determined to interview all those available who knew Primo Levi or Eligio Perucca. He ultimately communicated with Levi's son Renzo, his sister, Anna Maria, Levi classmate, Giovanna Rava, and Levi's coworker at the SIVA varnish and paint factory, Renato Portesi as well as Perucca's nephew, Giovanni, and Perucca's assistant, Radicati di Brozolo. By all accounts, Perucca was *un personaggio bizzarro*, but he was not anti-semitic. Quite the opposite in fact. According to Portesi, Levi would secretly attend Perucca's physics lectures because of their clarity, [34] and because Perucca was a "crypto-anti-fascist" [35] peppering his remarks with sarcasm toward Mussolini. Perucca could rally his barbed wit to humiliate students or the government. Perucca, as rector of the *Politecnico,* was tasked with hosting Mussolini's visit to campus. He is remembered by a former student, Angiola Maria Sassi Perino, to have arrived at this meeting in Piedmontese folk costume, not a requisite black shirt, a bold and dangerous statement of protest. She said of Perucca, "He was not a member of the Party" [36].

Afterward, BK and Viterbo discussed translating and analyzing Levi's undergraduate thesis, *L'inversione di Walden* (1941), his *Tesi di Laurea* in Pure Chemistry, a literature analysis supervised by Professor Giacomo Ponzio (1870–1945), an organic chemist with an interest in stereochemistry (Fig. 8.2B). Ponzio, a liberal, was nevertheless constrained by the Italian race laws and would not supervise an experimental thesis by a Jewish student [37]. No matter Ponzio's shortcomings, Levi appreciated his mentorship: "I still remember Professor Ponzio's first chemistry lesson, from which I got clear, precise, verifiable information, without useless words, expressed in a language I liked enormously, also from a literary point of view: a definite language, essential" [38]. "Zinc" [CW p 775–782], in *The Periodic Table*, however, paints a more equivocal portrait of "Professore P.", not named fully. Levi entered the University of Turin in 1937, and Jewish second-year students and beyond were allowed to finish their studies despite the 1938 legislation that prohibited Jews from enrolling in "mixed" educational institutions [39].

In 2015, WW Norton issued the CW which we became aware of two years later [13]. Many stories and essays were newly translated or retranslated, but there was no translation of *L'inversione di Walden*. Finally resolved to translate and make an analysis of the thesis, BK tried to contact Viterbo but discovered that his friend had just passed away [40]. Then, BK, while visiting author MR at Vassar College, noticed several books by Levi on her office shelves. MR is a native Italian speaker and a longstanding friend of Viterbo from the Italian X-ray crystallography community. She smoothly took the baton from Davide, and we set out to make sense of the document together.

Five scholars have previously commented on the *L'inversione.* First is Enzo Borello [41], an Italian physical chemist and Dean of the Science Faculty at the University of Turin. Then comes Naso [37] an organic chemist from the University of Bari currently at Synchimia Srl with an active interest in chemical history [42]. Next [43] are collaborators [44, 45] Giancarlo Cravotto, an organic chemist of the University of Turin with an Levi-like interest in translational research to industry and Pedro Cintas, a Spanish Levi-like stereochemist and historian of the 3rd dimension in chemistry [46], among other subjects. Lastly is Maria Vittoria Barbarulo [14, 47],[3] a chemistry instructor in Rome and a scholar of the Jewish race laws and the fascist period in Italy [48]. Their comments on the thesis are more comprehensible than those of Levi's biographers and literary critics (as we will explore below) because they are written by scholars trained in the language of chemistry. We will comment on their observations following the translation.

The questions we were interested in resolving by studying *L'inversione* were the following:

- Did Levi succeed in summarizing the vexing mechanistic chemistry of Walden inversion given his time and place and the urgencies in Italy in 1941?
- Is there any evidence that Levi was gifted writer and communicator as a student?
- In what way does the thesis echo through the nearly 3,000 pages of the otherwise CW [13]? Perhaps we now can say more? According to science historian Antonio Di Meo, "the deepest links between [Levi's] scientific education and his literary work are still waiting a full investigation" [49].

In the next section, we give an English translation of the thesis. In the final section, we will wrestle with its echoes in Levi's work and the work of others. We do not take on this job lightly. This is Primo Levi's writing, after all. And he had strong opinions about translation: "The author who finds before him a page of his own work translated into a language that he understands will, variously, or all at once, feel that he has been flattered, betrayed, ennobled, X-rayed, castrated, planed smooth, raped, embellished, or murdered." [CW p 2123] Needless to say, Levi won't see this. Nevertheless, we are scared, but not to inaction.

8.3.2 Translation

ROYAL UNIVERSITY OF TURIN

THESIS IN PURE CHEMISTRY

[3] Barbarulo's essay, *La pergamena nel cassetto* ("The parchment in the drawer"), was recently posted on the website of the *Centro Internazionale di Studi Primo Levi*, with an unparalleled searchable bibliography. It is also richly documented with Levi's school records (https://www.primolevi.it/it/pergamena-cassetto). Accessed 1 September 2025.

WALDEN INVERSION
PRIMO LEVI

[Signed] Prof. GIACOMO PONZIO.

Year 1941.

I—Significance of Walden Inversion

When replacing one of the groups linked to an asymmetric carbon atom with another group, so as not to symmetrize the molecule, a new active compound is usually obtained (that is, if you have no total or partial racemization), whose rotatory power may differ in value and in sign to that of the initial compound. This behavior was initially framed within mechanical and steric considerations under the assumption that such reactions would affect the smallest number of bonds, i.e., by simple substitution, and even conceding the absence of a simple relationship between chemical constitution and rotatory power.

While this last assertion is widely validated in materials by subsequent experiments—except for a small number of irregularities found empirically—the principle of minimum steric modification remains valid. However, it is at odds with many reactions, the first of which was accidentally discovered in 1893 [1] by Paul Walden.[4]

Firstly, Walden was able to obtain from optically active hydroxy compounds, halogenated compounds that are still optically active, in cases where his predecessors only obtained racemates. This outcome was due to experimental measures he adopted (low temperatures, accurate choice of solvent, short reaction times).

Walden wanted to prepare optically active bromosuccinic acid directly from optically active malic acid but did not get the desired result. He thought that he might succeed starting from asparagine [50]. It was known that (−)-asparagine, treated with nitrous acid, is transformed into (−)-malic acid. On the other hand, Walden prepared (+)-bromosuccinic acid from (−)-malic acid with PBr_5. It then would be expected that (−)-asparagine, treated for example together with gases of nitric oxide (NO) and bromine (Br_2) giving nitrosyl bromide (NOBr) would yield dextrorotatory bromosuccinic acid. Walden got levorotatory bromosuccinic acid instead (Fig. 8.3).[5] Therefore, depending on the reagents employed, the same asparagine can give rise to one or the other antipode of bromosuccinic acid, that is to states of opposing optical purity.

[4] [Some of Levi's citations were in his bibliography at the end of his document. Some were given as footnotes. They are included as sequential endnotes here, as they appear. Some citations were just indicated parenthetically as a name and/or year. If we were unsure of the exact reference, we have left these unspecified. Footnotes that we inserted, like this one, are enclosed between square braces. Otherwise, they are Levi's notes].

[5] [Levi did not number his figures, provide captions, nor call them out in the text. We have numbered them 8.3–8.26 and indicate when they are considered in the text. We have given them brief titles rather than invented captions].

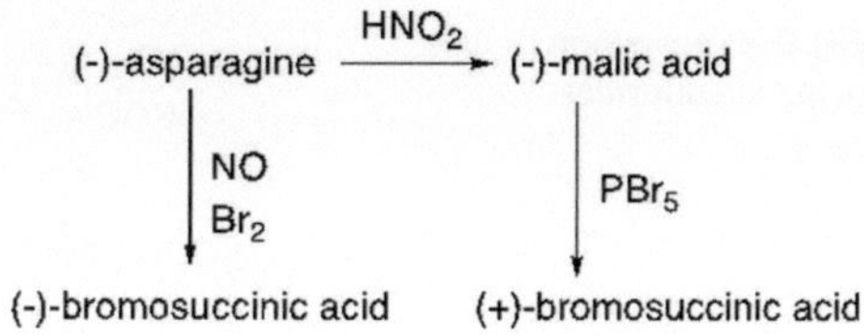

Fig. 8.3 Walden's initial observations

Walden, to whom the phenomenon was "truly unprecedented," stated: "Starting from a given optically active compound, containing only one asymmetric carbon atom, it is possible to obtain, at temperatures sufficiently low, two different substitution products, both optically active, which are antipodes of each other" [50].

The scope of Walden's discovery is remarkable. In the early years of stereochemistry, it was believed one could deduce immediately the configurational relationships among different optically active compounds through substitution reactions on the asymmetric carbon. We are now forced to doubt "static" representations of the molecular models that have proved suitable for deducing the number of possible geometric isomers of certain compounds. Such models cannot be applied to the interpretation of dynamic phenomena, such as substitution reactions and addition reactions to double or triple bonds, phenomena for which large disturbances of molecular structure cannot be excluded.

Zincke (1879), Baeyer (1888), Michael (1888)[6] and others had already suspected that this generalization was not prudent. As examples of "sterically anomalous"[7] reactions multiplied, it became evident that they are not exceptional. Even when reagents give substitutions on an asymmetric carbon and yield the starting compound, we cannot presume a "simple substitution". We emphasize that optical activity of the substances considered here is merely a requirement for the observability of the course of the reactions. It seems extremely unlikely that it is a necessary condition for the occurrence of sterically anomalous reactions. In fact, none of the many proposed interpretations applies exclusively to asymmetric molecular models.

The fundamental theoretical importance of the Walden Inversion (Emil Fischer proposed the name) [51] is that for all the reactions that permit the quantitative transformation to the antipode, it must be assumed that in many reactions, [52] substitutions among others, whether involving symmetrical or asymmetrical molecules, are carried out according to mechanisms up to now poorly known. But these processes are certainly less simple than is required by the principle of minimum structural modification, and likely to be different from case to case.

Furthermore, through laboratory technique Walden inversion can be conveniently guided toward the transformation of a naturally occurring optically active

6 [No citations are provided].

7 It seems today that the number of reactions with inversion is approximately equal to the number of similar reactions without inversion.

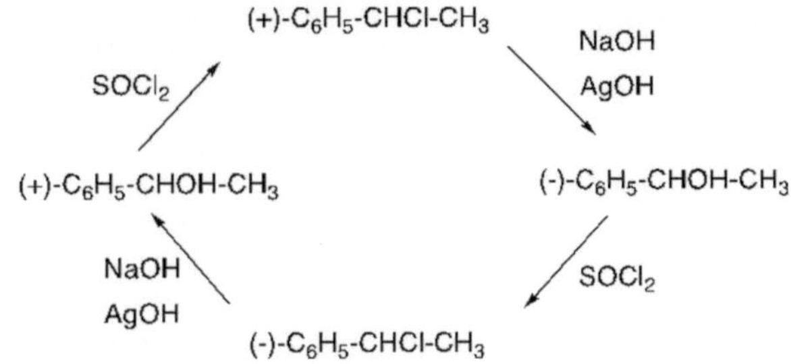

Fig. 8.4 Inversion of at benzylic chlorides

compound into its enantiomer, avoiding racemization. For example, Fischer and Warburg [53] obtained levorotatory alanine starting from natural dextrorotatory alanine with the reagents NOBr and NH_3.

II—Chemical correlates of the inversion phenomena

The existing experimental material on the subject is undoubtedly broad, but given the puzzling phenomenon, and the variety of guidelines and working assumptions that inform the individual studies, the results reported in the literature are rather incomplete, and lack consistency and coordination. The obvious question as to which compounds undergo Walden inversion cannot be yet answered definitively. It is known that many inversions of substituted acids and their derivatives (esters, amides, and acid chlorides), constitute most of the aforementioned reactions. On the other hand, inversions are also observed in compounds whose molecular structure does not contain a carboxyl group nor a carbonyl (Fig. 8.4).[8]

This last example also demonstrates that it is not necessary for the asymmetric carbon atom to be linked to an exchangeable hydrogen atom and in fact inversions of α-substituted acids (Fig. 8.5) have also been observed. Most of the observations of this phenomenon involve compounds that are not diverse, namely carboxylic acids and their derivatives. This does not imply that these groups as opposed to others prefer these inversion phenomena, but only that they compose the most common optically active substances known, that they are easier to obtain in pure form and have the simplest and most common chemical behavior.

The substitution reagent has a decisive influence on the outcome of the inversion but is vexingly irregular. PCl_5 and $SOCl_2$, reagents that usually replace hydroxyl groups with chlorine atoms, behave in an equivalent manner (i.e. they give rise to optical chloro-derivatives of equal sign) toward aliphatic hydroxy derivatives, while transforming the aromatic hydroxyacids to enantiomers (compare to the example above, *β*-phenyl-*β*-hydroxypropionic acid, Fig. 8.5).

Silver oxide (Ag_2O) and the alkaline hydroxides, which replace halogens with OH, act in a different way on chloropropionic, chlorosuccinic, and chlorophenylacetic acids. They behave equivalently elsewhere, for example, on 1,1-chlorophenylethane. PBr_5 and PCl_5 in general give substitution products of

[8] [Levi's product at the bottom of the figure is in error in the thesis.].

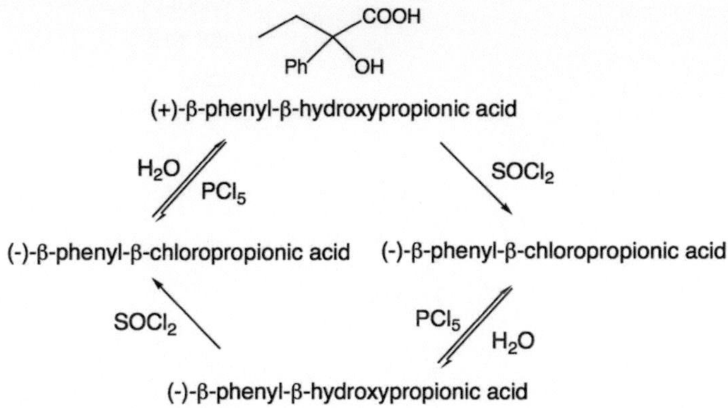

Fig. 8.5 Inversion of at benzylic hydroxides

the same optical sign (not surprising given the great similarity between the two halogens) but (−)-α-hydroxyisobutane[9] with PCl_5 gives a chloro-substituted levorotatory derivative and a dextrorotatory bromo derivative with PBr_5.

The same reagent usually acts in the same way on an acid and its esters; however, nitrosylbromide changes the α-amino acids into the corresponding α-bromoacid with opposite optical signs, while it causes no sign changes if it acts on esterified acids (Fig. 8.6). Since, as we shall see, one has good reason to exclude the possibility that a change of configuration occurs in the hydrolysis of carboxylic esters, the two resulting enantiomeric compounds demonstrate that one and only one of the two reactions caused by NOBr occurred with configuration change. The "sterically abnormal" reaction (I) was illustrated arbitrarily in the accompanying diagram. We will say more later on the possibility of executing this choice in analogous cases.

Finally, we have an example of the same reagent acting differently on the various members of the same homologous series. While $SOCl_2$ reacts without change of sign on methyl-, ethyl-, propyl-, isopropyl-, and *n*-butylphenylcarbinol, HBr, on the other hand, reacts with the first two compounds with a change of sign of optical activity; no sign change accompanies the reaction with the others.

To render explanation even more subtle, there are many examples in which the solvent, and in general the substances which ostensibly do not participate in the reaction, strongly influence the course of the inversion. By treating bromophenylacetic acid with ammonia, the product is right- or left-handed phenylglycine depending on the solvent used (Fig. 8.7). Instead, the higher homologue of bromophenylacetic acid, i.e. *α*-bromo-*β*-phenylpropionic acid, yields the same phenylalanine when treated with NH_3 under all experimental conditions [54–56]. Certain tertiary amine bases, in particular, pyridines, quinolines, and isoquinolines act in this manner [57].

[9] [This appears to be a misnomer. Perhaps Levi meant (-)-2-hydroxybutane].

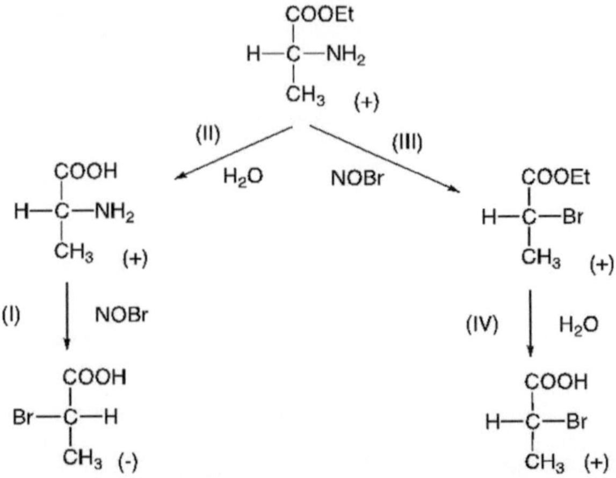

Fig. 8.6 Esters versus carboxylic acids

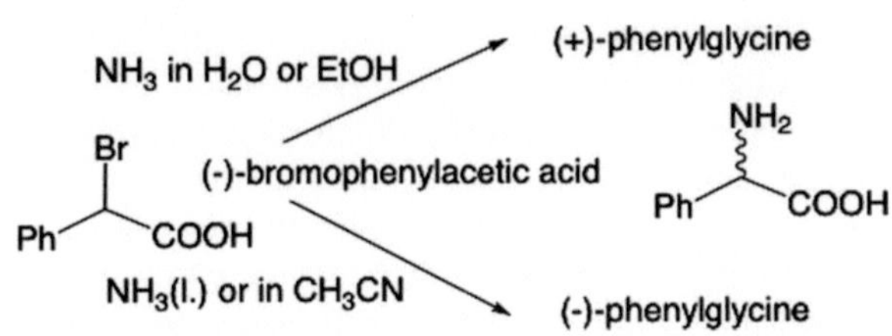

Fig. 8.7 Solvent effects

The presence of pyridine also affects the reaction between $SOCl_2$ and phenylmethylcarbinol (Fig. 8.8). On the other hand, the analogous reaction on aliphatic hydroxy derivatives do not show any modification. Most of the interpretations of the solvent effect are based on the formation of labile intermediate complexes. The matter has not yet been persuasively confirmed by experiment, and (although it is not easy to see a different explanation) like many questions that we will examine, remains open.

In the chemistry of sugars, there are examples of reactions that may give a particular epimer or another depending on whether there are adjunct reagents present (Fig. 8.9). There is no denying that this is a genuine Walden inversion, limited to only one of the asymmetric centers in the molecule. Thus, we must remember that the alternative epimers are not antipodes, and hence their formation energies are

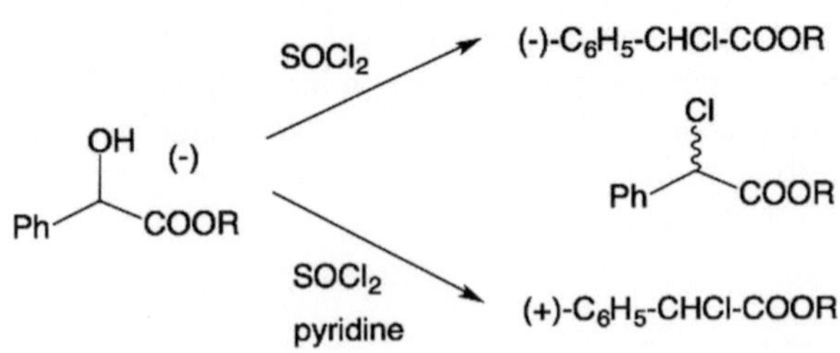

Fig. 8.8 Another example of solvent effects

different. One or the other will be more likely to form depending on experimental conditions. This is akin to the equilibrium α-glucopyranose ⇌ β-glucopyranose, which is displaceable toward the left by crystallization from water or alcohol or to the right by crystallization from pyridine or glacial acetic acid. Nothing of the sort accompanies a typical Walden inversion between antipodes as they have identical energies of formation. The reason one enantiomer is formed over the other should not be sought in energies but rather in the reaction mechanism.

Noteworthy is the hydrolysis of the malic acid lactone, since it is possible to quantitatively pass from *d*-malic acid to its antipode (Fig. 8.10). These reactions are called Walden inversions, although they are not substitutions. One explanation may be as follows: in one of two reactions the lactone ring would open in **a**, that is, at some distance from the asymmetric center rather than breaking at **b**, thus directly affecting the asymmetric carbon, with the possibility of a change of configuration [58]. Other cases are known in which a compound with an oxide bridge results in two epimers (antipodes) by acidic or basic hydrolysis, respectively. The formation of a cyclic lactone intermediate has been demonstrated [59].

Reaction of an optically active halosuccinic acid (-COOH) or xanthogenate (-CSSH) with hydrogen sulfide or its equivalent, generally yields a mixture of the two antipodes in unequal proportions, depending on the temperature, the concentration, and the presence of neutral salts, among other conditions (Fig. 8.11). There exist two distinct mechanisms for the two substitution products: Mechanism I is a direct, slow, bimolecular substitution. Mechanism II consists of two

Fig. 8.9 Epimerization

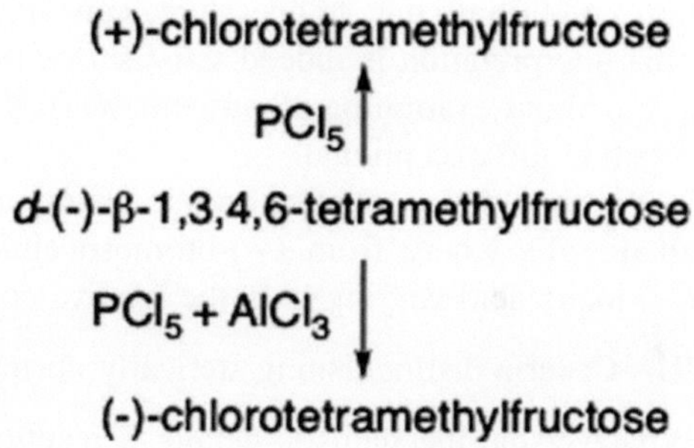

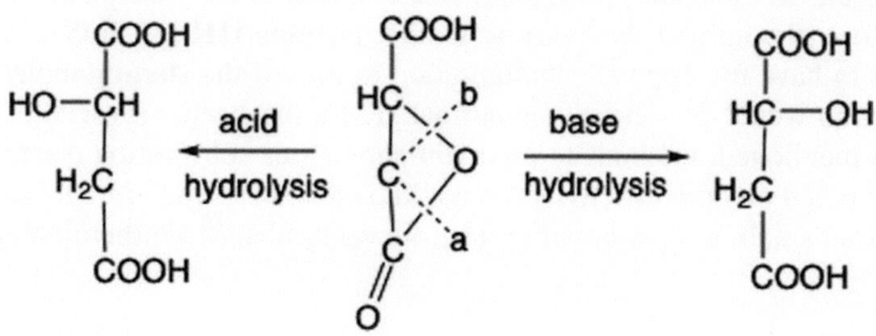

Fig. 8.10 Selectivity of malic acid lactone hydrolysis

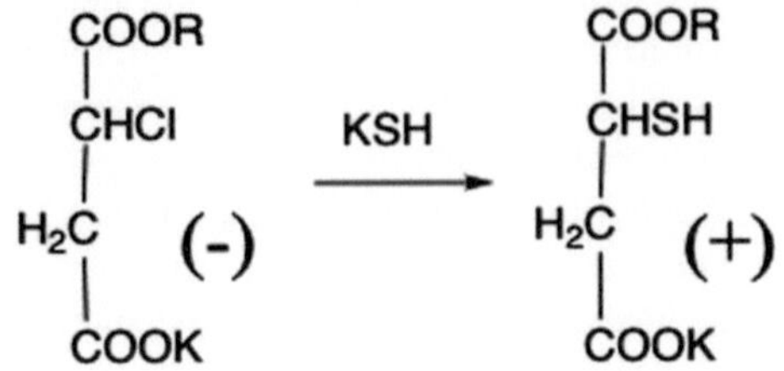

Fig. 8.11 Bisulfide nucleophile

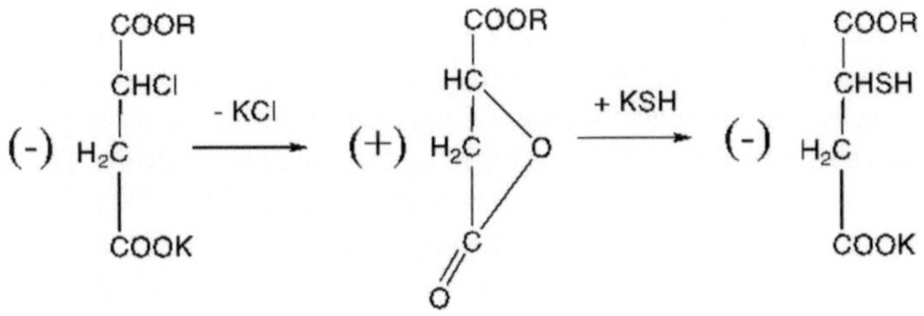

Fig. 8.12 Double inversion through lactone

steps: (1) formation of the lactone by slow, unimolecular addition, followed by (2) rapid bimolecular reaction with potassium thiol. Slight differences in the reaction conditions, e.g. temperature, can alter the product distribution (Fig. 8.12). This interpretation is indeed satisfactory, but it certainly cannot be applied generally without examining all cases of sterically abnormal reactions. This example is peculiar and exceptional.

We should also mention an interesting observation using either Ag_2O or caustic alkali salts where from (−)-bromosuccinic acid one can obtain (+) as well as (−)-malic acid varying only the relative concentration of the base [60].

III—Criteria distinguishing sterically normal and abnormal reactions

Whatever the mechanistic details, a reaction "with inversion" can be captured by Fig. 8.13. Replacement of the atomic group **e** in the molecule **(I)** Cabde with a new group **x** results in molecule (**II**) Cabdx, where **x**, may occupy the position in which it was before (the normal reaction) or another position **(III)**, such that the [Cabd] group has to have the opposite configuration to that of the starting molecule. The latter reaction would be sterically abnormal. There has been a reversal of configuration. As mentioned, methods to ascertain anomalous substitution reactions other than those based on optical activity are not known and it is therefore necessary that the five groups **a, b, d, e, x,** be different from each other.[10] Furthermore, since we

[10] [This is not strictly true as in the case of some of the stereoisomers of the tri-2,3,4-hydroxyglutaric acids].

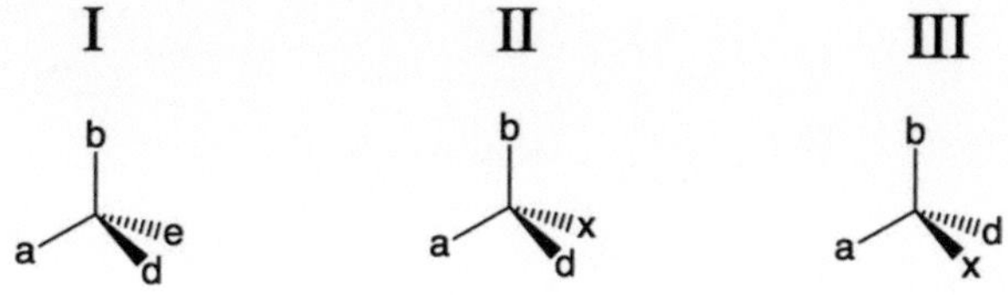

Fig. 8.13 Substitution of **e** (I) by **x** with retention of configuration (II) or inversion (III)

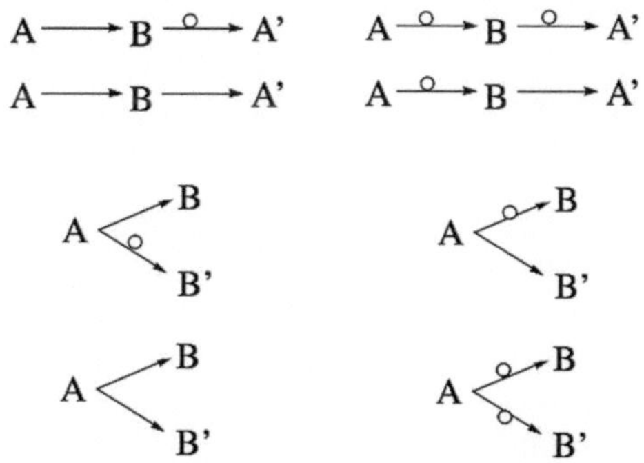

Fig. 8.14 Arrows with circles indicate transformation with inversions

do not know the relationship between chemical composition and rotatory power, the comparison between the configurations of two optically active compounds will be possible only in the case where they are antipodes to confirm whether the mechanism was normal or abnormal. The optical activity of the product must be compared with the starting material of established configuration.

According to the first mechanism, A → B → A' where A and A 'are identical and the substitutions are normal. Abnormal reactions give antipodes. Using the second procedure, we have: B ← A → B' (Fig. 8.14). However, the hitherto known experimental facts only allow us to assert in few cases that a certain reaction is normal or abnormal. Indicating → with normal reactions and $\overset{o}{\rightarrow}$ with inversion reactions merely masks the choice between mechanisms.[11] This is a great obstacle to interpreting sterically anomalous reactions, forcing many scholars to formulate criteria that allows the comparison of the configurations of active compounds of different series, thus making the steric course of any reaction knowable. We will summarize these results, with some consideration of the logical foundations to which we must adhere.

Ordinary examples of Walden inversions (in which the substitutions occur directly at an asymmetric center) prove to be in contradiction with the principle of minimum structural modification. One might wonder if this principle applies to those substitution reactions that do not directly affect bonds to asymmetric carbon atoms. Thus, one must also consider the possibility of inverting the configuration of a carbon atom merely because of a substitution elsewhere on a molecule. If this

[11] [Obscurantism, the greatest sin for Levi as a mature writer, permeates this section].

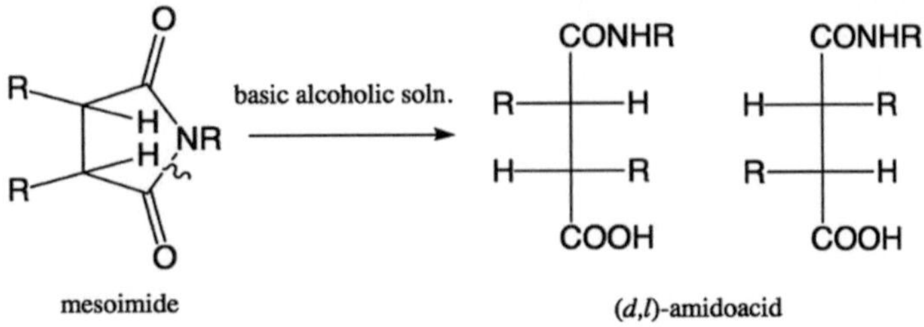

Fig. 8.15 Regioselective hydrolysis to give enantiomers

can be demonstrated, the foundations on which rests a good part of stereochemistry would be considerably weakened. Think of the research on the series of the natural amino acids with the same configuration, and the typical structural relationships among the carbohydrates. But, up to now, this never occurred. It has not been possible to demonstrate a change of configuration produced by a "distant" substitution (Fig. 8.15).

Auwers [61] interpreted this example as a partial Walden inversion due to the opening of the C–N bond that is far from both asymmetric centers. In fact, it seems far more correct (and proof of the contrary) to conclude that it is nothing more than a trivial hydrolysis followed by a complete transformation of an inactive form into a (*d, l*)-racemate that is more stable under the conditions of the experiment. This phenomenon is similar to common racemizations.

Conversely, a brilliant (although not definitive) proof of the lack of configuration changes caused by "distant" substitutions was obtained by Fischer [62]. He indirectly transformed a compound into its antipode without any change in the position of the four asymmetric carbon substituents (Fig. 8.16). In fact, none of the reported reactions directly affects the center of asymmetry. In conclusion, it is not yet reasonable to assume that changes in configuration occur, except by direct substitution on the asymmetric carbon. This principle has proven fruitful.

Without the need for any new experiments, and using only known facts, it was possible to resolve the question if the hydrolysis of a carboxylic acid ester occurs

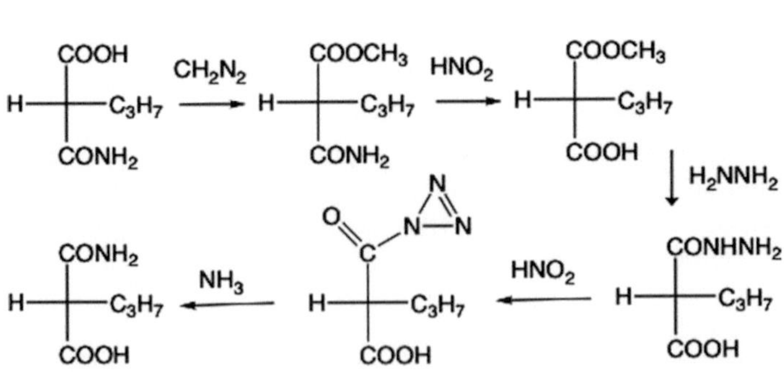

Fig. 8.16 Fischer's sleight of hand. Inversion without involving a stereogenic center

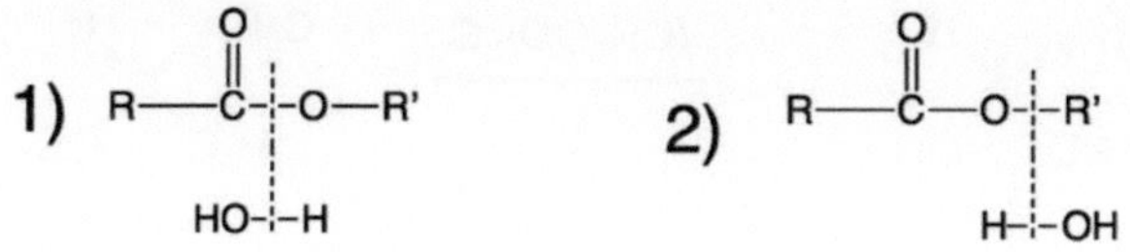

Fig. 8.17 Regioselective ester hydrolysis

according to scheme (1) or (2) in Fig. 8.17. Since it has never been possible to transform an optically active alcohol (wherein the asymmetric carbon is hydroxylated) into its antipode through simple esterification with a carboxylic acid and subsequent saponification, it is concluded that in these two phenomena the C–O bond in the alcohol molecule is not involved. Consequently, the preferable scheme is the first above.[12] Polanyi recently came to the same conclusion using the method (indeed more general and more reliable) based on the use of the oxygen-18 isotope [63].

Whenever it is possible to achieve the passage from an active compound to its antipode by means of two or more reactions, some of which do not affect the asymmetric carbon valencies, it is possible in some cases to state based on established principles which of these reactions occurred with inversion. A similar cyclic process was studied [57, 64, 65].

Of the three reactions shown in Fig. 8.18, the second consists of the elimination of a HCl molecule between the reagent and the alcoholic hydrogen, and therefore does not touch the asymmetric carbon. The same can be said for the first, since as we have seen, the configuration of an active alcohol is retained in its carboxylic esters. It is thus certain that the real inversion has occurred in the scission of a sulfonic ester group by potassium acetate in this way.

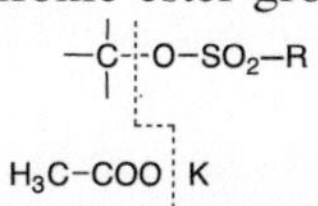

not this way,

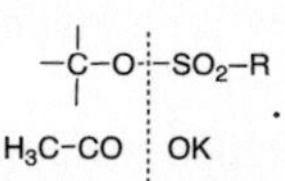

. And so, an interesting difference in behavior between sulfonyl esters and carboxylic esters is thus highlighted. Reasoning in an analogous manner, we see in that in Fig. 8.19 the configuration change must take place [66].

These examples show that the way to establish the process of a configuration change is quite limited and cannot be extended to the most common and simplest examples of Walden inversion. It therefore does not allow us to derive the configurational relationships which tie together, for example, an active α-hydroxy acid with the corresponding amino or halogenic acid.

The method developed by Freudenberg and collaborators [67–70] is susceptible to much broader application. It is based on the observation that similar compounds from different chemical series, when they have the same configuration,

[12] With a similar procedure it was also demonstrated that even in the formation and splitting of acetals and hemiacetals, the bond between the alcoholic hydroxyl and the carbon atom is never broken.

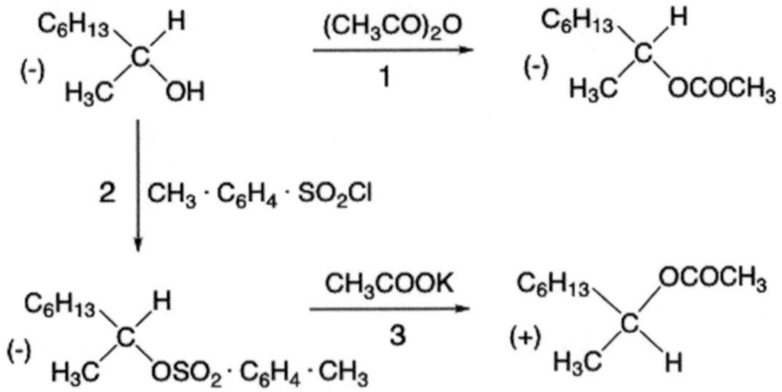

Fig. 8.18 Transformation of leaving group

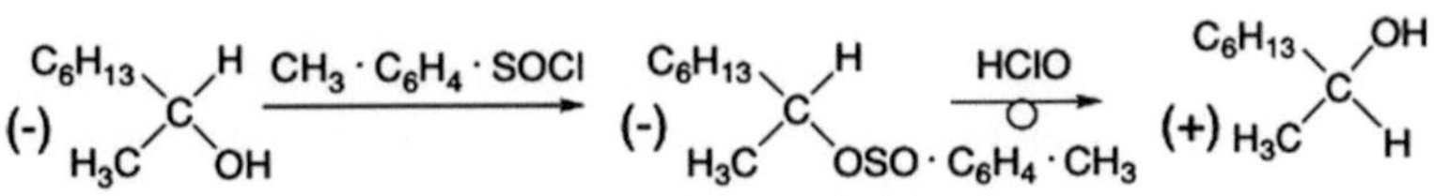

Fig. 8.19 Oxidation of sulfonyl esters

undergo variations of the same sign in their rotational power if one of the constituent groups common to both series undergoes the same substitution in the two compounds. Thus, if an ethyl ester is made from an active chloro acid and the corresponding amino acid of the same configuration, the optical sign is increased (or respectively decreased) in both cases. The time does not seem distant when an adequate theoretical interpretation will be given to this empirical rule (which is called the "Law of Optical Displacements").

Meanwhile, examining the most Walden inversion reactant cycles has led to the results that we have in Table 8.1 [67–70].

Perhaps more certain but less general is a method of investigation that was previously applied to compare (+)-bromo and chlorosuccinic acids [71]. From the examination of the melting points of the various mixtures of the two compounds, it was found that they form mixed crystals in all proportions without melting diagram anomalies. Conversely, the melting point diagram of mixtures of (+)-chlorosuccinic acid and (−)-bromosuccinic acid has similar characteristics to

Table 8.1 Summary of stereochemical outcomes

reagent	lactic acid malic acid	chloro(bromo)propionic acid chloro(bromo)succinic acid	alanine aspartic acid	alanine ester aspartic acid ester
Ag_2O				
N_2O_3				
NOCl, NOBr				
$SOCl_2$, PCl_5, PBr_5				
NH_3				
KOH, NaOH				

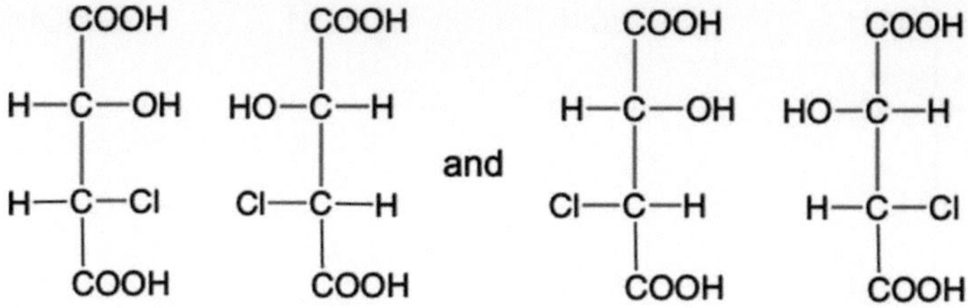

I (meso-like) II (racemic)

Fig. 8.20 Fischer projections of chloro substituted malic acids

that of the melting point diagram of the mixtures, for example, of (+) and (−)-chlorosuccinic acids. It was deduced that, in this case, compounds of same optical chirality possess the same configuration. Of course, much more difficult would be the interpretation of the diagrams if the substituents were less similar than the halogens.

Noteworthy are the results of Kuhn and Wagner-Jauregg, not because of their intrinsic importance, but rather because of the novelty of the principle from which they are deduced [58]. We will refer, for simplicity, to an example. There are, as is known, four chloromalic acid stereoisomers, one pair (I) with opposite configurations at C2 and C3 and the other (II) with the same configurations at those positions.[13] The meso-like (*d, l*)-chloromalic acids must, in their physical properties (Fig. 8.20), be intermediate between inactive dichlorosuccinic acid and inactive tartaric acid. The physical properties of the racemic (*d,1*)-chloromalic acids likewise must be intermediate of the acids (*d,l*)-dichlorosuccinic and (*d,l*)-tartaric. These predictions are in accordance with the measures of the dissociation constants of the various acids. In this way, we can know which of the two (*d,l*)-chloromalic acids is meso-like, and which are racemates.

[13] [Here, Levi uses his own shorthand representation of the shorthand that is the Fischer projection. It is not clarifying and is shown here.

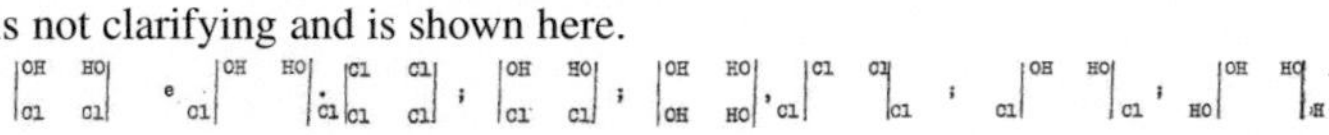

].

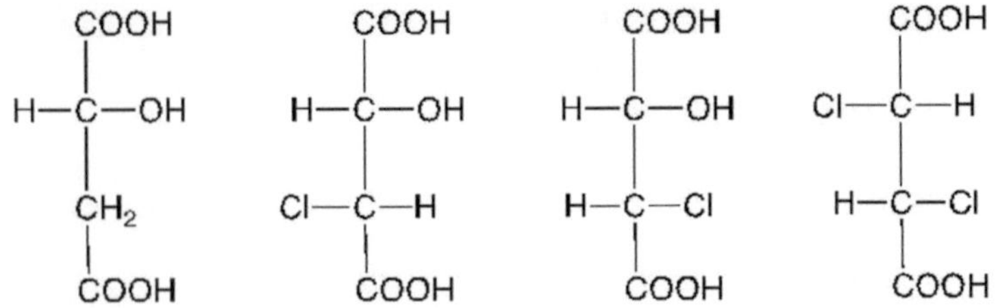

III (+)-malic acid meso-like | IV (+)-chloromalic acid racemic | V (-)-chloromalic acid meso-like | VI (-)-dichlorosuccinic acid

Fig. 8.21 Malic acid and derivative stereoisomers

One now can reduce the dextrorotatory component of the racemate with hydrogen gas producing *d*-(+)-malic acid (III) (Fig. 8.21). Since this step did not directly affect the asymmetric hydroxylated carbon, the dextrorotatory component of the racemate corresponds to configuration (IV). The levorotatory component of the meso-like chloromalic acid has configuration V, as it also gives *d*-malic acid by reduction. If it is treated with thionyl chloride, the (−)-dichlorosuccinic acid is obtained, with configuration VI. If the substitution of hydroxyl by chloro would have occurred normally, an inactive dichlorosuccinic acid should have been obtained. It is deduced that in the last reaction, there has been a change in configuration.

As for the stereochemical relation between (−)-dichlorosuccinic acid and (−)-chlorosuccinic acid, physical requirements based on the principle of optical superposition show that they have the same configuration. Following these steps, the series of α-hydroxy acids and α-chloroacids have been placed in a stereochemical relationship. For the "inversion" of malic acid, Fig. 8.22 agrees with other authors summarized in the previous table yet disagrees with Gadamer [72].

The validity of Kuhn's and Wagner-Jauregg's [58] hypothesis on the configuration of two non-equivalent asymmetric carbon atoms is, however, linked to the

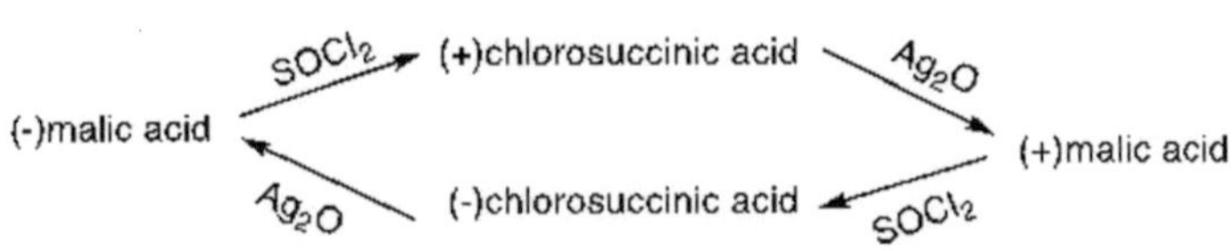

Fig. 8.22 Walden inversion summarized from Table 8.1

condition (Hückel) that the polarities of the groups through which the asymmetric centers are differentiated are not much different.[14]

IV. Mechanism of Walden inversion

Current data on the Walden inversion are inconsistent. It is not strange therefore that theories are numerous, more than twenty, and often proposed tentatively. For the most part, these theories deal with the construction of mechanical models, whose operation is described step by step, sometimes with excessive precision. Recent interpretations have also been proposed based on the new concepts regarding the structure of the atom and on the nature of chemical affinity. But nothing definitive has been put forward, generating a distrust of the overwhelming use of mechanical models. Nevertheless, such models can be useful, even if they compromise much. We report some of the latest theories.

The ideas of Meisenheimer [and Link, 72] and Kossel,[15] as well as Gadamer [73] can be applied to the inversion of Werner [74, 75]. The direction of the valence bond is determined by specific classical forces. Consider, for example, the reaction $CH_4 + Cl_2 \rightarrow CH_3Cl + HCl$[16] [76] and imagine that initially a molecule of chlorine is added from the far side of a hydrogen atom (Scheme I, Fig. 8.23). The formation of this complex pictured is such that part of the "affinity" unbinds one of the carbon atoms, now engaged in the labile H complexed to Cl. The C–H bond is weakened, and the bond length is increased. The methane molecule therefore loses its tetrahedral symmetry. The three other hydrogen atoms (H_2, H_3, H_4) assume new positions since the H_1 repulsion is decreased, such that the three faces of the trigonal pyramid with base $H_2H_3H_4$ increase. In the middle of the trihedron in the direction opposite to $C \rightarrow H_1$, the screening action of the hydrogen atoms is weakened. Therefore, under suitable experimental conditions the "residual affinity" of the carbon atom expands with the coordination of a second chlorine atom, which will be diametrically opposed to the first (Scheme II, Fig. 8.23). I and II are the possible intermediate substitution processes/products. Which of the two is formed, depends on even small changes in the parameters. The reaction continues in both cases. Process I runs normally. II, as we see, occurs with inversion. Due to their decidedly qualitative character, such considerations do not allow predictions as to whether inversion will occur under given conditions. They may, however, explain quite well, for example, the actions of the solvent as mentioned above,

---

14 [Here, Levi tries to puzzle out the conformations of compounds with adjacent tetrahedral stereogenic centers. The accompanying figure features undefined labels which renders the associated discussion uninterpretable].

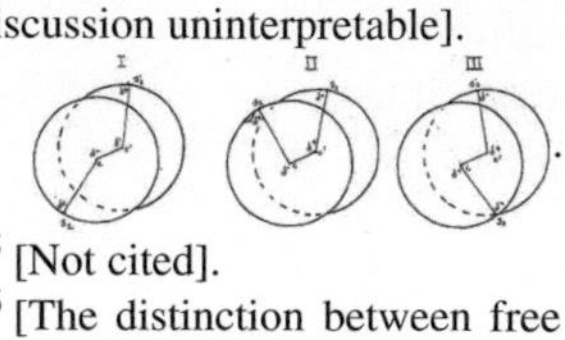

15 [Not cited].

16 [The distinction between free radical substitution and nucleophilic substitution started to be clarified in 1937, beyond Levi's sources. See [76].

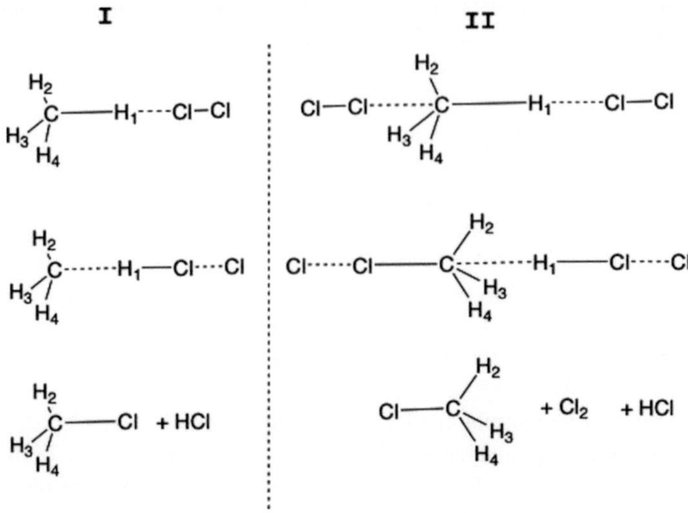

Fig. 8.23 Halogenation of methane

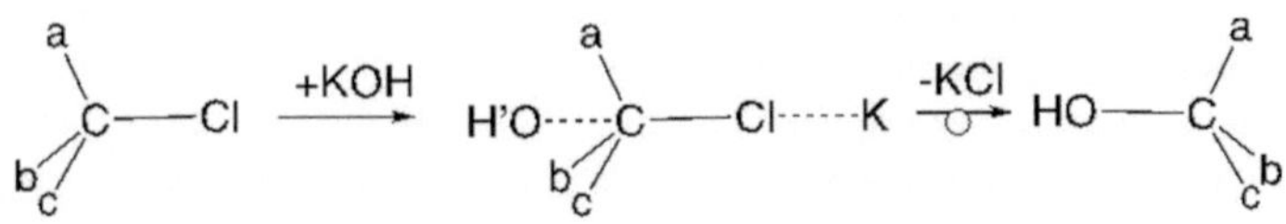

Fig. 8.24 Backside attack

and in some cases, the theory explains the anomalous steric addition to ethylene bonds, as well as reactions such as the following in Fig. 8.24.

Similar processes have been observed in large numbers (note the aforementioned anomalous reaction of halogenic succinic acids with caustic alkali salts). Even Holmberg [77] admits the prior formation of a labile complex, but also introduces the concept of "reaction distance."[17] Consider the reaction Cabdx + ny → Cabdy + nx. Imagine that the association takes place so that the n-y axis is parallel to, say, the C-x axis as pictured in Fig. 8.25. The labile compound that is formed will be represented by one or the other of the two schemes depending on whether the C-X distance will be greater or less than the distance n-y. If the complex Cabdy is formed, 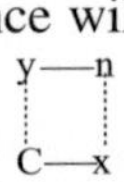 then transforms to 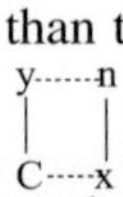 of the other antipode. The distances between the atoms in the molecules are known in many cases. Thus, it could be determined a priori whether a certain substitution occurs or not with inversion. But reaction distance predictions contradict experiment.

[17] [That is, the distance between the two reacting parts of the reagent acting upon the optically active substrate].

Fig. 8.25 Holmberg's reaction distance

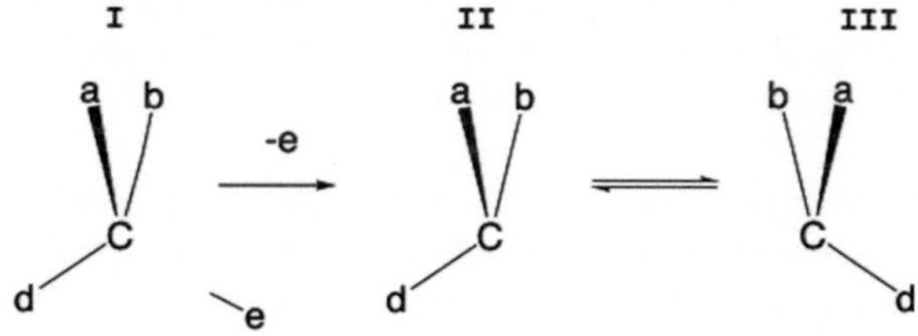

Fig. 8.26 Rørdam's equilibrium/oscillation

Rørdam [78–80] developed more credible concepts drawing from ideas of Le Bel in 1896[18] utilizing dynamic models. Given that the first step of a substitution reaction consists in simply removing the group to be replaced (**e**), the resulting radical [Cabd] I, as long as it remains free, will oscillate between the two enantiomeric configurations until the saturation of the free valence by the new substituent occurs while the initial configuration is preserved, or when it has already been reversed; there is a normal or anomalous reaction, respectively (Fig. 8.26).

Rørdam considers two distinct types of substitution reactions: (1)The reaction occurs with (or without) inversion for all molecules. (2) Inversion exists only for a certain percentage of the molecules. The interpretation is as follows:

(1) Before substitution, molecule I Cabde gives up one valence, -**e**. Thanks to symmetry, the addition to radicals II and III always requires the same amount of time. Assuming that the oscillations of the radical are sinusoidal, and by labeling the time t between the detachment of **e** and the replacing **f** [not pictured in Fig. 8.26], there will be inversion (or not) according to the sign of the function $\cos 2\pi(t/T)$ where T is the oscillation period of the radical.
(2) Cabde molecules and reactant molecules have varying distances between them over time. There is racemization if distances are distributed randomly. When labile complexes are formed, there is a limit on the variability of distances (and hence t). Then, partial inversion may occur. If we assume that $t < T$ (which is equivalent to the assumption that substitution with inversion takes place during the first radical reversal), it will be inferred that the partial inversion in non-inverted molecules will increase by all factors which accelerate the rate of the substitution reaction, as well as those which lengthen the period of oscillation of the radical. It would then be possible to establish the relationship between

[18] [Not cited].

> the configuration of the antipodes of the product and the starting compound. The configuration of the antipode that increases as the concentration of the reagent increases must correspond to that of the starting compound. Rørdam attempted a quantitative formulation of the dependence between concentration of reactant and percent yield, even though he needed simplifying assumptions because of the difficulty of a rigorous mathematical treatment. The theoretical equation that he finds is in excellent agreement with the experimental equation in the case of the reaction of chlorosuccinic acid; however, further confidence evades us since, as we have seen, that reaction differs somewhat from the norm.

Bonino [81] proposed an explanation of Walden's inversion which ultimately is a quantum mechanical description of Rørdam's theory. Based on the studies of Slater [82] and Pauling [83] on possible electron states for the carbon atom, Bonino considers the possibility that in a CHabc molecule, the whole CH group oscillates between two equilibrium positions located symmetrically between the two sides of the abc plane (similar to what is allowed for the nitrogen atom in the NH_3 molecule). When $a \neq b \neq c \neq H$, the two minimum potential energy configurations correspond to the two antipodes. Then the steric and energetic conditions that make oscillation possible would also be those that cause racemization.

If the following are then assumed: (1) For a CHabc molecule, oscillation is prevented due to a large "potential barrier" between the two stable positions and (2) Substitution of one of the a, b, c, substituents with another group may momentarily lower the barrier that regains a high value when the reaction is over. A half-oscillation takes place when the barrier is low, and the subsequent rise of the barrier prevents the return of the CH group to the starting position. As you can see, here is the relationship described earlier between T (oscillation period) and t (time of substitution).

From the foregoing, the main purpose of the study of inversion phenomena has not yet been achieved. The chemist is not able to predict the outcome of a substitution reaction on an asymmetric carbon atom nor do any of the theories described justify the hope of an immediate resolution of the matter (except for current developments in knowledge of chemical kinetics).

Rørdam's views are superior. It seems difficult to reconcile the views of Meisenheimer [73] and Holmberg [77] with recent concepts of modern physics while Rørdam's theory has gained reliability due to the recent explanation from Bonino [81]. Moreover, only the latter explains satisfactorily some of the observed phenomena. Recall the different behavior toward NOBr of alanine and its esters, and the irregularities in the action of $SOCl_2$ on the homologous series of alkylphenylcarbinols. The transition from a compound to higher homologues (or an acid into alkyl esters) results in an increase in the mass of the oscillating parts, hence an increase in the oscillation period. The variation of T can then cause a change of sign in the $\cos 2\pi\,(t/T)$ function (Rørdam's theory), and therefore the transition from a normal type of reaction to one that is abnormal.

A crucial experiment could be designed based on these considerations. The Rørdam–Bonino theory would find valid confirmation if, for any given compound that undergoes Walden inversion under certain reaction conditions, a higher homologue underwent normal substitution with the same reactants and reaction conditions. This confirmation would be even more convincing in the case of compounds (such as alanine, see table above) which are not inverted under conditions where the first (lowest molecular weight) of the homologue series is inverted. For homologues with greater molecular weights, the substitution should return to normal. In fact, according to Rørdam's theory, the starting compound shows no inversion because the radical (the CH group) has time to perform a whole oscillation, while in the case of intermediate molecular weight homologues the compound may undergo half oscillation (and inversion). It may remain in the initial configuration for a sufficiently high molecular weight making $T > t/4$, and therefore $\cos 2\pi\, t/T > \cos \pi/2 > 0$.

8.4 Analysis

We asked three questions at the outset. First: Did Levi succeed in summarizing the vexing mechanistic chemistry of Walden inversion given the urgencies in Italy of 1941?

The previous analysts of Levi's *tesi*, Borello [41], Naso, [37] Cravotto and Cintas, [43] as well as Barbarulo, [14, 87] are all appreciative of Levi's efforts. Borello tried to gives an overview of the problem of Walden inversion for non-chemists. Naso, as well as Cravotto and Cintas emphasized the time-dependent Rørdam–Bonino double inversion which came into focus at the end of *L'inversione*, and congratulated Levi for the stirrings of quantum mechanics that might not have impressed themselves on the average chemistry student of the time. Barbarulo [14] discusses many of the sources used by Levi and gives passages of more than a few of them. However, they are presented without prejudice, whereas Naso, Cravatto, and Cintas, stress those features that hint at the S_N2 mechanism, unheard of by Levi but familiar today.

Levi seized upon Rørdam's hypothesis, summarized in Fig. 8.26, in which, following dissociation of **e**, what we would describe today as the leaving group, the pyramidal moiety left behind would vibrate at a natural frequency in a double-well potential and the stereochemical outcome would depend on the relative vibrational frequency of the reactive intermediate and association rate of the nucleophile in contemporary parlance.

Ultimately, Levi was attracted to the pioneering work of G. B. Bonino [84] of the University of Bologna who aspired to apply Raman spectroscopy, discovered only in 1928 [85], to problems in organic chemistry. Bonino tried to adopt Fermi's umbrella inversion mechanism for ammonia, derived from rotational spectroscopy, [86] to Walden inversion at carbon atoms, imaging that bonded C–H atoms could behave like the lone electron pair in ammonia. "The possibility of CH passing from the $+x_0$ to the $-x_0$ equilibrium position will depend on the height of the 'potential

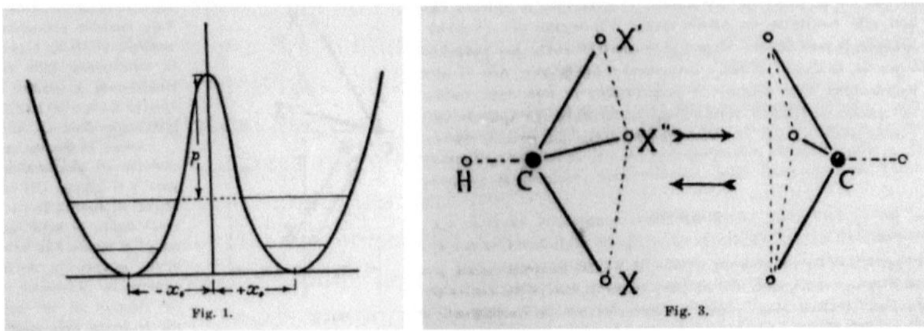

Fig. 8.27 Bonino's interpretation of Walden inversion. A double well potential ('Fig. 1', left) consistent with an oscillatory motion of CH ('Fig. 3', right) through the plane of the other ligands to C, X, X', and X". From ref. [73]

barrier'" according to Bonino. Refer here to Fig. 8.27 of Bonino's subfigures [81]. Bonino's assertion was based on Raman spectra consistent with a soft pyramidal oscillation, but that does not resolve how the C–H bond changes its sense.

Levi was forward thinking in recognizing that quantum theory predicted frequencies for molecular dynamics. He thought that the difference between normal and abnormal substitution may only depend on the intrinsic timing of the oscillation at the stereogenic carbon, that could be reduced to a formula that depended on the masses of the substituents. Thompson described Levi's attempts to reduce this mechanism to practice in the laboratory, experiments which Levi hints at in the final paragraphs of his thesis, despite institutional prohibitions. Thompson suggested that Levi tried to control the attached masses of oleic acid derivatives via Hoffman degradations, thereby systematically varying the kinetics of substitution at a stereogenic carbon. It is hard to puzzle backward what precisely he was aiming at from the fragmentary clues given, but apparently he lost way when, struggling with surfactants he "filled the lab with foam" [87]. Another student reported, "We could have shaved with the stuff" [88].

In our view, the value of the thesis is in the courage that Levi showed in searching for a physical, almost universal, explanation for Walden inversion in the dynamic processes of what today we would call "reactive intermediates." Levi strove beyond the account of Walden inversion given by his teacher, Ponzio, whose explanation in his organic chemistry textbook, [89] published in 1943, merely gave some reaction cycles that appear in *L'inversione* [10].

It is also rewarding to us to review those ideas in *L'inversione* that have been left behind by successive generations, such as Holmberg's concept of the reaction distance (Fig. 8.25). Obsolete scientific mechanisms are valuable since all mechanisms may ultimately be falsified. The rise and fall of credible ideas give all scientists the courage to be confused.

Levi's point of departure is Walden's 1895 paper [50], although Walden published many papers in quick succession [90–94]. Not all the chemical schemes in Levi's thesis make sense from a modern perspective. There are errors in judgment in the original sources from which he drew, errors in his interpretation of these results, and errors that arise in presumptions of relationships between configuration and optical rotation for which there was little basis, as Levi repeatedly reminds his reader.

Statements from Walden, for example that the oxidation of levorotatory asparagine with nitrous acid (HNO_2) gives levorotatory malic acid contains no mechanistic information a priori. Levi himself conceded that matters have "not yet been confirmed by experiment…and many questions that we will examine remain open." (See above.) Moreover, Levi acknowledged that "the results reported in the literature are rather incomplete and lack consistency and coordination." (See above.) It would take a mountain of annotation and a hundred additional citations to bring Levi's *tesi* in line with all that we know about nucleophilic substitution reaction mechanisms today. There seemed to us little reason for this labor. The 1941 thesis is not to be judged by the chemical facts that Levi and his sources got right or wrong by our standards.

Not Borello [81], Naso, [37] Cravotto and Cintas, [83] nor Barbarulo [14, 87] stress that Levi's most recent sources in 1941 were Bonino [81] and Olson [52], both from 1933. Levi's examiners did not quarrel either; he received 110/110 points from each of the eleven examiners, for a perfect score of 100%; Levi became a *dottore* with merit [95]. Borello surmised that the liberal though compliant faculty may have put their thumbs on the scale in Levi's favor, given his disadvantageous circumstances, and the opportunity to make a judgement that risked little outside of state control [81].

The Olson paper was a proposition, but supporting data unexamined by Levi, came from Olson with Long, in just the next year, 1934, and which established a relationship between some substitutions and inversion [96] that another contemporary reviewer (1945) described as "causal" [97]. The classic experiments showing that inversion takes place when a halogen is substituted by another radioactive atom of the same type and that the velocity of racemization is directly proportional to the acquisition of radioactivity could have been known to Levi [98–100]. The missing sources from 1934–1941 undoubtedly impacted the scope and texture of *L'inversione.*

Much of the classic mechanistic work of Ingold, Hughes and coworkers, correlating kinetics and stereochemical outcomes of substitution reactions was from the period that Levi did *not* consider in *L'inversione* [101–106]. Albeit, in the last chapter of his thesis Levi does allude to "current developments in knowledge of chemical kinetics" (see above), Levi may have heard about the developments in England without having access to the original publications or having advanced to the literature after 1933. In any case, Ingold, Hughes and coworkers, clarified many of the earlier propositions by showing that nucleophilic substitution sometimes involved concomitant association and disassociation with inversion and sometimes involved a pre-ionization step [107]. And much to the consternations of millions of organic chemistry students that followed, the mechanism(s) depends on many soft factors and reactions conditions.

Commenting on the work of Hughes [108], presented at the 1938 *Transactions of the Faraday Society* discussion, Hammett was congratulatory: "Hughes has reported on a beautifully complete and thorough and an extremely important investigation. It demonstrates the possibility of determining relative configurations of optically active compounds by reaction kinetic measurements and breaks a path through the jungle of confusion which has always surrounded the Walden inversion" [109]. In principle, Levi had a compass to sort his way out of the "jungle" but did not seem to have one in practice. In the words of Ingold in his classic text of 1953, "[W]hen, why, and how a Walden inversion occurs…remained unsolved for 40 years. It is easy to see now that it could not have been solved before the discovery in 1933–35 of the two mechanisms of nucleophilic aliphatic substitution" [107].

One might surmise that during Mussolini's reign, the University of Turin library was not receiving the current journals. However, the librarians at the *Biblioteca Ponzio* (named for Professore Giacomo Ponzio who approved the thesis), assured the authors that the university was receiving international journals through 1939 and 1940. It is likely that Levi was proceeding through the literature available to him chronologically and merely ran out of time. Turin, home to Fiat, was subjected to heavy bombing raids by the Royal Air Force beginning in June 1940. Most of the soldiers assembled from Turin were killed retreating from the Soviet Union. [110] Finding food in wartime Turin in 1941 [111] surely interfered with reading the latest paper from the *Journal of the Chemical Society*.

It is unfortunate and ironic that young Levi did not access Henry Gilman's classic work in English, *Organic Chemistry, An Advanced Treatise*, 2 vol. from 1938 [112]. It is unfortunate because it gives two concise descriptions of Walden's mechanistic puzzle (in Chap. 3, "Stereoisomerism" by Shriner, Adams, and Marvel, and in Chapter 21, "Rotatory Dispersion," by Levene and Rothen) that cover five critical years that postdated Levi's last sources, and ironic, because in 1956, Levi and Anglesio translated Gilman to Italian [113]. We stress that you must know *a lot* of organic chemistry merely to translate Gilman. Curiously, as pointed out by Barbarulo, [14] the forward to the translation was written by the beacon of Levi's thesis, Bonino. Levi and Bonino ultimately worked together in some fashion. Even though Bonino was a touchstone in Levi's search for a mechanism underlying the stereodynamics of nucleophilic substitution, he likely would have repelled Levi later in life. During the war, Bonino was an enthusiastic fascist eager to bolster scientific cooperation along the Rome-Berlin axis. "Science and fascism," he said, "must become an inseparable combination in the world, a source of new successes for Human Activity" [114]. Perhaps, this is why Levi named the character in "Uranium" [CW 909–921, *In the Periodic Table*] Bonino, a dishonest raconteur who claimed, at war's end, to have been given a lump of uranium metal by passing "Nazi aeronaut[s]" as a gift for pointing them in the direction of Switzerland. Maybe Bonino was the real name of a liar, or maybe it was chosen by Levi as a private satisfaction. Bonino, the character, constructs the past "that most pleases him." Perhaps that is what Bonino did in the 1950s in the face of what he is recorded to have said in the 1930s, a change in outlook common in a nation that struggled to come to terms with prewar ideologies. We do not know whether Levi followed the career of Walden. If he had, his discoveries might also have been dispiriting.

In 1919, Walden resigned his position in Riga and moved to Rostock as Professor of Inorganic Chemistry. Opportunities to return to the Soviet Union were declined. Maier [115] has provided evidence that Walden's embrace of German culture and institutions, in hindsight, may have been too enthusiastic. Walden joined the Nazi party in 1934 and provided financial support to the SS subsequently. According to Maier, in 1936, Walden as the history chair of the Association of German Chemists, "embodied the highly political character of the history of chemistry in the Third Reich like no other" [116]. His work as a historian emphasized distorted national and racial characteristics of the science of chemistry that were consistent with his elevated positions in the German chemical establishment during this time.

Walden's home and extensive library in Rostock were bombed in 1942. Walden himself was bombed and wounded in Berlin in 1943. After the war, he was disconnected from his retirement savings held in Rostock, now in the Soviet zone. He lived in penury until 1957. It was speculated that his wartime sympathies may have doomed some of his Nobel Prize nominations, particular those after 1945 [117].

While Levi's thesis may have been incomplete in 1941, he continued to work on it in a fashion. Many years later, in an essay, "Asymmetry and life" [CW pp 2656–2657], Levi returned to stereochemistry and offered an apology: "I know I have strayed into a field that is not (any longer) mine. It was, however, the subject of my dissertation. I have returned to it with reverence, some regret, and the fear of having made mistakes; you pay a price for years of retirement." Here, Levi acknowledged that stereochemistry is hard to write about, and that he may not know all that he needs to know, but that he loves it, and once loved it. And, indeed, there are some mistakes in "Asymmetry and life." In describing seven possible mechanisms for the origin of biomolecular handedness, he addresses the actions of magnetic fields. He says that "The Earth's magnetic field...could have driven the same asymmetrical synthesis to the advantage of only one of the two opposites throughout the surface of the Earth." Many before Levi had made this mistake. A magnetic field is a pseudovector not a polar vector, and therefore, it cannot be a so-called true chiral influence [118]. One celebrated instance of asymmetric synthesis in a magnetic field turned out to be fraudulent [119]. On the other hand, Levi's suspicions were aroused: "As far as I know (but I have never had—nor do I have now, after so many years without practicing chemistry—any competence in this field; if anyone knows more, I will be glad to recant), there are no organic reactions sensitive to a magnetic field of reasonable intensity." [CW p 2664] That said, the last word about magnetic field effects may not yet be known [120]. In any case, Levi was critically apprehensive of what he didn't know and could sense from experience.

Echoes of Levi's long-standing interest in the origin of biomolecular homochirality can also be found in his story "Adam's clay," [CW 2759–2763] a review of the inventive crystals-as-genes hypothesis of Cairns-Smith, [121] which has continued to inspire us [122].

Levi read Giulio Natta's book on *Stereochemistry* (1972) [123] which he characterized as "beautiful" and from which deduced that if a "polymerization is carried out in the presence of small quantities of an inert but highly asymmetrical substance, then the resulting polymer is also asymmetrical for its whole length. The inert substance acts as a sort of mold; from an asymmetrical mold we can obtain asymmetrical pieces in virtually unlimited quantities." [CW p 2665] This is a foreshadowing of the so-called sergeant's and soldiers process in supramolecular and polymer chemistry that influenced a variety of subfields in chemistry [124]. Levi's scientific training led him to see clearly, rather far into the future.

8.5 Levi as a Young Writer

The second question we asked at the outset was this: Is there any evidence that Levi was a gifted writer and communicator as a student? The thesis is not a joy to read. Throughout, Levi relied on ordinary words such as *interessante*, *molto*, *forse* (interesting, very, perhaps) to emphasize observations or insights while cautioning against things unsure. In a number of instances, we eliminated synonyms of the word "very," honoring Levi's admonition in "A Tranquil Star": "How many times higher than a high tower is a very high tower?" [CW p 1420] An author trying to convey something that is difficult by nature and unclear by virtue of the immaturity of the subject, can place the reader in the position not unlike the passenger in a car driven by a traveler relying too heavily on the gas and the brakes at the same time.

Of course, no one should be judged by a first draft. And, we have conceded that conditions in Turin were severe, and Levi had a hard graduation deadline, one that could have been endangered by additional race prohibitions. Levi's manuscript is lightly corrected and only features corrected spelling errors with the typewriter or pen.

We can compare this treatment to Levi's typical writing habit. For decades, Levi's livelihood was not dependent on his writing. He therefore could be deliberate. Belpoliti examined original texts preserved in the archive of the publisher, Einaudi. After examining the manuscript of Levi's novel, *If Not Now, When?* (CW 1547–1853), Belpoliti wrote:

> Among the first 159 pages of yellowing paper, one can see additional sheets of whiter paper with revisions and rewrites of passages of text. There are also several strips of lighter paper stuck onto various sheets of the original typescript. In addition, there are numerous handwritten corrections, where Levi used white-out for his edits. Levi did not retype the whole manuscript. He made partial changes, rewriting passages or single pages as he went along. Some of the pages were punched for a file holder and were numbered differently. Levi also made handwritten notes on the back of the sheets, as if he wanted to check the correct order of the events he was narrating in the novel. He crossed text out with a black pen, stuck white strips onto previously typed text and wrote in pen on top [125].

L'inversione was incomplete and imperfect—not carefully crafted or worked over—but it was his passage to an advanced degree and paid employment during a precarious time.

L'inversione di Walden is the earliest surviving writing of Primo Levi of which we are aware. He did write some poetry as a student but dismissed these efforts as part of a youthful phase; "We were all writing poetry," he said [126]. An earlier story, *Uomo*, was never published. It is apparently in the possession of an associate of Levi's and was shown to a biographer who described it [127]. We cannot say anything about it.

According to Lang, by the time Levi returned to Turin in October of 1945 he "had published nothing and ventured only a few forays into writing" [128]. He had settled on a profession in chemistry, and there is no indication that he considered writing a vocation. Decades later, Levi remarked that students (among others) frequently ask whether he would have become a writer, or the same kind of writer, without his wartime experiences and/or his training as a chemist. He answered that that experiment could not be done without a fresh Primo Levi [129]. Yet, there are traces of Levi before and after the *Lager* that give evidence.

The earliest of Levi's writing in the CW is a poem, *Crescenzago* [CW p 1879–1880], dated February 1943. Crescenzago was a grim suburb of Milan where Levi worked at Wander, a Swiss Drug Company, as a "semi-clandestine" [130] chemist after his university training because he was officially unemployable. It is a haunting poem *before* the *Lager* about miserable people engaged in miserable work in a windy, sunless wasteland that ends with this verse:

From morning till night they make
The menacing black steamroller heave,
Or spend the whole day eyeing
The hand that quivers on the dial.
Saturday nights they make love in the ditch
Beside the crossing keeper's house.

This stanza smacks with a breathtaking absence of romance.

Levi's poetry was first collected in *Ad ora incerta* ("At Uncertain Hour") [131]. In an analysis accompanying the second 1990 edition Raboni wrote, "[F]rom the start, it seems to me that Levi's poetic writings have the same solemn moral acumen, the same power of memory, warning, and piety that makes his prose so substantial, so to the point, and so naturally memorable" [132]. And, from the start, Raboni must mean from the first poem, "Crescenzago" (February 1943). It is an apt characterization. Lang, in his intelligent analysis of Levi and his work observed about "Crescenzago" that "[T]he backstory of Levi as a writer is more substantial than the impression he conveys. At least some of what he wrote as a student proved to be more than diversionary or casual jottings" [133]. These are hints. We can be more particular by evaluating the second poem, "Buna," December 28,

1945. [CW p 1881] It is separated from the first poem by a biographical chasm described in *If This Is a Man*[19] [CW pp 2–205], Levi's account of Auschwitz. Buna refers to synthetic rubber made from butadiene and sodium (Na) which gave its chemical name to the Monowitz-Buna (Auschwtiz III) *Arbeitslager* (work camp). This second poem is also about dull work in a chemical factory, and it is astonishingly similar in tone to *Crescenzago*. (Numbered stanzas a/b are paired out of sequence for best comparison between "Crescenzago" (a) and "Buna" (b)).

1a. From Crescenzago (February 1943):

The wind rushes out of the mountains,
Running free across the endless plain,
But when it notices the smokestack here
It suddenly turns tail and flees
Because the smoke's so black and poisonous

1b. From Buna (December 1945):

Wounded feet and curse earth,
The line long in the gray mornings.
Buna's thousand chimneys smoke.
A day like every other day awaits us.

2a. Crescenzago:

When the siren sounds at dawn
They slip out of their rumpled beds.
They come into the street with their mouths full,
With dull dyes and ringing ears,

2b. Buna:

The sirens are terrific in the dawn:
"You, multitude with wasted faces,
Another day of suffering begins
On the monotonous horror of the mud."

[19] Translated from the 1958 edition. First edition was 1947.

3a. Crescenzago:

In Crescenzago there's a window,
And at it stands a girl who's going pale.
She always holds her needle and her thread,
She sews and mends and always eyes the clock.
And when the whistle blows at the day's end
She sighs and weeps;

3b. Buna:

The last courage has been broken in you.
Gray companion, you were a strong man,
A woman traveled next to you.
Empty comrade who has no more name,
A desert who has no more tears,

4a. Crescenzago:

The faces of the children are the color
Of the dead dust in the streets

4b. Buna:

Up in the sweet world under the sun,
With what face would we confront each other?

The subjects of the poems before and after the *Lager* are the same: Work in chemical factories. The tones are indistinguishable. The images are identical: Smokestacks, sirens, and suffering workers with gray faces predate *and* postdate the *Lager*. As a writer, it must be true that Levi brought with him to Poland many ingrained capacities, images, and wells of feeling. Rowland emphasized the commonalities of the two poems:

> The 'monotonous horror' and robotic slaves…of 'Buna', where the dehumanization of the prisoners is registered in synecdoches of suffering. Repetitive labour – a 'day like every other day' – is refigured from the last poem Levi wrote before 'Buna' (nearly three years earlier) about factory life: 'Crescenzago', the first piece in the Collected Poems, has a sewing girl who 'never stops looking at the clock'; men keep 'The grim black stonecrusher panting' [134].

According to Lang, "[T]he question of whether Levi would have become a writer if not for the Auschwitz months becomes more complex than it might initially seem. Even if one views his wartime experience as a sufficient condition for his becoming a writer, this would not mean that without it he would not have followed the fork in the road which led to that destination" [135]. In his essay, the "Non-Writer Writer" Levi says "I arrived at the designation of writer because, captured as a partisan, I ended up in the Lager as a Jew" [CW p 1214] but also something more foundational, "I write because I am a chemist." [CW, p 1219]. Both can be true.

If we had only *L'inversione* the path from chemist to writer would be harder to trace, most likely because the substitution mechanism as of 1933 remained so vague. In 1989, Levi explained that he is much more successful at writing about things that are definite, like a "two-lira coin," but less successful about writing about things that are indefinite, "for example a human personality" [129]. Even if we concede that Levi had innate gifts as a writer, an essay about stereochemistry would hardly be the best showcase for them. This is because the author needs to master one of the many "languages of chemistry," whose limitations Levi identified with care, within the context of a traditional language by a gifted stylist. Moreover, a synthesis of the language [CW pp 2129 and 2134] of chemistry and a narrative language needs to reach the eyes and ears of readers who have likewise mastered the same pair of languages. Many things need to conspire to make the process of Walden inversion leap from the page.

It became obvious why *L'inversione* was not translated in the CW [13]. To translate the thesis, the translator needs to be a chemist. It is impossible to negotiate the evidence that would enable the translation of Italian into English unless one understands what Levi calls the third symbolic language of chemistry, structural formulas. [CW 2132] *L'inversione* was probably excluded from the CW because none of the writers who made the CW [13] vivid in English were chemists like Levi. It would have been impossible for them to translate. The Italian OC [12] reproduces the thesis photographically, simply inserted as is, without need for a translation or interpretation. [OC II pp 1721–1762].

Levi does succeed elsewhere in making chemical mechanisms come to life but by stand-ins for real chemical entities. Speaking of a paint polymerization in batch gone awry at great expense in "A molecule's defiance":

> [E]verything is fine as long as one molecule connects to another as if each had only two hands: they're not supposed to make more than a chain, or a rosary of molecules—it can be long, but only a chain. And you have to keep in mind that, among the many molecules, some have three hands, and there's the rub. In fact, they are inserted on purpose: the third hand is the one that is supposed to catch hold later—when we decide, not when they do. If the third hands grip too soon, every rosary joins with two or three other rosaries, and in the end they've formed a single molecule, a monster molecule as big as the whole reactor, and then you're in a fix [CW p 1506]

The molecule is meta-anthropomorphized, given a third hand. All concessions to composition, constitution, and configuration are sacrificed for the general reader.

8.6 Mirrors Real and Metaphorical

Lastly, we asked: In what way does the thesis echo through the OC [12] and CW [13]? Foremost, Levi may have written more about mirror manufacturing that any other author ever (albeit this would be a hard conjecture to prove). Nevertheless, mirror manufacturing appears in four instances in Levi's oeuvre.

The story "Lead" [CW p 819–830] from *The Periodic Table* features an ancient family, the Rodmunds: The narrator recounts that, "My father and all of us Rodmunds in the paternal line have always followed this trade, which consists in recognizing a particular heavy rock, finding it in distant lands, heating it in a particular way we know, and digging out the black lead."[CW p 819] The Rodmunds were hunters of galena, PbS, the black crystal that is a principal source of lead metal by smelting. The narrator traveled far for sources of black lead. A fishing village he visited was populated with...

> strange people, who came by ship from various distant lands...They knew glass but not mirrors: that is, they had only cheap mirrors of polished bronze, the type that are easily damaged and distort colors. If you have lead, it's not very hard to make a glass mirror, but I revealed the secret as if doing them a great favor, telling them it's an art that only we Rodmunds know, that a goddess named Frigga taught it to us, and other nonsense that they lapped right up. I needed money: I looked around; near the port I found a glazier who had a fairly intelligent look, and I made an agreement with him.... I taught him that you can pour molten lead on a hot sheet of glass and obtain a mirror that's not too big but is luminous, without flaws, and will last for many years.... All summer and fall we sold mirrors to the merchants, who paid us well. [CW p 819]

Mirror making is mentioned in "Tin" [CW p 903–908 also from *The Periodic Table*]: "There are friendly metals and hostile metals. Tin was a friend: not only because, for some months, Emilio and I had been living on it, converting it into stannous chloride to sell to mirror makers, but for other, less overt reasons."[CW p 826] There is another mention in *The Wrench* [CW pp 948–1117]: "I knew, oh yes, the feminine suppleness of copper, metal of mirrors, metal of Venus."[CW p 1020].

In the "The mirror maker," a story that was collected in a volume of short stories by the same name, [136] Levi returns to describe an ancient family of mirror makers. "Timoteo, his father, and all his ancestors, going back to the remotest times, had always made mirrors," [CW p 2298] begins Levi. Why is reflection, for Levi, something ancient? Perhaps because he knew the choice among chemical and reflected chemical, that which ultimately gave rise to the origin of biomolecular handedness, was something that happened even longer ago, long before the remotest times of Timoteo's ancestors. But, Timoteo was a rebel. Behind the backs of his father and grandfather and their guild, when he finished making "boring" "flat, transparent, colorless" "truthful" mirrors, Timoteo invented the full range of what a mirror could do:

> Normal mirrors obey a simple, inexorable physical law; they reflect like a mind that is rigid and obsessed, that claims to grasp *the* truth about the world. As if there were only one! Timoteo's secret mirrors were more versatile.... [CW p 2299] A contrivance that had cost Timoteo weeks of work inverted top and bottom and reversed left and right. [CW p 2300]

Here, Levi is acknowledging the magic of ordinary mirrors that don't reflect left and right as they seem to in Fig. 8.1a, and not top and bottom in Fig. 8.1b,c, but rather, they reflect front and back, something that never happens in reality. But a "truthful" mirror can still be magical as in the Walden inversion. And Timoteo's fantasy mirror is fantastic especially because it does something that only a fictional mirror can do, exchanging top and bottom. Unlike complex organisms such as humans, molecules with stereogenic carbon atoms are simple enough that they do not need mirrors to turn inside out. Miraculously, they can invert, exchanging front and back.

Timoteo was making untruthful mirrors, even advanced metaphysical mirrors (*specchio metafisico* or "Spemet"), reflectors that reveal the image of the observer as seen from the perspective of the person facing you. This invention was not very popular. Who wants to see ourselves as we are truly seen by others? Timoteo was ruined financially in trying to maintain his patent, then "resigned himself, and went on making plain mirrors…until he reached the age of retirement." [CW p 2302] A truthful mirror is not a metaphor for Auschwitz. A Spemet is closer.

In "Order at a good price" [CW p 451–459], Levi imagined a kind of contemporary 3D printer with unimaginable chemical versatility. Even though every photocopy machine delivers a direct copy, that is a document with retention of configuration without the necessity of holding it up to a mirror, Levi recognized that this was a double inversion, a reflection of light from a document onto a charged drum that attracts toner particles which are then transferred to paper as a second mirror image. Levi imagined that a fantasy machine, his imaginary 3D printer, must work likewise. "This was truly a technical revolution: organic synthesis at low temperature and pressure, order from disorder, silently, quickly, and cheaply. It was the dream of four generations of chemists," Levi wrote. A machine that could reproduce chemical objects in all their dimensionality. "This wasn't easy for them, you know" he continues.

> From what they tell me, the forty technicians assigned specifically to the Mimete project, having already brilliantly resolved the fundamental problem of directed synthesis, didn't obtain anything for two years but mirror images, by which I mean reversed copies, which were useless. NATCA's management was ready to put the machine into production anyway, even though it would have to be operated twice for every duplication, incurring twice the expense and twice the time. The first actual direct copy happened by chance, thanks to a providential error in assembly. [CW p 904]

Has there ever been a writer who dwelt so insistently on looking glasses, real or imagined? Thus, Ferrero's remark "all of Levi as a thinker and writer stems from the nutshell of his thesis" [137] might seem obvious. Levi loved mirrors as a subject in fiction, and enantiomers are mirror images of one another. But "*all of*

Levi" [emphasis added] as Ferrero remarked? He wasn't limiting himself to Levi's mirrors in fiction. Levi is best known for his mirrorless memoires about the second world war and its aftermath, as for his philosophical masterpiece, the *The Drowned and the Saved* [CW pp 2407–2574]. The Levi scholars Robert Gordon and Marco Belpoliti give some insight. They wrote, "At the most general levels, Auschwitz is for Levi an *inverted* world, a mirror of the civilized world or of human nature; both in the sense of showing the latter for what it is at its essence, and in the sense of its being turned against itself, literally *inverted*." [138], [emphasis added]. But, this is not precisely Levi's language, and Levi's is honored foremost for his precision (see below). When Belpoliti asks in the introduction of *L'asimmetria e la vita* [139], "Isn't it true that in the extermination camp, reason and non-reason are enantiomorphic?" The answer isn't *yes* or *no* but rather, *huh*? "Enantiomer" is not a synonym for all oppositional qualities in the human experience. Enantiomers are metrically identical and have indistinguishable physical properties (except under a dissymmetric influence in which they lose their identities as mirror images). Reason and madness or good and evil are pairings for which enantiomers are an inapt analogy.

Angier was looser still: "[Levi's] intellectual life, throughout his boyhood and youth, had itself been a Walden's Inversion—an oscillation between confidence in his abilities, and fear they were not enough for what he wanted to do" [37]. This is a silly way of speaking. Enantiomers are opposites of a specific kind. "This" and "that" don't make a racemate, and anyone's battle with imposter syndrome and a good day at the laboratory don't make for a Walden Inversion, a specific chemical mechanism. Further cementing her confusion about mirrors, albeit widely shared, Angier continues, "Only a very few times in his life would he become sufficiently unstable to flip over like the carbon atom…his impossible, longed for opposite." *Ribaltare* would be reserved in Italian for something flipping over, like a car in an accident. In any case, nothing flips over or turns upside-down in a Walden inversion, nor in a mirror.

In fact, in translation, *inverted* was once placed on Levi's pen to contrast good and evil. On reencountering his chemistry supervisor at the Buna works, Levi's translator wrote, "Years ago, a strange, even provocative, fate put me on the track of 'someone from the other side': certainly not one of the great evildoers, perhaps not even a legitimately evil man, nevertheless a specimen and a witness…He was an almost-me, *a me inverted*." [CW p 2278] However, in his native Italian Levi used *ribaltato*, overturned [OC II p 1036], not *invertito*. Mirrors exchange backs and fronts, they don't overturn. Enantiomers can be inverted by Walden inversion, they don't turn over. Levi, criticizing the obscurantism that he sensed in the holocaust inspired poetry of Paul Celan insisted, "We have a responsibility, as long as we are alive: we must answer for what we write, word for word, and ensure that every word hits its target."[CW p 2065].

Widespread misunderstandings about how mirrors operate may have given too broad a license to some in the interpretation of Levi's work. Such misunderstandings are in evidence, as described in the introduction; university students commonly are confused about how looking glasses work during annual introductions to molecular chirality.

In fairness to Belpoliti, who has worked on Levi studies for many years and made invaluable contributions, we should look at his most recent writing in his native Italian. Here, he says, "*Il Lager, Levi lo ha spiegato a più riprese, è un. Mondo rovesciato*..." [140] ("The Lager Levi explained several times, was a world upside-down" [141]). Levi rarely used *rovesciare* and its conjugations (see below). Here, *rovesciato* is sometimes translated as *upside-down,* but sometimes it is translated as *inverted,* or like a latex glove removed, turned inside-out. Translation is subtle, and distinctions that may not mean much to the average reader, would not be lost on a stereochemist like Levi obsessed with choosing the right word.

Auschwitz and fascist Italy might flip the world over, but Walden inversions and mirrors don't flip anything. Turning something upside-down is *precisely what a mirror doesn't do* and Levi knew this in his bones. Levi's translators said the following about the upside-down world of 1943–45. Of course, what is crucial is the Italian word Levi used. The Italian in each case is inserted in the quotes in square braces.

> ...It may seem surprising that one of the most frequent states of mind in the Lager was curiosity. And yet we were not only frightened, humiliated, and hopeless but also curious: starved for bread and also for understanding. The world around us seemed **upside down** [*capovolto*], therefore someone must have turned it **upside down** [*capovolto*], and therefore must be **upside down** [*capovolto*] himself. [CW pp 2277–2278 OC II p 1036]

> ...the man who can explain to us how, in the heart of our Europe and our century, the commandment "Thou shalt not kill" was turned **upside down** [*capovolto*]. [CW p 1190, OC II p 1358]

> Russians...did not recognize in those gaunt yet determined faces the zhid of their tradition, the foreigner in their home, who speaks Russian to swindle you but thinks in his own strange language, does not know Christ but instead follows his own ridiculous and incomprehensible precepts, who is wealthy and cowardly, his only strength that of his cleverness. The world was turned **upside down** [*capovolto*]: these Jews were armed allies. [CW p 1671, OC II p 518]

> ...I do not mean to imply that they were made of a perverse human substance that was different from our own (there were sadists and psychopaths among them, but they were few). Put simply, for several years they had been subjected to a schooling that turned the reigning morality **upside down** [*capovolta*]. [CW p 2499, OC II p 1222]

> In countries where human rights and basic freedoms are suffocated, history, too, belongs to the ruling class: not just current history but also past history. It can be obliterated, turned **upside down** [*capovolta*], made up; what is true today can become false tomorrow, encyclopedia entries and schoolbooks are rewritten, and the individual memories of loyal citizens are required to be equally submissive. [CW p 2603, OC II p 1693]

> ...we returned in the long, windy evenings of the Polish summer to find the camp **upside down** [*sconvolto* = upset], no water to drink or wash in, no soup for our empty bellies, no light by which to defend our piece of bread against someone else's hunger...[CW p 112, OC I p 231]

> A giant Polish common criminal punched him in the face over nothing. Améry hit him back as hard as he could, not out of an animal reaction but rather out of a reasoned revolt against the **upside-down** [*stravolto*=upset] world of the Lager. [CW pp 2509–2510, OC II p 1231]

> ...the good guys are the ones who should win, otherwise the world would be turned **upside down** [*sovvertito* = subverted]. [CW p 2431, OC II p 1164]

> It's the Lager, the KZ: new words for him, never heard before. In a way, it's an **upside-down** [*rovescia* = overturned] world, where honesty and gentleness are punished, while violence, violence, betrayal, and deception are rewarded. [CW p 2783, OC II p 1667]

Seven times Levi used *capovolto/a*, a straightforward Italian word meaning upside-down. He used *sconvolto* once and *stravolto* once. Both mean "upset" or "disturbed." He used *sovvertito* once meaning "subverted," and once *rovescia,* meaning "reversed." He once combined inversion and reversal when writing for middle school children: "[The 1938 the racial laws] separated the Jews from the rest of the population...Absurd, unjust, and oppressive laws followed. Every day the newspapers were full of lies and insults. It was an **inversion** [*inversione*], a ridiculous and cruel **reversal** [*capovolgimento*] of the truth." [CW p 2742, OC I p 1380] Here, he mixes a function of real mirrors and fantasy mirrors. This is a minor inconsistency. Levi is talking to middle school children, with familiar language, not connecting fascism to enantiomers or Walden inversion. He never used *riflesso* (reflected) or *riflettere* (to reflect) to convey the opposite world of the *Arbeitslager*.

Franco Baldasso, a Levi scholar at Bard College, described it correctly when he said, "Levi never concealed the fact that the *Lager* seemed to him first of all like a world turned upside down." This clarity, however, is quickly reversed when he continues: "The asymmetry and the consequent 'enantiomorphism'...constitute the sources of the vocabularies from which Levi draws the languages employed to narrate Auschwitz" [142]. This inconsistency and those highlighted above are peccadillos, however. The average reader, unlike chemists, do not make important distinctions between symmetry operations of the 1st and 2nd kind, proper rotations and improper rotations, respectively. To those who have not shared Levi's training in stereochemistry, *reflected*, *upside-down*, *inverted*, *backwards*, and *reversed*, are all synonyms for *the other way round*. Levi was more particular about words, and especially more particular about words in this general category.

We close this essay, written in 2024 in the USA, by quoting Levi without additional comment:

> Every era has its fascism; the warning signs can be seen wherever the concentration of power denies a citizen the opportunity and the capacity to express and carry out his will. There are many ways of getting there.Not necessarily through the terror of police intimidation but also by censoring and distorting information, polluting justice, paralyzing schools, disseminating in many subtle ways nostalgia for a world where order reigned supreme, and where the security of the privileged few rested on the forced labour and forced silence of the many. (CW p 1199).

Acknowledgements We are grateful for the assistance and correspondence of Paola Novaria of the Archivio Storico dell'Università di Torino, Barbara Martin of the Biblioteca Ponzio of the Università di Torino, Professor Sergio Luzzatto of the University of Connecticut, and Professor Stefano Sacanna of New York University. Permission for the translation of the thesis was provided by Ms. Valeria Zito (Foreign & Film Rights, Giulio Einaudi Editore, S.p.A.) in consultation with the Levi family. We thank Professors Roald and Arthur Greenberg for calling to our attention Maier's *Chemiker im der Dritte Reich* [115].

References

1. Walden P (1893) Ber dtsch Chem Ges 26:210–215
2. Frankland PF (1913) Presidential address. The Walden inversion. J Chem Soc Trans 103:713–742
3. Campbell IG (1964) Stereochemistry of carbon compounds. Elsevier, In Rodd's chemistry of carbon compounds, pp 201–259
4. Lewis DE (2022) Addition, elimination and substitution: Markovnikov, Hofmann, Zaitsev XE "Zaitsev" and Walden: discovery and development. Elsevier, Amsterdam Chapter 6
5. Walden P, Bischoff CA (1894) Handbuch der Stereochemie. Verlag von H, Bechhold, Frankfurt am Main
6. Boeck G (2019) Paul Walden (1863–1957): the man behind the Walden inversion, the Walden rule, the Ostwald-Walden-Bredig rule and ionic liquids. ChemTexts 5:6
7. Walden P (1919) Optische Umkehrerscheinungen (Waldensche Umkehrung). Wissenschaft, vol 64. In: Wiedemann E (ed). Springer, Wiesbaden
8. Adams CC (1994) The knot book. WH Freeman and Company, New York Ch. 10
9. Cervantes M de (1605) Don Quixote de la Mancha, (trans: Grossman E) https://www.publiconsulting.com/wordpress/donquixoteoflamancha/front-matter/introduction/. Accessed 5 June 2024
10. Levi P L'inversione di Walden. www.primolevi.it/sites/default/files/inline-files/levi%20inversione%20walden.pdf. Accessed 5 June 2024
11. *Centro Internazionale di Studi Primo Levi.,* https://www.primolevi.it. Accessed 31 August 2025
12. Levi P (1997) Opere complete, 2 volumi. In: Belpoliti M (ed). Einaudi, Turin
13. Levi P (2015) The complete works of Primo Levi, 3 volumes. In: Goldstein A (ed) (trans: Woolf S, McPhee J, Goldstein A, Shugaar A, Galassi J, Appel AM, Moore MF, Bastagli A, Bastagli F). WW Norton, New York
14. Barbarulo MV (2024) La pergamena nel cassetto, 17 January 2024. https://www.primolevi.it/it/pergamena-cassetto. Accessed 31 August 2025
15. Stille A (2009) The double bind of Italian Jews: Acceptance and assimilation. In: Zimmerman JD (ed). Jew in Italy under Fascist and Nazi Rule. 1922–1945. Cambridge University Press, Cambridge, p 19
16. Wood J (2015) The art of witness: How Primo Levi survived, The New Yorker, 21 September. https://www.newyorker.com/magazine/2015/09/28/the-art-of-witness. Accessed 1 September 2025
17. As quoted by Geras N, Hope, shame, and resentment: Primo Levi and Jean Améry. In: Vuohelainen M, Chapman A (eds) Interpreting Primo Levi. Palgrave MacMillan, p 7
18. Gordon RSC (ed) (2007) The Cambridge Companion to Primo Levi. Cambridge University Press, Cambridge, p xvii
19. Calvino I (1988) Six memos for the next millennium. Harvard University Press, p 26
20. Perucca E (1919) Nuove osservazioni e misure su cristalli otticamente attivi ($NaClO_3$). Il Nuovo Cimento 18:112–154

21. Bing Y, Kaminsky W, Kahr B, Viterbo D (2009) Turinese stereochemistry: Eligio Perucca's enantioselectivity and Primo Levi's asymmetry. Angew Chem Int Ed 48:2–8
22. Bing Y, Selassie D, Paradise RH, Isborn C, Kramer N, Sadilek M, Kaminsky W, Kahr B (2010) Circular dichroism tensor of a triarylmethyl propeller in sodium chlorate crystals. J Am Chem Soc 132:7454–7465
23. Eligo Perucca. https://en.wikipedia.org/wiki/Eligio_Perucca. Accessed 29 Dec 2023
24. See also, Wataghin G (1966) Eligio Perucca, Accademia Nazionale dei Lincei, Roma
25. de Alfaro V (1999) La Facoltà di Scienze Matematiche, Fisiche e Naturali dell Università di Torino, 1848–1980, vol I. In: Roero CS (ed) Deputazione Subalpina di Storia Patria, Torino
26. Dallaporta N (1998) La mia esperienza di studi e di vita, Annali di storia dell'educazione e delle istituzioni scolastiche, 5
27. Levi P Comportamento dielettrico della miscela ternaria C_6H_6, $CHCl_3$, C_6H_5Cl is reproduced in manuscript form in OC vol II, pp 1707–1720
28. Viterbo D (2013) Primo Levi, William Henry Bragg and the atomic theory of matter. Acta Crystallogr A 69:452–456
29. Anissimov M (2000) Primo Levi: Tragedy of an Optimist (trans: Cox P). Overlook Press, New York
30. Anissimov M (2007) Email to Kahr B. 27 January
31. Jacoby M (2008) Recognizing a pioneer, Chemist-historians uncover the stereochemical contiributions–and idiosyncracies of a little-known Italian scientist 86(33). Chem Eng News. https://cendev.acs.org/articles/86/i33/Recognizing-Pioneer.html. Accessed 1 September 2025.
32. Sanderson K (2008) New account of Primo Levi racism row. Nature. https://doi.org/10.1038/news.2008.870. Accessed 2 August 2024.
33. Viterbo D. (2008) Eligio Perucca: Serendipity across the Atlantic. IUCr Newsletter 16(2). September 5, https://www.iucr.org/news/newsletter/volume-16/number-2/eligio-perucca. Accessed 2 Aug 2024
34. Anissimov (2000) p 64
35. Levi P, Regge T (1989) Dialogo. Princeton, Princeton University Press, p 19
36. Sassi Perino AM (1991) Un docente affascinante e terribile: ricordo di un'allieva. In: Demichelis F (ed) Mezzo secolo di fisica per gli ingegneri: l'insegnamento di Eligio Perucca al Politecnico di Torino. Politecnico di Torino, Torino, pp 39–43
37. Naso F (2015) Chimica, Stereochimica e Fantachimica Negli Scritti di Primo Levi: Oltre il Sistema Periodico. Parte 2: Stereochimica. Chim Ind 4. https://www.soc.chim.it/sites/default/files/chimind/pdf/2015_4_56_ca.pdf. Accessed 5 June 2024
38. Levi P, Regge T, p 21
39. Lang B (2013) Primo Levi. The Matter of a Life. Yale University Press, New Haven
40. Mealli C, Milanesio M (2017) Davide Viterbo (1939–2017). Acta Crystallogr A 73:375–376
41. Borello E (1997) Primo Levi, studente di chimica. L'Ateneo. Notiziario dell'Università degli Studi di Torino XIII(3):6–10. https://www.primolevi.it/sites/default/files/inline-files/borello%20.pdf. Accessed 1 September 2025
42. Naso F (2017) Mario Betti: A giant in the chemistry scenario of the twentieth century. Substantia 1(2):111–121
43. Cravotto G, Cintas P (2021) Primo Levi: a frustrated stereochemist?—insight in hindsight. Chirality 33:72–80
44. Martínez RF, Cravotto G, Cintas P (2021) Organic sonochemistry: A chemist's timely perspective on mechanisms and reactivity. J Org Chem 86(20):13833–13856
45. Cravotto G, Cintas P (2006) Power ultrasound in organic synthesis: moving cavitational chemistry from academia to innovative and large-scale applications. Chem Soc Rev 35(2):180–196
46. Cintas P (2002) On the origin of tetrahedral carbon: A case for philosophy of chemistry? Found. Chem 4:149–161
47. Reference [14] is an expanded and richly illustrated version of the following: Barbarulo MV (2022) Le cinque lire dell'aquila ei fiordalisi del commendatore: Primo Levi dal liceo alla

laurea in chimica, ai primi lavori negli anni 1935–1942. Chimica nella Scuola Sep 20(Speciale):55–61
48. Barbarulo MV, Calascibetta F (2019) Per la purezza della cattedral Le Leggi antiebraiche del '38 e le vicende di cinque professori universitari chimici di religione ebraica, Rendiconti Accademia Nazionale delle Scienze detta dei XL, Memorie di Scienze Fisiche e Naturali, 137° Vol. XLIII, Parte II, Tomo I:21–36
49. Di Meo A (2013) Primo Levi's chemistry: languages and literary style. Trauma Memory 1:23–33
50. Walden P (1895) Weiteres über optisch active Halogenverbindungen. Ber dtsch Chem Ges 28(3):2766–2773
51. Fischer E (1906) Synthese von Polypeptideen. XV. Ber dtsch Chem Ges. 39:2893–2931
52. Olson AR (1933) The mechanism of substitution reactions. J Chem Phys 1:418–423
53. Fischer E (1906) Untersuchungen über Aminosäuren, Polypeptide und Proteïne, Verlag von Julius Springer, Berlin
54. Senter G, Drew HDK (1915) LXV.–Studies on the Walden inversion. Part I. The influence of the solvent on the sign of the product in the conversion of phenylchloroacetic acid to phenylaminoacetic acid. J Chem Soc Trans 107:638–642
55. Senter G, Drew HDK (1916) LXXVIII.–Experiments on the Walden inversion. Part IV. The influence of the solvent on the sign of the product in the conversion of phenylchloroacetic acid to phenylaminoacetic acid—(continued). J Chem Soc Trans 109:1091–1106
56. Senter G, Drew HD, Martin GH (1918) XVIII.–Studies on the Walden inversion. Part VII. The influence of the solvent on the sign of the product in the conversion of α-bromo-β-phenylpropionic acid to α-amino-β-phenylpropionic acid (phenylalanine). Iminodiphenyldipropionic acid. J Chem Soc 113:151–163
57. Lipscomb AG, Phillips H (1930) The interaction of ethyl *l*-mandelate and thionyl chloride in the presence of pyridine. The mechanism of the replacement of hydroxyl by chlorine by means of thionyl chloride. J Chem Soc 415–423
58. Kuhn R, Wagner-Jauregg T (1928) In welcher Reaktionsphase findet bei der Waldenschen Umkehrung die Umgruppierung der Substituenten statt? (Zur Stereochemie des tetraedrischen Kohlenstoffatoms, IX.). Ber dtsch Chem Ges 61(3):504–521
59. McKenzie A, Clough GW (1913) LXXVII.–Experiments on the Walden inversion. Part IX. The interconversion of the optically active phenylmethylcarbinols. J Chem Soc Trans 103:687–699
60. Holmberg B (1913) Stereochemie der halogensubstituierten Bernsteinsäuren. J prak Chem 88(1):553–603
61. Auwers K (1899) Ueber Aminsäuren und Imide aliphatischer Dicarbonsäuren. Justus Liebigs Ann Chem 309(3):316–347
62. Fischer E (1914) Über Phosphorsäureester des Methyl-glucosids und Theophyllin-glucosids. Ber dtsch Chem Ges 47(3):3193–3205
63. Polanyi M, Szabo AL (1934) On the mechanism of hydrolysis. The alkaline saponifications of amyl acetate. Trans Farad Soc. 30:508–512
64. Kenyon J, Phillips H, Turley HG (1925) LXI.–Investigations on the dependence of rotatory power on chemical constitution. Part XXIV. Further experiments on the Walden inversion. J Chem Soc London 127:399–417
65. Kenyon J, Phillips H (1930) Some recent developments in the study of the Walden inversion. Trans Faraday Soc 26:451–458
66. Houssa AJ, Kenyon J, Phillips H (1929) CCXXV.—The relative configurations of d-β-octanol and its dextrorotatory halides. The interconversion of the optically active β-octanols by a new method. J Chem Soc 1929:1700–1711 [Not the article cited by Levi.]
67. Freudenberg K, Rhino F (1924) Die Konflguration des Alanins.(4. Mitteilung über sterische Reihen). Ber dtsch Chem Ges, 57(8):1547–1557
68. Freudenberg K, Noë A (1925) Die Konfiguration der Asparaginsäure.(VII. Mitteilung über sterische Reihen). Ber dtsch Chem Ges 58(10):2399–2408

69. Freudenberg K, Markert L (1927) Die Konfiguration der α-Brom-propionsäure (8. Mitteilung über sterische Reihen). Ber dtsch Chem Ges 60(11):2447–2458
70. Freudenberg K, Lux A (1928) Die Konfiguration der monosubstituierten Propion- und Bernsteinsaeuren. (9. Mitteilung über sterische Reihen) Ber dtsch Chem Ges 61:1083–1089
71. Timmermans MJ (1931) Quelques propriétés physiques des molécules organiques en relations avec leur consitutiton et leur configuration, in Rapports et Discussions Relatifs a la Constitution et a la Configuration des Molécules Organiques, Inst. Internat. Chim. Solvay, Gauthier-Villars, Paris, pp 191–227
72. Gadamer J (1913) Über die Racemisation der Tropasäure und ihrer Ester; zugleich eine Theorie der Racemisation, Substitution und Waldenschen Umkehrung. J prak Chem 87(1):312–392
73. Meisenheimer J, Link J (1930) Über die Verschiebung in der Allyl-Gruppe. 3. Mitteilung über Substitution und Addition. Justus Liebigs Ann Chem 479:211–277
74. Werner A (1911) Über den räumlichen Stellungswechsel bei Umsetzungen von Raumisomeren Verbindungen. Ber dtsch Chem Ges 44:873–882
75. Werner A (1912) Über die raumisomeren Kobaltverbindungen. Justus Liebigs Ann Chem 386:1–272
76. Kharasch MS, Engelmann H, Mayo FR (1937) The peroxide effect in the addition of reagents to unsaturated compounds. XV. The addition of hydrogen bromide to 1-and 2-bromo-and chloropropenes. J Org Chem 2:288–302
77. Holmberg B (1926) Stereochemische Studien, XI.: Die Reaktions-Distanzen der Anionen. Ber dtsch Chem Ges 59(2):125–134
78. Rørdam HN (1928) CCCXXIV.–The Walden inversion. J Chem Soc 2447–2453
79. Rørdam HN (1929) CLXVI.–The Walden inversion. Part II. J Chem Soc 1282–1290
80. Rørdam HN (1930) CCLXI.–The Walden inversion. Part III. J Chem Soc 2017–2026
81. Bonino GB (1933) Su un possibile meccanismo della racemizzazione e dell'inversione di Walden. Gazz Chim Ital 63:448–452
82. Slater JC (1931) Directed valence in polyatomic molecules. Phys Rev 37:481–489
83. Pauling L (1931) The nature of the chemical bond. II. The one-electron bond and the three-electron bond. J Am Chem Soc 53:3225–3237
84. Long DA (1988) Early history of the Raman effect. Int Rev Phys Chem 7:317–349
85. Raman CV (1928) A new radiation. J Ind J Phys 2:387–398
86. Fermi E (1932) Sulle bande di oscillazione e rotazione dell'ammoniaca. Il Nuovo Cimento 9:277–283
87. Angier C (2002) The Double Bond: Primo Levi, Farar, Straus, Giroux, p 159
88. Thompson I (2002) Primo Levi: A Life. Metropolitan Books, New York, p 100
89. Ponzio G (1942) Chimica Organica, Seconda Edizione, Unione Tipografico-Editrice Torinese, Torino, pp 96–98.
90. Walden P (1895) Ueber optisch active Halogenverbindungen. Ber dtsch Chem Ges 28:1287–1297
91. Walden P (1896) Ueber die gegenseitige Umwandlung optischer Antipoden. Ber dtsch Chem Ges 29:133–138
92. Walden P, Lutz O (1897) Ueber die gegenseitige Umwandlung optischer Antipoden. Ber Dtsch Chem Ges 30:2795–2798
93. Walden P (1898) Ueber Autoracemisirung. Ber dtsch Chem Ges 31:1416–1422
94. Walden P (1899) Ueber die gegenseitige Umwandlung optischer Antipoden. Ber dtsch Chem Ges 32:1855–1864
95. Thompson, p 103
96. Olson AR, Long FA (1934) The mechanism of substitution reactions. J Am Chem Soc 56:1294–1299
97. Deasy CL (1945) The Walden inversion in nucleophilic aliphatic substitution reactions. J Chem Ed 22:82–83
98. Hughes ED, Juliusburger F, Masterman S, Topley B, Weiss J (1935) 362. Aliphatic substitution and the Walden inversion. Part I. J Chem Soc 1525–1529

99. Hughes ED, Juliusburger F, Scott AD, Topley B, Weiss J (1936) 250. Aliphatic substitution and the Walden inversion. Part II. J Chem Soc 1173–1175
100. Cowdrey WA, Hughes ED, Nevell TP, Wilson CL (1938) 39. Aliphatic substitution and the Walden inversion. Part III. Comparison, using radioactive bromine, of the rates of inversion and substitution in the reaction of bromide ions with α-bromopropionic acid. J Chem Soc 209–211
101. Hughes ED, Ingold CK (1935) 55. Mechanism of substitution at a saturated carbon atom. Part IV. A discussion of constitutional and solvent effects on the mechanism, kinetics, velocity, and orientation of substitution. J Chem Soc 244–255
102. Hughes ED (1938) Rate of ionisation. Part II. Ionisation of C—hal. bonds. Trans Farad Soc 34:185–202
103. Hughes ED (1938) Aliphatic substitution and the Walden Inversion. Trans Farad Soc 202–221
104. Bateman LC, Church MG, Hughes ED, Ingold CK, Taher NA (1940) 188. Mechanism of substitution at a saturated carbon atom. Part XXIII. A kinetic demonstration of the unimolecular solvolysis of alkyl halides.(Section E) A general discussion. J Chem Soc 979–1011
105. Bateman LC, Hughes ED, Ingold CK (1940) 184. Mechanism of substitution at a saturated carbon atom. Part XIX. A kinetic demonstration of the unimolecular solvolysis of alkyl halides.(Section A) Kinetics of, and salt effects in, the hydrolysis of tert-butyl bromide in aqueous acetone. J Chem Soc 960–966
106. Bateman LC, Hughes ED, Ingold CK (1940) 187. Mechanism of substitution at a saturated carbon atom. Part XXII. A kinetic demonstration of the unimolecular solvolysis of alkyl halides.(Section D) Kinetics of, and salt effects in, the hydrolysis of *pp*'-dimethylbenzhydryl chloride in aqueous acetone. J Chem Soc 974–978
107. Ingold CK (1953) Structure and Mechanism in Organic Chemistry. Cornell University Press, Ithaca Chapter 7
108. Hughes ED (1938) Aliphatic substitution and the Walden Inversion. Trans Farad Soc 34:202–221
109. Hammett LP (1938) Discussion. Trans Farad Soc 34:221
110. Gooch J (2020) Mussolini's War. Fascist Italy from Triumph to Collapse. 1935–1943. Penguin Books, Dublin. p 294.
111. Luzzatto S (2016) Primo Levi's Resistance, (trans: Randall F). Metropolitan Books, New York
112. Gilman H (ed) (1938) Organic Chemistry. An Advanced Treatise, 2 vol. John Wiley, New York
113. Gilman H (ed) (1955–56) Chimica organica superiore, volumi 1 e 2 (trans: Levi P, Anglesio G) Edizioni Scientifiche Einaudi, Torino
114. Karachalios A (2022) Chimica e fascismo: Il contributo epistemologico e culturale di Giovanni Battista Bonino (1899–1985) durante gli anni del fascismo. Chimica nella Scuola. 20(Speciale):80–90
115. Maier H (2015) Chemiker im "Dritten Reich." Die Deutsche Chemische Gesellschaft und der Verein Deutscher Chemiker im NS-Herrschaftsapparat; VCH: Weinheim, pp 434–436
116. Maier, p 329
117. Lewis, p 156
118. Barron LD (2021) Symmetry and chirality: where physics shakes hands with chemistry and biology. Isr J Chem 61:517–529
119. Kaupp G, Marquardt T (1994) Absolute asymmetric synthesis solely under the influence of a static homogeneous magnetic field? Ang Chem Inter Ed Engl. 33:1459–1461
120. Bloom BP, Paltiel Y, Naaman R, Waldeck DH (2024) Chiral induced spin selectivity. Chem Rev 124:1950–1991
121. Cairns-Smith AG (1990) Seven Clues to the Origin of Life: A scientific Detective Story. Cambridge University Press, Cambridge
122. Bullard T, Freudenthal J, Avagyan S, Kahr B (2007) Test of Cairns-Smith's 'crystals-as-genes' hypothesis. Farad Disc 136:231–245
123. Natta G, Farina M (1972) Stereochemistry, Longman, London.

124. Green MM, Park JW, Sato T, Teramoto A, Lifson S, Selinger RL, Selinger JV (1999) The macromolecular route to chiral amplification. Angew Chem Int Ed 38:3138–3154
125. Belpoliti M (2022) Primo Levi. An Identikit. (trans: Botsford C) Seagull Books, Kolkata, p 471
126. Lang, p 52
127. Angier, p 685–686
128. Lang, p 47
129. Levi P, Regge T. p 60
130. Belpoliti M (ed) (2001) The Voice of Memory. Interviews 1961–1987. Primo Levi (trans: Gordon RSC), p 65
131. Levi P (1984) Ad ora incerta, 1st edn. Garzati, Milan
132. Raboni G (1990) Primo Levi un poeta vero ad ora incerta. Ad ora incerta. Garzanti, Milan
133. Lang, p 56
134. Rowland A (2008) Poetry as testimony: Primo Levi's collected poems. Textual Pract 22:487–505
135. Lang, p 57
136. Levi P (1986) The Mirror Maker, (trans: Rosenthal R). Schocken Books, New York
137. Ferrero E (2007) Primo Levi: La vita, le opere. Einaudi, Turin, p 9
138. Gordon, p 58
139. Levi P (2002) L'asimmetria e la vita. In: Belpoliti M (ed) Einaudi. Turin, p XI
140. Belpoliti M (2015) Primo Levi di fronte e di profilo. Guanda, Milan, p 159
141. Belpoliti (2022) p 174
142. Baldasso F Weightless flight. Primo Levi and the "Break of Civilization". Nemla Italian Studies, XXXII/2009–2010, pp 62–90

How Sweet It is: Emil Fischer's Stereochemical Studies

9

David E. Lewis

Abstract

Oddly enough, the work of Emil Fischer (1852–1919) leading to the elucidation of the relative structures of the monosaccharides was only a part of his work that was recognized by the Nobel Prize in 1902. By 1889, Fischer had begun to consider that the monosaccharides were stereoisomers of each other, and he set out the sixteen isomers predicted by the van't Hoff model of the carbon atom. The arguments used by Fischer to derive a structure for glucose still stand, over 100 years later, as one of the most impressive feats of chemical logic in the history of the discipline. The whole structure of his logical derivation of a structure for glucose appeared in two papers, published in 1891. These papers also contain the first publication of Fischer's new projection formulas for what we now term stereogenic centers. His arguments stand on three foundations: (1) the interrelationships between the monosaccharides, established by oxidation–reduction chemistry; (2) the molecular symmetry of the saccharic acids (and the alditols); and (3) the application of the Kiliani-Fischer homologation of aldopentoses, and the symmetry of those pentoses. The prelude to his work and Fischer's analysis is presented.

It takes encouragement from his peers to persuade an organic chemist to pursue the study of the history of chemistry. This chapter is dedicated to the memory of one such individual, Dr. Paul R. Jones, with gratitude.

D. E. Lewis (✉)
Department of Chemistry and Biochemistry, University of Wisconsin-Eau Clare, Eau Claire, USA
e-mail: lewisd@uwec.edu

A. Greenberg and D. E. Lewis (eds.), *The Birth of the 3rd Dimension in Chemistry*, Perspectives on the History of Chemistry, https://doi.org/10.1007/978-3-031-97743-5_9

9.1 Introduction

Emil Fischer's elegant deduction of the relative configuration of glucose [1] was one of the most impressive accomplishments of nineteenth-century organic chemistry, as the Nobel Committee recognized when they awarded him the second Nobel Prize in Chemistry, in 1902. Strictly speaking, the Fischer "proof" of the structure of glucose, designated in 2004 as one of the ten most beautiful chemistry experiments, was neither a single experiment, nor had the deliberate aim of proposing a structure for glucose at the onset of what became almost a decade odyssey into the realm of stereochemistry. And yet, despite the fact that this is not, in fact, "an experiment," it richly deserves to be included in the ten most beautiful applications of the chemist's art.

9.2 Emil Fischer (1852–1919)

Emil Fischer [2] was born in Euskirchen, near Bonn, the son of a lumber merchant. After convincing his father that he had no aptitude for business, Fischer was permitted to enroll at the University of Bonn in 1871, and in two years there he attended lectures by Kekulé, Engelbach, and Zincke. Nevertheless, he was unimpressed by the instruction in analytical chemistry, and determined to study physics, instead. He left Bonn in 1872, accompanied by his cousin, Otto, who persuaded him to join him in attending the newly founded University of Strasbourg, and to give the study of chemistry a second chance. Strasbourg was one of the new universities founded by the Germans in the immediate aftermath of the Franco-Prussian war, and Germany had sent some of its finest young minds as professors to these universities. Among these was the young Adolf von Baeyer, under whom Fischer obtained his Ph.D. in chemistry in 1874 for a thesis on fluorescein and orcin-phthalein. Following his graduation, Fischer remained as an assistant to von Baeyer, who took him to Munich when he accepted the Chair there. For his first three years in Munich, Fischer had no official position at the University, becoming Privatdozent only in 1878, and *Ausserordentlicher* in analytical chemistry (succeeding Jakob Volhard) in 1879. This same year, he declined appointment to the Chair of Chemistry at Aix-la-Chapelle. In 1882 he moved to Erlangen, where he again succeeded Jacob Volhard as Professor of Chemistry, and three years later he accepted the call to Würzburg. His final move, as Professor of Chemistry, was to Berlin, in 1892, succeeding A.W. von Hofmann in that Chair. Fischer remained at Berlin until his death, by his own hand, in 1919. His suicide may well have been precipitated by an accumulation of events. The first was the loss of his wife, Agnes, née von Gerlach, whom he had married at the beginning of 1888, after only seven years of marriage. This left him with three young boys to raise alone: Hermann Otto Laurenz (1888–1960), Walter (1891–1916), and Alfred (1894–1917). The loss of two of his sons in World War I—Alfred, by typhus, and Walter, by suicide as a result of compulsory military training—was certainly a major contributing factor in his decision, as was the pain arising from his terminal cancer, diagnosed in 1919. In addition to his Nobel Prize, Fischer received numerous accolades for his scientific work, including the Davy Medal of the Royal Society (1890), the Faraday Medal of the Chemical Society (London), the Helmholtz medal (1909) and the Elliot Cresson Medal of the Franklin Institute (1913), as well as honorary doctorates from the Universities of Christiania, Cambridge, Manchester and Brussels. From his native land, he was made a Prussian Geheimrat (Excellenz) and a member of the Maximilian Order for Arts and Sciences (1898) and the Prussian Order *Pour le Mérite* for Sciences and the Arts (1904). Following his death, the Emil Fischer Memorial Medal was instituted by the German Chemical Society.

9.3 What Techniques Were Available to Fischer When He Started His Carbohydrate Research?

It is often difficult to assess the work of early chemists without having one's judgment shaded by modern knowledge, and so let us begin by looking at the techniques that Fischer did and did not have available to him at the time of his work (the last decades of the nineteenth century). Fischer was without some of the most powerful modern methods for purification and characterization: he did not have access to chromatography, nor did he have access to modern spectroscopy—NMR, IR, uv–visible, and mass spectrometry. What techniques he did have available to him were much more simple physical techniques: fractional crystallization, distillation, and melting and boiling point determination, along with optical microscopy and crystal habit (crystallography). Combustion analysis provided necessary analytical data. In 1878, François-Marie Raoult (1830–1901) [3] reported his law relating molecular weight and the colligative properties of solutions, [4] but it is worth noting that the molecular weight of glucose was only determined experimentally by Bernhard Christian Gottfried Tollens (1841–1918) [5] in 1888, [6] as part of an extensive series of molecular weight determinations of carbohydrates. Therefore, its molecular formula was not actually known until after Fischer had embarked on his work.

9.4 The Starting Line

When Fischer began his work on glucose in 1884, just three sugars—sucrose, glucose, and fructose—had been isolated in pure form, along with the alditol, mannitol. Sucrose, which is a disaccharide, had been known pure and crystalline since antiquity, having been isolated by Indians during the Gupta dynasty, around 350 C.E [7]. It took another millennium and a quarter before the first pure crystalline monosaccharide, glucose, was isolated—from raisins, by Andreas Sigismund Marggraf (1709–1782), [8, 9] from honey, by Johann Tobias Lowitz (Tovii Yegorovich Lovits, Ловиц Товий Егорович, 1717–1804), [10, 11] from grapes, by Joseph Louis Proust (1754–1826) [12, 13] and by the hydrolysis of starch by

Gustav Robert Kirchhoff (1824–1887) [14, 15]. Its mutarotation was reported by Augustin-Pierre Dubrunfaut (1797–1881) in 1847 [16, 17].

Fructose proved to be the most difficult of the three sugars to be isolated in crystalline form. It was isolated in pure form as an uncrystallizable syrup by Dubrunfaut in 1847, but it was not until 1880 that preliminary reports of its isolation from honey by Émile Clément Jungfleisch (1839–1916) [18] appeared in the *Extraits des Procès-Verbaux des Séances* of the Société chimique de France [19]. In 1881, a more complete report by Jungfleisch and his student, Lefranc was published in the *Comptes rendus* [20]. Proust is credited with the isolation of the alditol, mannitol, in 1806 [21]. In 1888–1889, Fischer and Hirschberger reported the preparation of mannose by the controlled oxidation of mannitol with 15% aqueous nitric acid at 40 °C [22].

The analytical values for glucose were reported in 1843 [23] by Jean-Baptiste André Dumas [24] (1800–1884) as C 36.8, H 7.0. and O 56.2%. These values correspond to the empirical formula $C_6H_{14}O_7$ [25]. However, the molecular weight of anhydrous glucose was not actually known until 1888, [6] when Tollens and Mayer determined it by the Raoult cryoscopic method. Thus, the molecular formula of glucose was not known with certainty until after Fischer had embarked on his work.

The reduction of alcohols with hydrogen iodide and phosphorus [26] in a sealed tube, reported by French chemist Marcellin Berthelot (1827–1907), [27] was important in establishing the length of the carbon chain of the aldoses. The reduction of the aldaric acids (**1**) from aldohexoses was studied by Alexander Crum Brown (1838–1922), [28] who demonstrated the reduction of mucic acid from galactose (galactaric acid) to adipic acid (**2**), [29] and by de la Motte, who demonstrated the reduction of saccharic acid from glucose to the same diacid (Fig. 9.1) [30].

The conversion of 3-iodopropionic acid (**3**) to adipic acid by heating with silver powder, reported by Johannes Wislicenus [31] (1835–1902) in 1868, [32] demonstrated that adipic acid was an unbranched molecule with six carbon atoms in the chain (Fig. 9.1). In 1881, the Croatian chemist Julije (Julius) Domac (1853–1928) [33] reported the structure he had deduced for the hexene (**6**) he obtained by

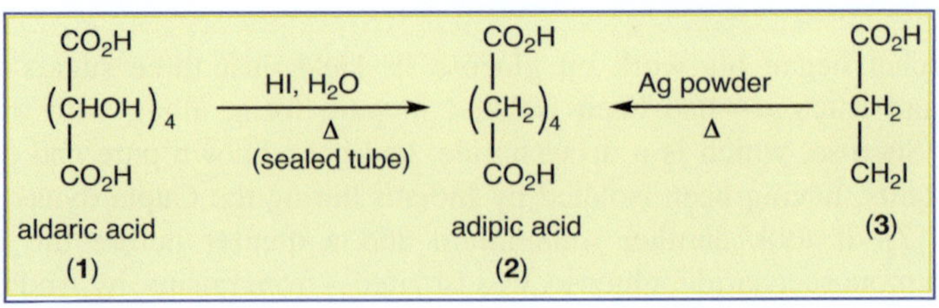

Fig. 9.1 Early studies of the Berthelot reduction of aldose derivatives to adipic acid and Wislicenus' synthesis of adipic acid

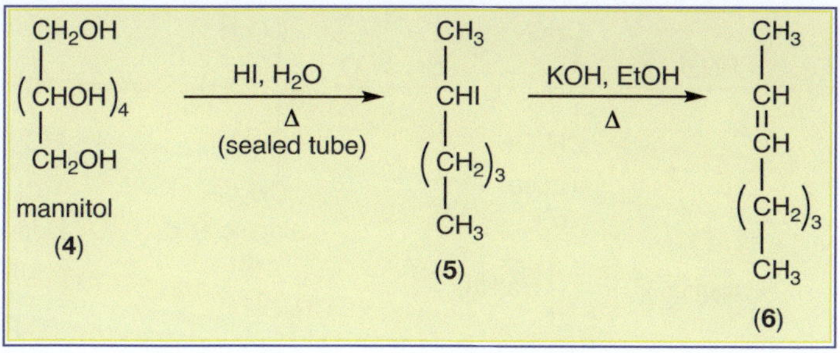

Fig. 9.2 Berthelot reduction of mannitol and conversion of the product to 2-hexene

base elimination of the iodohexane (**5**). He had prepared compound **5** by Berthelot reduction of mannitol (**4**) [34]. These reactions (Fig. 9.5) were confirmed by Fischer himself [35].

With the pure material in hand, data began to accumulate about glucose, in particular. The presence of a six-carbon unbranched chain in the glucose molecule was established by the oxidation of galactose to mucic (galactaric) acid by Franz Hugo Schwanert (1828–1902), [36] who heated the carbohydrate with commercially available (presumably 70%) nitric acid, [37] the reduction of saccharic acid to adipic acid three years later by Crum Brown, [22] and the reduction of glucose and gluconic acid to hexane derivatives by Krusemann [38] and Scheibler [39]. Even before the presence of the unbranched six-carbon chain had been established, glucose had been shown to be a pentahydroxy compound by Berthelot, [40] by Schützenberger and Naudin, [41] and by Colley [42]; in 1883, Tollens proposed a cyclic structure corresponding to the modern furanose form [43].

Although Fischer had a great deal of information available to him when he began his work, the field of carbohydrate chemistry was still disorganized, and nobody had considered applying the stereochemical theories of van't Hoff and Le Bel to the problem.

Once the sugars had been isolated in pure form, they were subjected to oxidation with concentrated nitric acid, a reagent which oxidizes aldehydes to carboxylic acids, and selectively oxidizes primary alcohol groups of polyols to carboxylic acids. The reagent thus permitted the conversion of aldoses (**7**) to aldaric acids (**1**). The milder oxidizing agent, bromine water, was found to permit the selective oxidation of aldoses to aldonic acids (**8**) or aldonolactones (**9**). During the same period, the reduction of glucose to mannitol (**4**), a representative alditol (**10**), was reported by Dewar [44] and Bouchardat [45]. In 1862, Linnemann [46] reported the reduction of glucose to mannitol with sodium amalgam. Nearly two decades later, Krusemann reported the more facile reduction of fructose (**11**) to mannitol (**12**) [38]. These results are summarized in Fig. 9.3.

Based on oxidations with silver (I) and copper (II) reagents under basic conditions, carbohydrates were characterized as reducing or non-reducing. Ammoniacal silver oxide [$Ag(NH_3)_2NO_3$], which was developed by Tollens, and carries his

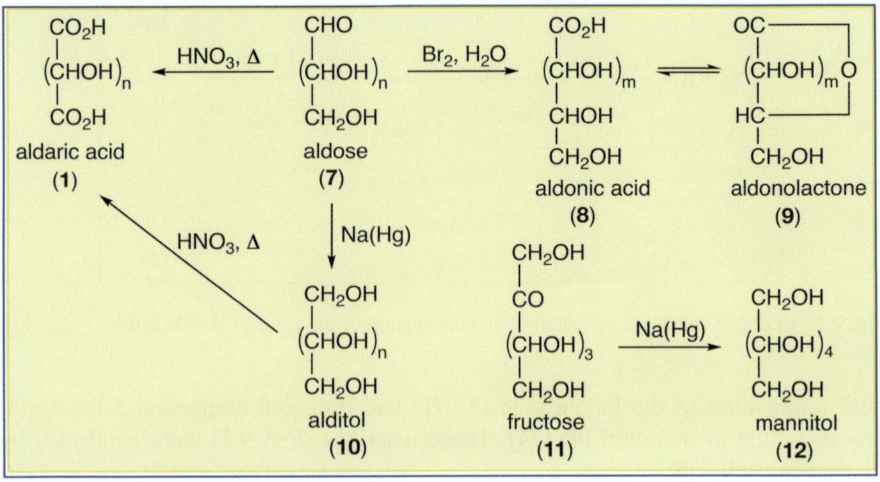

Fig. 9.3 Studies of the selective redox reactions of aldoses

name [47]. Tollens' reagent oxidizes aldoses with a free hemiacetal group to the corresponding lactones, generating a silver mirror on the glass surface of the reaction vessel. There are two analogous reagents based on copper (II) that oxidize reducing sugars with the development of copper (I) oxide: (1) Fehling's reagent, [48] used by mixing Fehling's solution A (a solution of cupper (II) sulfate) and Fehling's solution B (a strongly basic solution of sodium potassium tatrate and sodium hydroxide), and reported by German chemist, Hermann von Fehling (1812–1885), [49] and (2) Benedict's reagent, [50] a solution of copper (II) sulfate, sodium citrate and sodium carbonate developed by American chemist Stanley Rossiter Benedict (1884–1936) [51].

In 1888, Sohst and Tollens reported the results of their studies of the sugar (aldaric) acids and their lactones [52]. Although the configurations shown in Fig. 9.4 are not correct, the important results to come from this work was the observation that mannaric acid forms a di-γ-lactone (**13**), and that glucaric (saccharic) acid forms a monolactonic acid (**14**), which we now know to exist preferentially as the δ-lactone.

Saccharic acid monolactone (**14**) provided an important piece of information about the structure of glucose. In a short series of papers beginning in 1886, Hans Thierfelder [53] (1858–1930) reported the isolation and chemistry of glucuronic acid (**15**) [54]. In the first issue of the *Berichte* for 1891, Fischer and his student, Oskar Piloty [55] (1866–1915) published their synthesis of the same compound, which they named gulonic acid, and the corresponding aldose, gulose, from saccharic acid (Fig. 9.5) [56]. Their synthesis involves the reduction of the lactonic acid **14** with sodium amalgam in neutral solution to give glucuronic acid (**16**), which was then reduced further to gulonic acid (**17**) by allowing the pH to rise.

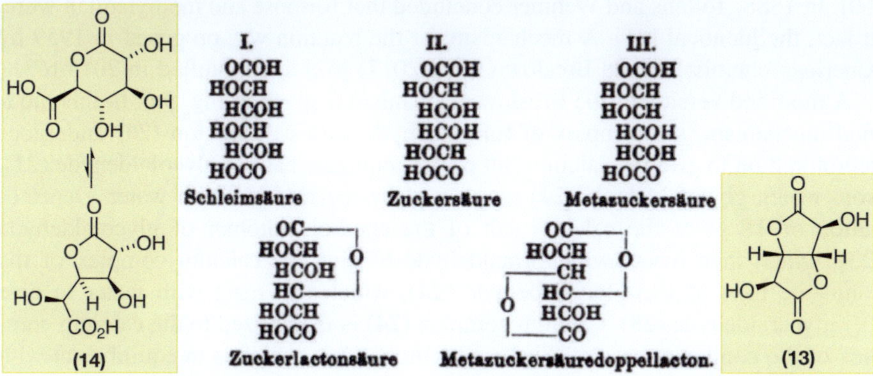

Fig. 9.4 Sugar diacids and their lactones as proposed by Sohst and Tollens [52]. The modern structures of the lactones are given in the colored highlight boxes

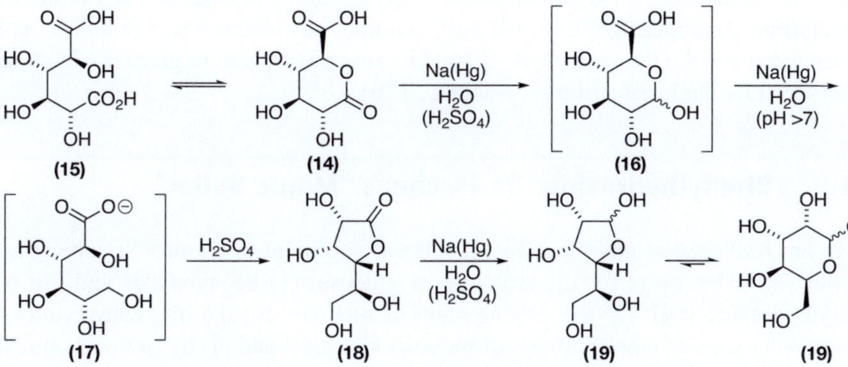

Fig. 9.5 The formation of gulose derivatives **17–19** from saccharic acid (**15**)

Acidification of the solution gave gulonolactone (**18**), which was then reduced to gulose (**19**).

The importance of the sequence in Fig. 9.5 is that it permitted the interchange of the terminal functionalities of the aldose molecule.

9.5 Formose: Preparation of a Saccharic Substance

The apparent simplicity of the carbohydrate molecule invariably led to attempts to effect synthesis from a simple precursor. In 1861, the Russian chemist, Aleksandr Mikhailovich Butlerov (1828–1886), [57] reported that the reaction between formaldehyde and limewater gave a material that exhibited carbohydrate properties [58]. He called his material, "methylenitan," but it is known today as, "formose," a name given to it by agricultural chemist Oscar Loew (1835–1902) [59] in 1886

[60]. In 1888, Tollens and Wehmer concluded that formose and methylenitan were, in fact, the identical [61]. A mechanism for the reaction was proposed in 1959 by American chemist Ronald Breslow (1931–2017) [62] and modified in 2014 [63].

A modified version of the Breslow mechanism is given in Fig. 9.6. In this modified mechanism, the complex of formaldehyde with calcium ion (**20**) undergoes deprotonation to give the calcium salt of the conjugate base of glycolaldehyde (**21**), from which glycolaldehyde (**22**) is released by the reaction with water. Deprotonation of **18** gives the calcium salt of the enediol tautomer of glycolaldehyde (**23**), which then reacts with formaldehyde to give the calcium complex of the conjugate base of (±)-glyceraldehyde (**24**), which can react with water to give (±)-glyceraldehyde (**25**). Calcium complex (**24**) is isomerized to the calcium complex of the conjugate base of dihydroxyacetone (**26**), which is in equilibrium with its alkoxide isomer (**27**). The calcium complex **26** reacts further with formaldehyde to give the calcium complex (**28**) of the conjugate base of erythrulose (**29**). The subsequent base-promoted isomerization of the erythrulose gives a mixture of threose and erythrose (**30**). Further elaboration by sequences of aldol addition reactions with formaldehyde, glycolaldehyde or glyceraldehyde leads to the mixture of carbohydrate analogues (**31**); in particular, compounds where $n = 5$ (ribulose and ribose) are formed. Glyceraldehyde, (±)-(**25**), reacts with (**26**) to give (±)-fructose (α-acrose, **11a**) and (±)-sorbose (β-acrose, **11b**).

9.6 Phenylhydrazine: Dr. Fischer's "Magic Bullet"

There are four factors that played vital roles at one point or another in generating Fischer's final proposal for the structure of glucose: (1) his synthesis and use of phenylhydrazine [64]; (2) the development of methods for the interconversion of monosaccharides; (3) the synthesis of monosaccharides; and (4) the stereochemical theories of van't Hoff and Le Bel.

9.6.1 Discovery of Phenylhydrazine

Fischer's first independent work, carried out while he was without an official position in Munich, was concerned with phenylhydrazine, a reagent that permeated his life's work. In 1875, Fischer reported the syntheses of phenylhydrazine by reduction of phenyldiazonium nitrate (**32**) with potassium hydrogen sulfite, and established its structure. [lxv] The reaction gives two different sulfonate products, depending on the reaction conditions. With neutral potassium hydrogen sulfite at low temperature a red crystalline salt with a composition corresponding to the potassium salt of *N*-(phenylimino)sulfamic acid (**33**) was obtained. At higher temperatures under acidic conditions, a new crystalline salt was formed, which had the composition of the potassium salt of *N*-(phenylamino)sulfamic acid (**34**) (Fig. 9.7).

Fischer showed that this compound was a selective reagent for aldehydes and ketones (**35**) to give phenylhydrazones (**36**) (Fig. 9.8), [65] a reaction he later

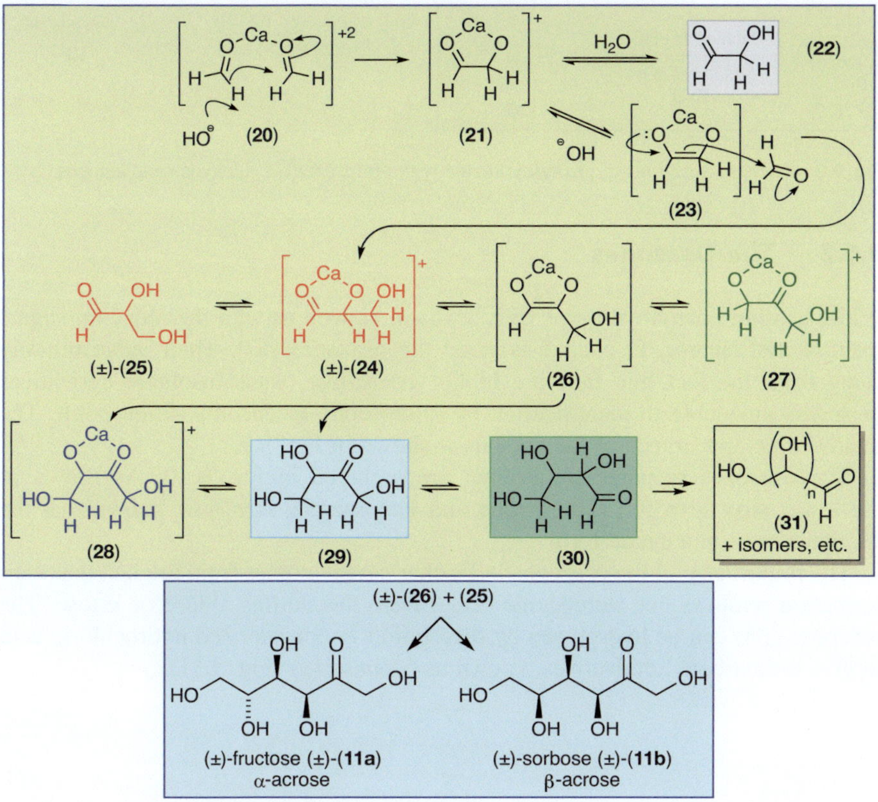

Fig. 9.6 A modified version of the Breslow mechanism for the formose reaction and the formation of acrose

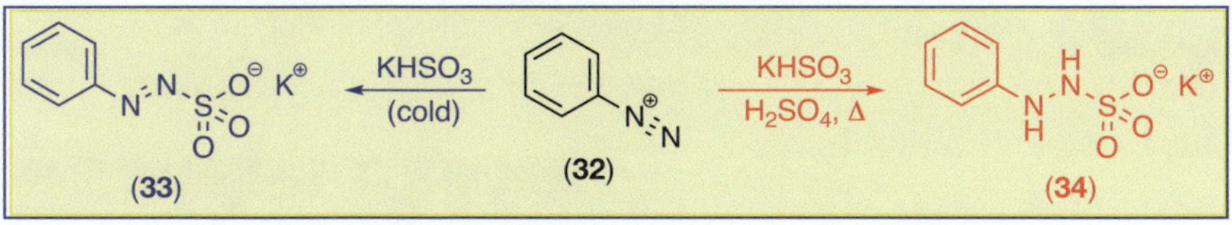

Fig. 9.7 Fischer's reactions of phenyldiazonium nitrate with potassium sulfite

compared with the reaction between hydroxylamine and aldehydes and ketones to form oximes [66].

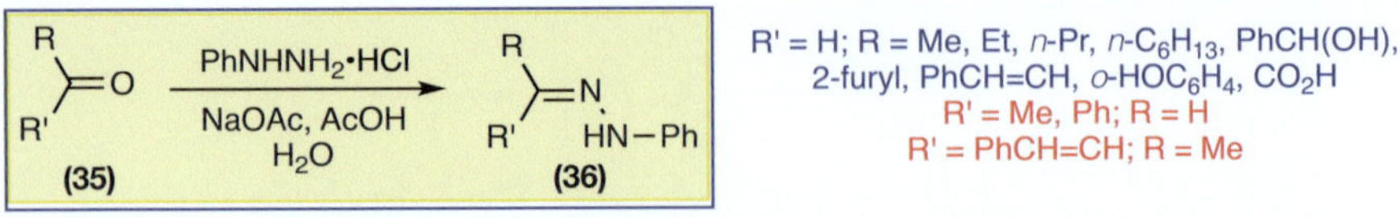

Fig. 9.8 Fischer's reactions of phenylhydrazine with representative aldehydes and ketones

9.6.2 The Osazones

While heating phenylhydrazine with hexoses, as well as with the reducing sugars maltose and lactose, Fischer discovered the osazones [67]. Their value initially came from the fact that they are highly crystalline, water-insoluble derivatives of sugars amenable to identification by microscopy or decomposition point. The microscopic appearance of the crystals is shown in Fig. 9.9.

The osazones prepared by Fischer are gathered in Fig. 9.10. As shown in the lower structures for maltosazone and lactosazone, it is now known that the osazones exist in a cyclic form.

The importance of the osazones to Fischer's work comes from the fact that their formation removes one stereogenic center from the starting aldose or ketose. The osazones (**38**) can be hydrolyzed by dissolution in concentrated hydrochloric acid to give α-dicarbonyl derivatives, known as osones (**41**) (Fig. 9.11).

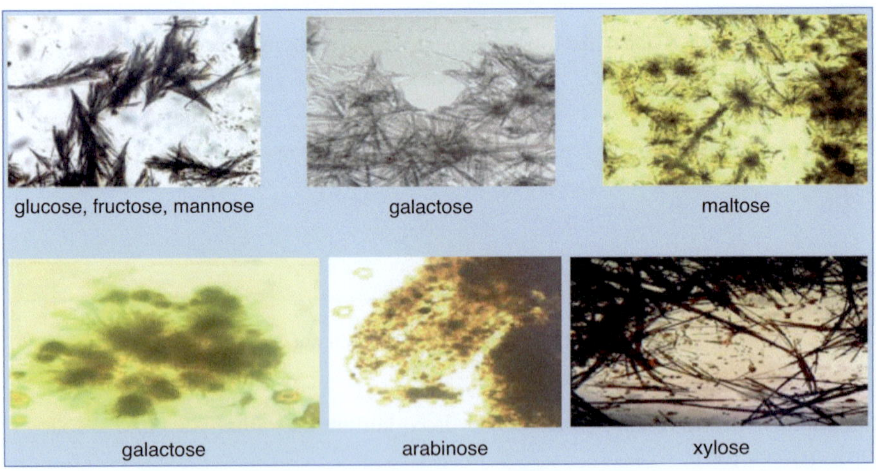

Fig. 9.9 Osazones of representative reducing sugars [68]

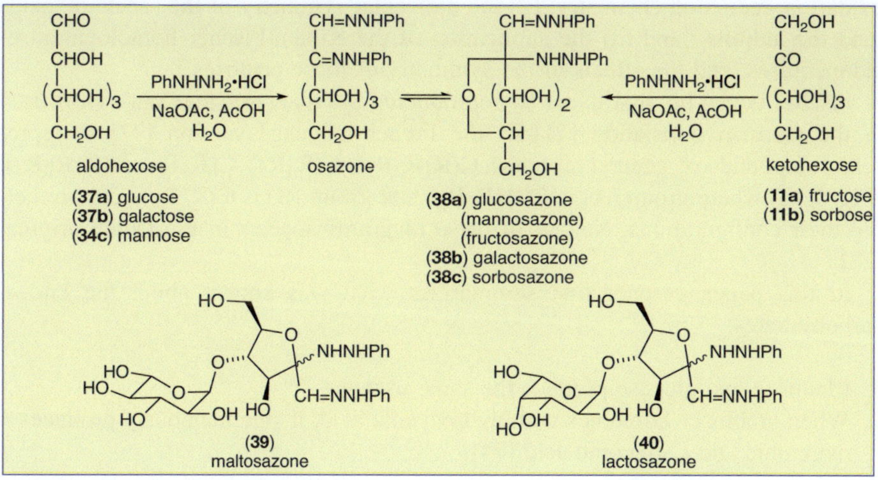

Fig. 9.10 Formation of osazones

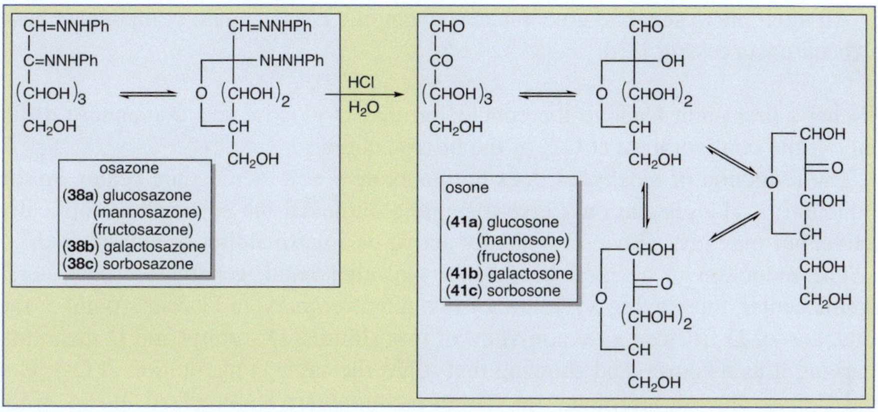

Fig. 9.11 Conversion of osazones to α-ketoaldoses (osones) [69]

9.7 Fischer's Deduction of the Stereochemistry of Glucose [70]

The arguments used by Fischer to derive a structure for glucose still stand, over 100 years later, as one of the most impressive feats of chemical logic in the history of the discipline. The whole structure of his logical derivation of a structure for glucose appeared in two papers, published in 1891 [77, 78]. As noted earlier, these papers also contain the first publication of Fischer's new projection formulas for what we now term stereogenic centers. The arguments stand on three foundations: (1) the interrelationships between the monosaccharides, established by oxidation–reduction chemistry; (2) the molecular symmetry of the saccharic acids (and the alditols); and (3) the application of the Kiliani-Fischer homologation of aldopentoses, and the effects of the symmetry of those pentoses.

Fischer began his first paper by reproducing the diagrams that van't Hoff used in the German translation [71] of the French-language version [72] to represent compounds of general structure $C(R_1R_2R_3)–C(R_4R_5)–C(R_4R_5)–C(R_1R_2R_3)$, (Fig. 9.12) where group (A) is $C(R_1R_2R_3)$ and group (B) is $C(R_4R_5)$, and predicting their configurations. Neither of these diagrams appears in the Dutch original [73].

In this paper, Fischer first summarizes what was known about the known carbohydrates:

1. Mannose and glucose produce the same osazone. [xxii]
2. When arabinose combines with hydrocyanic acid, it simultaneously produces *l.* mannonic and *l.* gluconic acid [74].
3. Fructose is converted into a mixture of mannitol and sorbitol by sodium amalgam [75].
4. Mannonic acid and gluconic acid can be converted into one another by heating with quinoline [76].
5. All attempts to separate gluconic and mannonic acid into two components have been unsuccessful [76].

Fischer's first point leads to the conclusion that D-glucose and D-mannose differ only at the configuration at C-2 of the hexose chain.

The reduction of aldehydes does not generate a new stereogenic center, so the reduction of aldoses can only give a single alditol. All the pentoses are optically active, but may give either an optically active or a *meso* alditol or aldaric acid.

The reduction of a chiral ketone, on the other hand, generates a new stereogenic center, thus giving a mixture of two diastereomers. In Fischer's point 3, the reduction of D-fructose gave a mixture of two alditols, D-sorbitol and D-mannitol, marking it as a ketose, and showing that it has the same configuration at C-3, C-4 and C-5 as glucose and mannose. These reactions are summarized in Fig. 9.13, where the header comes from Fischer's paper [77].

By 1889, Fischer had begun to consider that the monosaccharides were stereoisomers of each other, and he set out the sixteen isomers predicted by the van't Hoff model of the carbon atom. The possible structures are displayed in two ways: The original van't Hoff notation for an open-chain compound with four stereogenic centers, $R–(X–C–Y)_4–R'$, [77] are shown in the red highlight box in Fig. 9.14. The much more intelligible (to modern chemists) Fischer projections [78] of the same compounds are given in the blue highlight box.

At the time of Fischer's work, the absolute configurations of chiral substances were not known, so Fischer arbitrarily assigned the D configuration to (+)-glyceraldehyde (**33**), giving him a 50:50 chance of being correct. Fischer's decision was shown to be correct by Dutch chemist, Johannes Martin Bijvoet

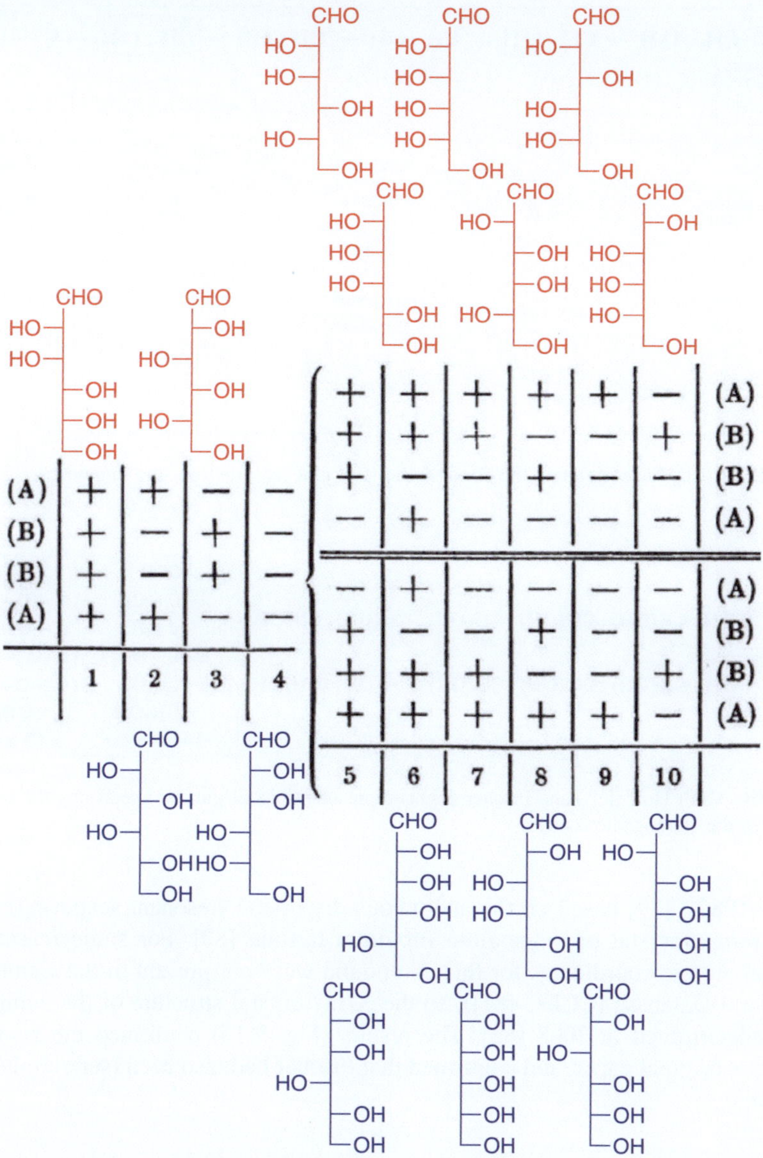

Fig. 9.12 Van't Hoff and Fischer representation of the stereochemistry of carbohydrates. The D-aldohexoses are blue and the L-aldohexoses are red

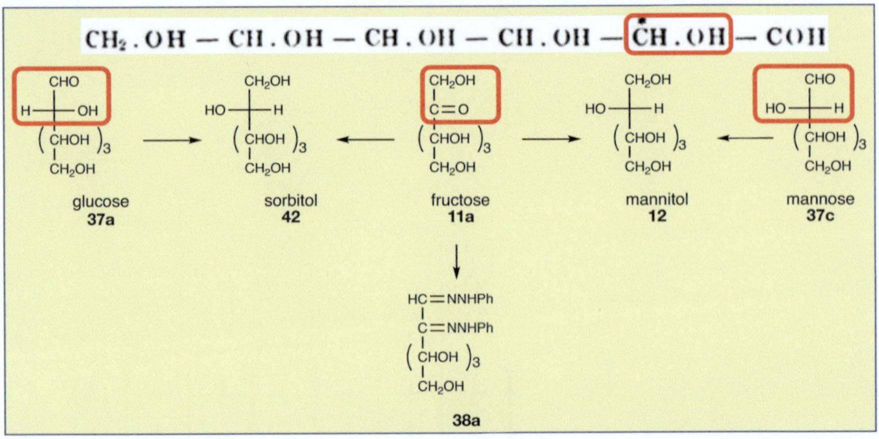

Fig. 9.13 Reactions correlating the configurations of glucose, fructose, and mannose

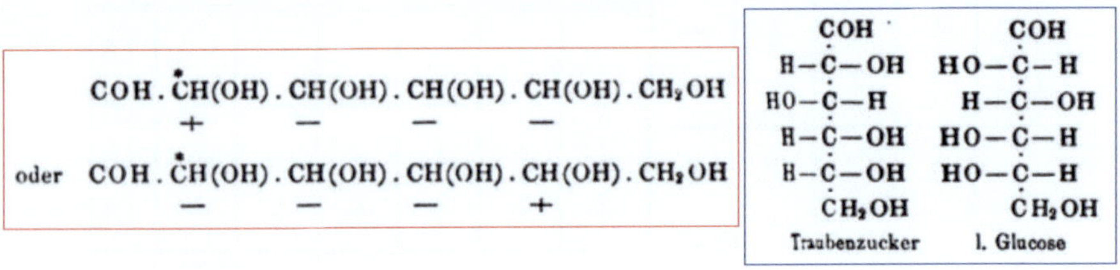

Fig. 9.14 Van't Hoff (left) and Fischer (right) representations of glucose specifying the configurations of the stereogenic centers

(1892–1980), [79] based on the anomalous dispersion (resonant scattering) of X-rays from a crystal of (+)-sodium rubidium tartrate [80]. For some reason, the original atomic coordinates for this compound were not present in the Cambridge Structural Database (CCDC [81]), so the X-ray crystal structure of the compound was redetermined in 2008 [82]. The results (Fig. 9.15) confirmed the results in Bijvoet's original paper, and confirmed that Fischer had also been (serendipitously) correct.

9.7.1 *The Role of* Meso *Compounds*

Carbohydrates derived from natural sources are optically active, so their reaction products are amenable to study by polarimetry; Fischer used the presence or absence of optical activity as a diagnostic tool. Although the aldoses themselves are all optically active (the groups at the two terminal carbons are different), their alditols and aldaric acids (where the groups at the terminal carbons are identical)

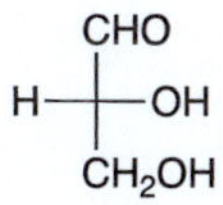

39

D-(+)-glyceraldehyde

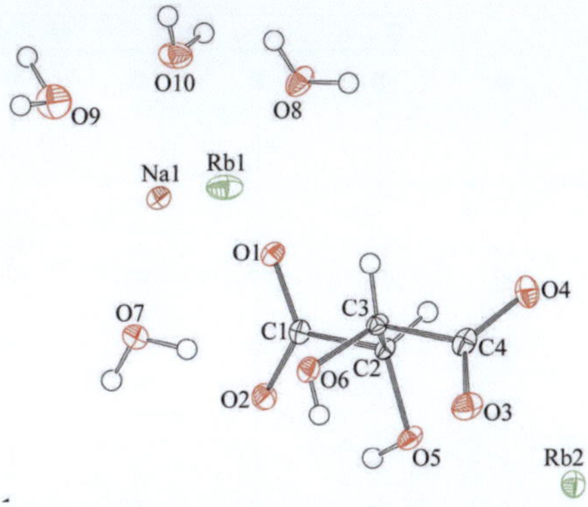

A redetermination of the structure of sodium rubidium tartrate tetrahydrate (Ref. 65)

Fig. 9.15 (l) Absolute configuration of D-glyceraldehyde as arbitrarily assigned by Fischer; (r) the modern X-ray crystal structure of sodium rubidium tartrate tetrahydrate

may, depending on the configurations of the individual stereogenic centers in the molecule, exist as *meso* diastereomers.

In a *meso* compound with an odd number of carbon atoms (Fig. 9.21), there is an atom at the mid-point of the chain, so the symmetry element (a mirror plane, center of symmetry, or an improper axis of rotation) must be located on that atom. In odd-membered aldaric acids and alditols, this atom carries a hydroxyl group, which makes it a stereogenic center; the configuration of stereogenic centers is specified in the *R/S* system using lower case letters. The configurations of the C_5 aldoses, alditols and aldaric acids are given in Fig. 9.16.

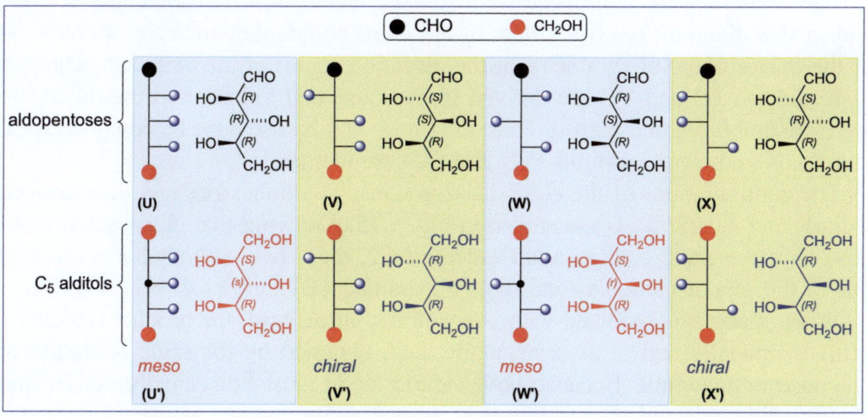

Fig. 9.16 Diastereomers of the aldopentoses and the alditols derived from them. The location of the mirror plane or center of symmetry in the *meso* compounds is specified by the black dot

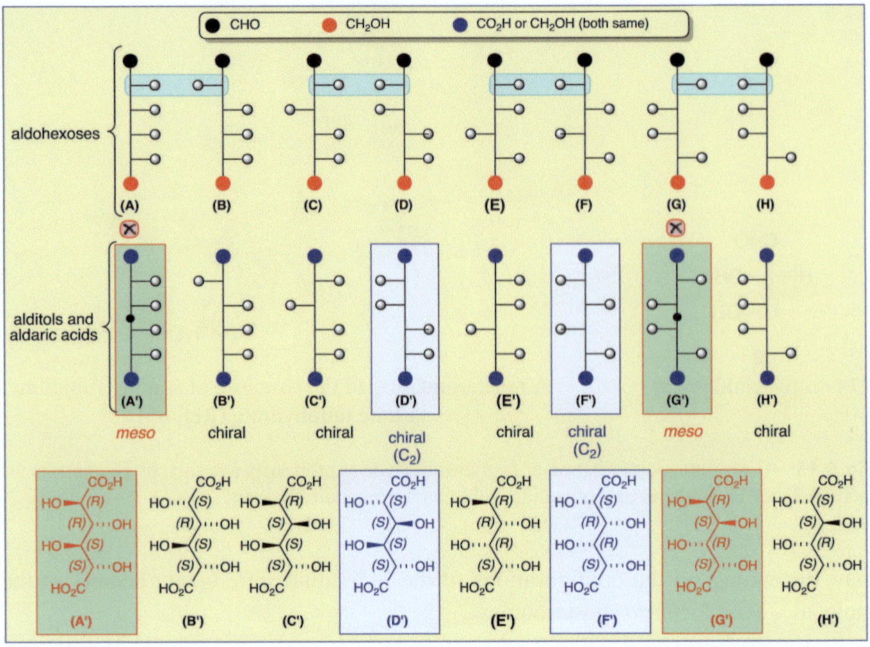

Fig. 9.17 Stereochemistry of the aldohexoses and their derived aldaric acids or alditols. The location of the mirror plane or center of symmetry in the *meso* compounds is specified by a black dot. The modern line structures and stereochemistry are given in the bottom row

In a *meso* compound with an even number of carbon atoms (Fig. 9.17), on the other hand, there is no atom at the mid-point of the chain, so the symmetry element cannot be located on an atom, but must be located at the center of a σbond.

The isomers of the alditols derived from aldopentoses were shown in Fig. 9.16, and in this diagram, the black dots in the *meso* compounds indicate the location of the mirror plane of symmetry or the inversion center of the molecule. The two *meso* alditols (**U** and **W**) are derived from ribose and xylose, and the other two are from arabinose and lyxose. Thus, arabinose and lyxose must be compounds (**V** and **X**). Note how the alditols (**V′**) and (**X′**) are identical.

The configurations of the eight diastereomeric D-aldohexoses and their derived alditols and aldaric acids are shown in Fig. 9.15. Knowing that glucose and mannose have the same configuration except at C2, these two carbohydrates must be one of the sets of pairs, (**A**) and (**B**), (**C**) and (**D**), (**E**) and (**F**), or (**G**) and (**H**).

When glucose is oxidized with concentrated nitric acid, the product (saccharic acid) is optically active, as is mannaric acid, obtained by the same oxidation of mannose or mannitol. Because both aldaric acids must be optically active, and aldaric acids (**A′**) and (**G′**) in Fig. 9.22 are *meso* compounds, neither (**A**) nor (**G**) can be glucose or mannose, which means that their C-2 epimers, structures (**B**) and (**H**), are also eliminated. To resolve which of the remaining pairs, (**C**) and

(**D**) or (**E**) and (**F**), correspond to glucose and mannose, Fischer resorted to the Kiliani-Fischer homologation of aldopentoses.

9.7.2 The Kiliani-Fischer Homologation

In a series of papers between 1885 and 1889, Heinrich Kiliani (1855–1945) [83] described a method for the homologation of aldoses to the next higher aldonic acid by means of the cyanohydrin, which he applied to the conversion of fructose, [84] glucose, [85] arabinose [86] and galactose [87] to the corresponding next-higher aldonolactones. The final step in the formal homologation of an aldose to the next-higher aldose was completed in 1889, when Fischer published a method for the reduction of an aldonolactone to an aldose with sodium amalgam [88]. Therefore, the homologation sequence carries the names of both chemists.

The Kiliani-Fischer homologation (Fig. 9.18) is carried out by first converting the aldose (**7**) to the diastereomeric cyanohydrins (**42**) by treatment with potassium ferricyanide and dilute sulfuric acid [89]. The acid hydrolysis of the cyanohydrins then gives the homologated aldonic acids (**43**), which are in equilibrium with the corresponding aldonolactones (**44**). The reduction of the lactones with 2.5% sodium amalgam in dilute sulfuric acid gives the aldose (**45**).

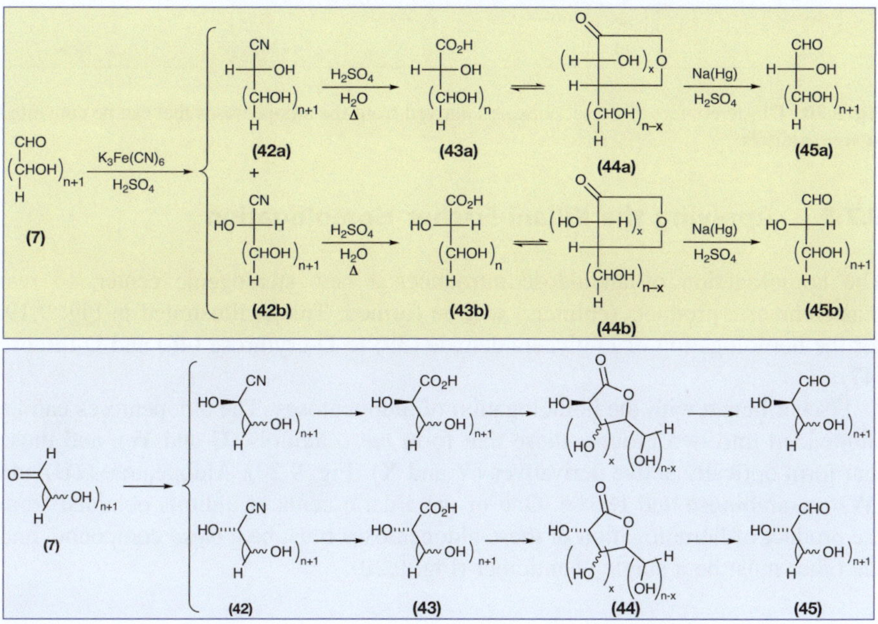

Fig. 9.18 Kiliani-Fischer homologation of aldoses to the next-higher aldoses

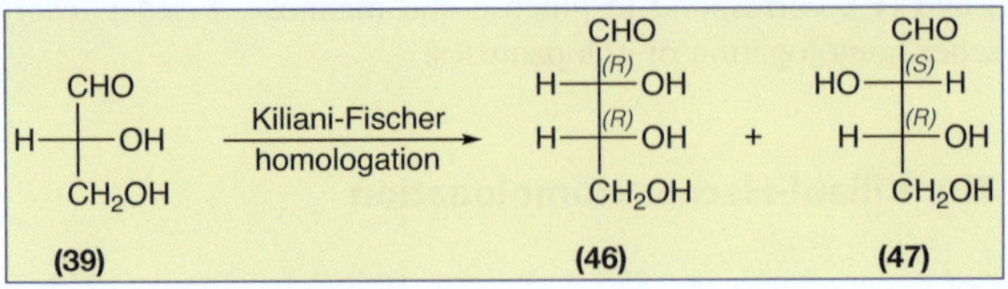

Fig. 9.19 Kiliani-Fischer homologation of D-glyceraldehyde

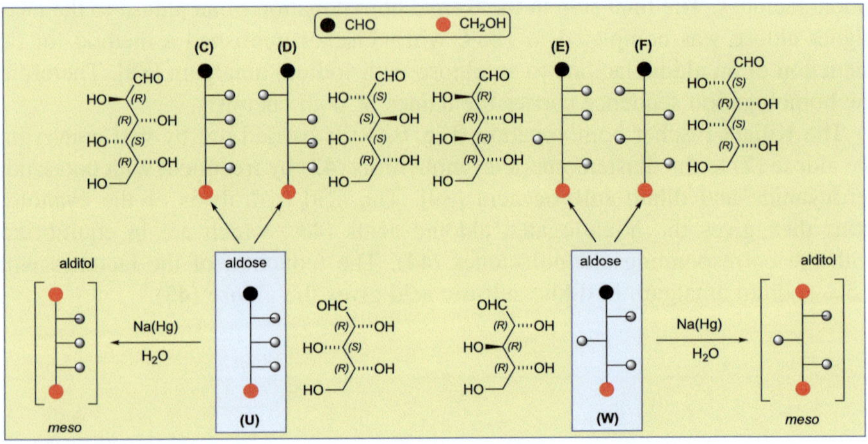

Fig. 9.20 Diastereomers of the aldohexoses derived from the aldopentoses that can be converted to *meso* alditols

9.7.3 Applying the Kiliani-Fischer Homologation

The homologation of an aldose introduces a new stereogenic center, so two diastereomeric products (epimers) will be formed. This is illustrated in Fig. 9.19, for the homologation of D-glyceraldehyde (**39**) to D-erythrose (**46**) and D-threose (**47**).

Fischer began with the homologation of aldopentoses. The aldopentoses can be subdivided into two groups: those that form *meso* alditols (**U** and **W**), and those that form optically active derivatives (**V** and **X**) (Fig. 9.20). Aldopentoses (**U**) and (**W**) are arabinose and lyxose. One of the aldaric acids or alditols obtained from the product of homologation of these aldopentoses must be a *meso* compound, and the other must be a single enantiomer (Fig. 9.20).

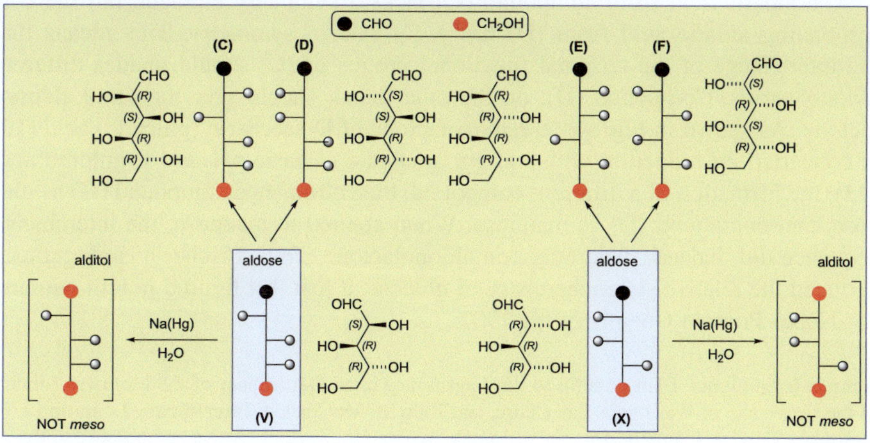

Fig. 9.21 Diastereomers of the aldohexoses derived from the aldopentoses that can be converted to chiral alditols

Aldopentoses (**V** and **X**) are ribose and xylose, and their homologation is summarized in Fig. 9.21.

The homologation of arabinose, as noted earlier, had been studied by Kiliani, who reported that the homologation of arabinose yielded a mixture of gluconolactone and mannonolactone [86]. The fact that the configuration of arabinose is L rather than D was fortuitous because it gave Fischer access to the L-aldohexoses. The oxidation of arabinose with nitric acid gave arabinaric acid, which was optically active, so this eliminated structures (**U**) and (**W**) from consideration for the structure of arabinose. The homologation of pentose (**V**) should give two aldonic acids, and the nitric acid oxidation should give two optically active aldaric acids. The same treatment of pentose (**X**) should lead to one optically active aldaric acid (corresponding to hexose **E**), and one inactive (*meso*) aldaric acid (corresponding to hexose **F**). Because both products of the sequence were optically active, arabinose had to have the structure (**V**), and thus glucose and mannose had to be compounds (**C**) and (**D**). This is where the interchange of the terminal functions of the aldonolactones acids became critically important.

The aldaric acid from compound (**C**) lacks a symmetry element, but the corresponding aldaric acid from (**D**) has a C_2 axis of symmetry. This means that the interchange of the terminal functional groups of (**C**) should yield a different aldonolactone. Compound (**D**), on the other hand, should give the same aldonolactone. As shown in Fig. 9.8, the monolactone of D-saccharic (glucaric) acid (**10**) is converted by reduction with sodium amalgam at basic pH into L-gulonic acid (**13**); the formation of a different compound, then eliminated compound (**C**) as glucose, and compound (**D**) as mannose. When applied to mannose, the interchange sequence did, indeed, yield only one aldonolactone. Thus, Fischer unambiguously deduced the relative stereochemistry of glucose, a feat that figured in him winning the Nobel Prize in Chemistry for 1902.

Acknowledgements I am grateful to the Interlibrary Loan Department of the McIntyre Library of the University of Wisconsin-Eau Claire, and John M. McCrackin, **Interlibrary Loan and UW Resource Sharing Coordinator**, for assistance in obtaining various elusive and obscure sources.

References

1. Lichtenthaler FW (1992) Emil Fischer's proof of the configuration of sugars: a centennial tribute. Angew Chem Int Ed 31:1541–1556
2. An extensive biography of Fischer in German was published as a Special Issue of the Berichte der Deutschen Chemischen Gesellschaft: (a) Hoesch K (1921). Emil Fischer. Sein Leben und sein Werk. Ber Dtsch Chem Ges 54:1–480. For other biographical information on Fischer, see: (b) Harrow B (1920) Eminent chemists of our time. D. Van Nostrand Company, New York, pp 217–239. (c) Forster MO (1920) Emil Fischer Memorial Lecture. J Chem Soc Trans 117:1157–1201. For an image of Fischer, see: https://commons.wikimedia.org/wiki/File:Hermann_Emil_Fischer_c1895.jpg
3. For biographies of Raoult, see: (a) van't Hoff J (1902). Raoult memorial lecture. J Chem Soc Trans 81:969–981. (b) Ramsay W (1901) Prof. François Marie Raoult. Nature No 1644 64:17–18. (c) Jones HC (1901) François Marie Raoult. Science 13:881–883. (d) Getman FH (1936) François-Marie Raoult, master cryoscopist. J Chem Educ 13(3):153–155. (e) Morachevskii AG (2005) François Marie Raoult (To 175th Anniversary of His Birthday). Russ J Appl Chem 78:856–858 [in English]. For an image of Raoult, see: https://en.wikipedia.org/wiki/Fran%C3%A7ois-Marie_Raoult#/media/File:Raoult.jpg
4. (a) Raoult F-M (1878) Loi de congélation des solutions benzéniques des substances neutres. Compt rend hebd Séances Acad Sci 5:187–189. (b) Raoult F-M (1878) Loi générale de congélation des dissolvants. Compt rend hebd Séances Acad Sci 5:1030–1033. (c) Raoult F-M (1878) Solutions benzéniques des substances neutres. Compt rend hebd Séances Acad Sci 5:187–189. (b) Raoult F-M (1878) Loi générale des tensions de vapeur des dissolvants. Compt rend hebd Séances Acad Sci 104:1430–1433
5. For biographies of Tollens, see: (a) Wallach O (1918) Nekrolog: Bernhard Tollens. Ber dtsch chem Ges 51(2):1539–1555. (b) Browne CA (1942). Tollens B (1841–1918) and some American students of his School of Agricultural Chemistry. J Chem Educ 19(6):253–259. For an image of Tollens. see: https://en.wikipedia.org/wiki/Bernhard_Tollens#/media/File:Tollens_Bernhard.jpg
6. Tollens B, Mayer F (1888) Ueber die Bestimmung der Moleculargrösse der Raffinose und des Formaldehydes mittelst Raoult's Gefriermethode. Ber dtsch chem Ges 21:1566–1572
7. Shaffer LN (2001) Southernization. In : Adas M (ed) Agricultural and pastoral societies in ancient and classical history. Temple University Press, pp 308–324

8. . For a biography of Marggraf, see: von Lippmann EO (1961) Andreas Sigismund Marrgraf. In: Farber E (ed) Great chemists. Interscience, New York, London, pp 193–200. For an image of Marggraf, see: https://en.wikipedia.org/wiki/Andreas_Sigismund_Marggraf#/media/File:Andreas_Sigismund_Marggaf.jpg
9. Marggraf [A] (1747) Expériences chymiques faites dans le dessein de tirer un véritable sucre de diverses plantes, qui croissant dans nos contrées. Mém Acad Roy Belles Lett Berlin III:79–90
10. For biographies of Lowitz (Ловиц), see: (a) Figurovslii N (1955) In Figurovskii N ed 919550. T. E.Lovits. Izbrannye trudy po khimii i khinlicheskoy tekhnologii [T. E. Lovits. Selected Works in Chemistry and Chemical Technology], Moscow, pp 405–514. (b) Leicester HM (1945) Tobias Lowitz, Discoverer of basic laboratory methods. J Chem Educ 22:149–151. For an image of Lowitz, see: https://en.wikipedia.org/wiki/Johann_Tobias_Lowitz#/media/File:LowitzT.jpg
11. (a) Lowitz T (1792) Anzeige neuer Bemerkungen über die Natur des Honigs, und die Darstellung seines zuckerartigen Bestandtheils in trockner Gestalt. Crells Ann I:218–224. (b) Lowitz T (1792) Anzeige neuer Bemerkungen über die Natur des Honigs, und die Darstellung seines zuckerartigen Bestandtheils in trockner Gestalt [Fortsetzung]. Crells Ann I:345–348
12. For biographical material on Proust, see: (a) Mauskopf SH (1969) Thompon before Dalton: Thomas Thomson's consideration of the issue of combining weight proportions prior to his acceptance of Dalton's chemical atomic theory. Ann Sci 25:229–242. (b) Mauskoph SH. Proust, Joseph Louis. https://www.encyclopedia.com/people/science-and-technology/chemistry-biographies/joseph-louis-proust [accessed June 12, 2024]; this is the online version of the article from the Complete Dictionary of Scientific Biography. For an image of Proust, see: https://commons.wikimedia.org/wiki/File:Portrait_of_Louis_Proust._Wellcome_L0006972.jpg
13. (a) Proust [J-L] (1802) Extrait d'une lettre du Professeur Proust à J.-C. Delamétherie. J Phys 54:199–200. (b) Proust [J-L] (1802) Lettre du Professeur Proust à J.-C. Delamétherie. Sur le sucre du raisin. J Phys 56:113. (c) Proust [J-L] (1805) Sur le sucre de raisin. J Phys 61:399–400. (d) Proust [J-L] (1806) Sur le sucre de raisin. Ann Chim 57:131–174. (e) Proust [J-L] (1806) Suite. Ann Chim 57:225–272
14. For biographies of Kirchhoff, see: (a) Boltzmann L (1905) Gustav Robert Kirchhoff. Populäre Schriften, Leipzig, 51–75. (b) v. Voit C (1889) Gustav Robert Kirchhoff. Sitzungsber. königl. bayer. Akad. Wiss. 181–186. (c) Warburg E (1925) Zur Erinnerung an Gustav Kirchhoff. Naturwiss 13:205–212. (d) Everest AS (1969) Kirchhoff–Gustav Robert (1824–1887). Physics Educ 4(6):341–343. For an image of Kirchhoff, see: https://en.wikipedia.org/wiki/Gustav_Kirchhoff#/media/File:Gustav_Robert_Kirchhoff.jpg
15. (a) Kirchhoff [CS] (1812) O prigotovlenii sakhara iz krakhmala [On the preparation of sugar from starch]. Tekhnol Zh 9:3–26. (b) Kirchhoff C (1815). Ueber die Zuckerbildung beim Malzen des Getreides, und beim Bebrühen seines Mehls mit kochenden Wasser. Schweiggers J Chem Phys 14:389–398.
16. For a biography of Dubrunfaut, see: Fletcher JG Jr (1940) Augustin-Pierre Dubrunfaut—an early sugar chemist. J Chem Educ 17(4):153–156. For an image of Dubrunfault, see: https://en.wikipedia.org/wiki/Augustin-Pierre_Dubrunfaut#/media/File:Le_chimiste_Augustin_Dubrunfaut_(1797-1881).jpg
17. (c) Dubrunfaut [A-P] (1846) Note sur quelques phénomènes rotatoires et sur quelques propriétés des sucres. Compt rend hebd Séances Acad Sci 23:38–44. (b) Dubrunfaut [A-P] (1846) Sur quelques phénomènes rotatoires et sur quelques propriétés des sucres. Ann Chim Phys 18:99–107. (c) Dubrunfaut [A-P] (1847) Note sur le glucose. Ann Chim 21:178–180
18. For a biography of Jungfleisch, see: Charle C, Telkès E (1988) Jungfleisch (*Émile* Clément). Publications de l'Institut national de recherche pédagogique 1988:113–115. For an image of Jungfleisch, see: https://commons.wikimedia.org/wiki/File:%C3%89mile_Jungfleisch.jpg
19. (a) Friedel C (1880) Bull Soc chim France [2] 34:674–675. (b) Friedel C (1881) Bull Soc chim France [2] 35:97–98

20. Jungfleisch [E], Lefranc (1881) Sur le levulose. Compt rend hebd Séances Acad Sci 93:547–550
21. Sonnedecker G (1976) Kremers and Urdang's History of Pharmacy, 4th ed. American Institute of the History of Pharmacy, Madison, WI, p 360
22. (a) Fischer E, Hirschberger J (1888) Ueber Mannose. Ber dtsch chem Ges 21:1805-1809. (b Fischer E, Hirschberger J (1889) Ber dtsch chem Ges 22:365–376
23. Dumas (1843) Traité de Chimie Appliquée aux Arts. Béchet jeune: Paris. vol 6, p 273 (misprinted as p. 375 in the original]
24. For biographies of Dumas, see (a) E. J. R (1884) Jean Baptiste André Dumas. Proc Roy Soc 37:x–xxvii; (b) Hofmann AW (1884) Jean Baptiste André Dumas. Ber dtsch chem Ges 947–949; (c) Urbain G (1961) Jean-Baptiste Dumas and Charles-Adolphe Wurtz. In Faber E (ed) Great chemists, Publisher, city, 1961; (d) Partington JR (19xx) Dumas; (e) Le Blanc F (1884) Le laboratoire et l'enseignement de J.-B. Dumas. Bull Soc chim Paris Nouv sér 41:549–559; (f) Cannizzaro S (1884) Discours à la mémoire de J.-B. Dumas. Bull Soc chim Paris Nouv sér 42:130–141. For an image of Dumas, see: https://en.wikipedia.org/wiki/Jean-Baptiste_Dumas#/media/File:Jean_Baptiste_Andr%C3%A9_Dumas.jpg
25. This formula is obtained from the analytical data referred to by Dumas. These data actually correspond to glucose monohydrate, $C_6H_{12}O_6{\bullet}H_2O$, which is now known to be the most stable crystalline form of glucose—anhydrous glucose is hygroscopic, and bsorbs water vapor to give the monohydrate. On p. 275 of the same book, Dumas reports that his glucose melts near 100 °C and loses approximately 9% of its weight (the loss of water from the monohydrate results in a 9.1% decrease in its mass)
26. (a) Berthelot M (1867) Nouvelles applications des méthodes de réduction en chimie organique. Bull Soc chim Paris [2] 7:53–65; (b) Berthelot M (1868) Méthode universelle pour réduire et saturer d'hydrogène les composées organiques. Bull Soc chim Paris [2] 9:8–31
27. On his birth certificate and officlal documents signed later in his life, Berthelot spells his name, "Marcelin," but for most of his life he used the spelling, "Marcellin." (a) Armstrong HE (1928) Marcelin Berthelot and Synthetic Chemistry. J Roy Soc Arts 76:145–171. (b) Ashdown AA (1921) Marcellin Berthelot. J Chem Educ 4:1217–1232. (c) Bredig G (1907) Marcelin Berthelot. Z angew Chem 20(1):689–694. (d) Boutaric A (1927) Marcellin Berthelot. Payot, Paris. (e) Graebe C (1908) Marcelin Berthelot. Ber dtsch chem Ges 41:4805–4872. (f) Jungfleisch E (1913) Notice sur la vie et les travaux de Marcellin Berthelot. Bull Soc chim France 13:i–cclx. For an image of Berthelot, see: https://en.wikipedia.org/wiki/Marcellin_Berthelot#/media/File:Marcellin_Berthelot.jpg
28. For biographies of Crum Brown, see: (a) Kendall J (1932–1935) Obituary notices Fell. Roy Soc 1:537–549. (b) Kendall JC (1927) The University of Edinburgh. J Chem Educ 4:565–569. (c) Walker J (1923) Alexander Crum Brown. J Chem Soc Trans 123:3422–3431. (d) Walker J (1923) Alexander Crum Brown 1838–1922. Proc Roy Soc A 105(734):i–v. For an image of Crum Brown, see: https://en.wikipedia.org/wiki/Alexander_Crum_Brown#/media/File:Brown_Alexander_Crum.jpg
29. Crum Brown A (1863) Ueber Umwandlung der Schleimsäure in Adipinsäure. Vorläufige Notiz. Ann Chem Pharm 125:19–21
30. de la Motte H (1879) Ueber die Einwirkung von Phosphorpentachlorid und Iodwasserstoffsäure auf Zuckersäure. Ber dtsch chem Ges 12:1571–1573
31. For biographies of Wislicenus, see: (a) Beckmann E (1905) Johannes Wislicenus. Ber dtsch chem Ges 37:4861–4946; (b) PFF (1907) Johannes Wislicenus 1835–1902. Proc Roy Soc A 78:iii–xii; (c) Perkin WH Jr (1905) Wislicenus memorial lecture. J Chem Soc Trans 87:501–534. For an image of Wislicenus, see: https://picryl.com/media/eth-bib-wislicenus-johannes-1835-1902-portr-13923-44b7d7
32. (a) Wislicenus J (1868) Vorläufige Mittheilungen aus dem Universitäts-Laboratorium Zürich. Z Chem 11:680–684. (b) Wislicenus J (1869) Synthetische Untersuchungen über die Säuren der Reihe $C_nH_{2n}(CO.OH)_2$. 1. Synthese der Adipinsäure. Ann Chem Pharm 149:220–224
33. For biographical material on Domac, see: (a) Inić S (2011) Julije Domac – Život i djelo [Julije Domac – Life and Achievements]. Doctoral Thesis, Zagreb (Croatian; Summary in

English; https://urn.nsk.hr/urn:nbn:hr:163:361666). (b) Inić S, Kujundužić N (2011) The first independent pharmacognosy institute in the world and its founder Julije Domac (1853–1928). Pharmazie 66:720–726. For an image of Domac, see: https://urn.nsk.hr/urn:nbn:hr:163:361666

34. (a) Domac J (1881) Über das hexylen aus Mannit. Monatsh Chem 2:309–322. (b) Domac J (1882) II Ueber die Einwirkung der Unterchlorsäure auf Hexylen. Justus Liebigs Ann Chem 213:124–132
35. Fischer E (1890) Synthesen in der Zuckergruppe. Ber dtsch chem Ges 23:2114–2141
36. For a biography of Schwanert, see: Friedrich C (2013) Schwanert, Franz Hugo (1828–1902) Apotheker, Chemiker, Universitätsprofessor. In : Alvermann D, Jörn N (eds) Biographisches Lexikon für Pommern. Böhlau Verlag, Köln Weimar Wien, vol 1:241–243 (in German). For an image of Schwanert, see: https://dcmny.org/do/c1223e15-3f9f-4ef8-807b-8700aca82a4b
37. Schwanert H (1860) Ueber einige Zersetzungsproducte der Schleimsäure. Ann Chem Pharm 116:257–287
38. Krusemann HD (1876) Ueber die Reduction der Levulose. Ber dtsch chem Ges 9:1465–1469
39. Scheibler C (1883) Ueber die Einwirkung des Natriumsamalgams auf Glusosen und die Saccharine. Ber dtsch chem Ges 16:3010–3011
40. Berthelot M (1860) Sur les combinaisons des sucres avec les acides ou saccarides. Ann Chim Phys [3] 60:95–104
41. (a) Schützenberger P, Naudin (1869) Sur les dérivés acétiques des substances hydrocarbonées. Compt rend hebd Séances Acad Sci 68:814–821. (b) Schützenberger P, Naudin (1869) Ueber die Acetylderivate der Kohlenhydrate. Z Chem 5:264–266
42. Colley A (1870) Action des haloïdes libres et de quelques chlorures sur la glucose. Compt rend hebd Séances Acad Sci 70:401–403
43. Tollens B (1883) Ueber das Verhalten der Dextrose zu ammonalkalischer Silberlösing. Ber dtsch chem Ges 16:921–924
44. Dewar J (1870) Note on inverted sugar. Philos Mag 1870(39):345–347
45. Bouchardat G (1872) Recherches sur la dulcite et les sucres en générale. Ann Chim Phys 27:68–109
46. (a) Linnemann E (1862) Umwandlung des Zuckers in Mannit. Ann Chem Pharm 123:136–140. (b) Linnemann E (1863) Umwandlung des Zuckers in Mannit. J prakt Chem 88:59–60
47. Tollens B (1882) Ueber ammon-alkalische Silberlösung als Reagens auf Aldehyd. Ber dtsch chem Ges 15(2):1635–1639
48. Fehling H (1849) Die quantitative Bestimmung von Zucker und Stärkmehl mittelst Kupfervitriol. Ann Chem Pharm 72(1):106–113
49. For biographies of Fehling, see: (a) Hofmann AW (1885) Dr. Hermann von Fehling, Vise-Präsident der Deutschen chemischen Gesellschaft. Ber dtsch chem Ges 18(2):1881–1818. (b) Müller H (1886) Annual General Meeting. J Chem Soc Trans 49:329–356 (esp. 346–347). For an image of Fehling, see: https://en.wikipedia.org/wiki/Hermann_von_Fehling#/media/File:Hermann_fehling.jpg
50. Benedict SR (1909) A reagent for the detection of reducing sugars. J Biol Chem 5(6):485–487
51. For a biography of Benedict, see: McCollum EV (1952) Stanley Rossiter Benedict 1884–1936. Natl Acad Biogr Mem 27:153–177. For an image of Benedict, see: https://commons.wikimedia.org/wiki/File:Stanley_Rossiter_Benedict.png
52. Sohst O, Tollens B (1888) Ueber kristallizirte Zuckersäure (Zuckerlantonsäure). Justus Liebigs Ann Chem 245:1–27
53. For a biography of Thierfelder, see: Klenk E (1931) Hans Thierfelder †. Z physiol Chem 203 (1–2):1–9. For an image of Thierfelder, see: https://www.sammlungen.hu-berlin.de/objekte/portraetsammlung-berliner-hochschullehrer/12738/
54. (a) Thierfelder H (1886) Über die Bildung von Glykuronsäure beim Hungerthier 10:163–169. (b) Thierfelder H (1887) Ueber die Untersuchungen der Glycuronsäure. Z Physiol Chem 11:388–409. (c) Thierfelder H (1889) Ueber die Untersuchungen der Glycuronsäure. 2 Mittheilung. Z Physiol Chem 13:275–284. (d) Thierfelder H (1891) Ueber die Reduction der Glycuronsäure durch Natriumamalgam. Z Physiol Chem 15:71–76

55. For a biography of Piloty, see: Harries C (1920) Oskar Piloty. Ber dtsch chem Ges 53:A153–A168. For an image of Piloty, see: https://commons.wikimedia.org/wiki/File:Oskar_Piloty_ca1914.jpg
56. Fischer E, Piloty O (1891) Reduction der Zuckersäure. Ber dtsch chem Ges 24(1):521–528
57. For biographies of Butlerov in English, see: (a) Leicester HM (1940) Alexander Mikhaĭlovich Butlerov. J Chem Educ 17:203–209. (b) Bykov GV (1970) Butlerov, Aleksandr Mikhailovich. In: Gillespie CC (ed) Dictionary of scientific biography. Charles Scribners and Sons New York, vol. 2:620. (c) Brooks NM (2008) Butlerov, Aleksandr Mikhailovich. Complete Dictionary of Scientific Biography, Scribner's Sons, Detroit, vol 19:465–468. (d) Lewis DE (1994) The University of Kazan—provincial cradle of Russian organic chemistry. Part I. Nikolai Zinin and the Butlerov School. J Chem Educ 71(1):39–42. (e) Lewis DE (2012) Early Russian organic chemists and their legacy. Springer Verlag, Heidelberg, pp 47–50. For an image of Butlerov, see: https://commons.wikimedia.org/wiki/File:Butlerov,_A._M._1828-1886.jpg
58. (a) Boutlerow A (1861) Formation synthétique d'une substance sucré. Compt rend hebd séances Acad Sci 53:145–147. (b) Butlerow A (1861) Bildung einer zuckerartigen Substanz durch Synthese. Ann Chem Pharm 120:295–298. (c) Fischer E (1888) Ueber die Verbindungen des Phenylhydrazins mit den Zuckerarten. III. Ber dtsch chem Ges 21:988–991
59. For a biography of Loew, see: Oesper RE (1930) Oscar Loew. J Chem Educ 7(2):314–315. For an image of Loew, see: https://en.wikipedia.org/wiki/Oscar_Loew#/media/File:Oscar_Loew.png
60. (a) Loew O (1886) Ueber Formaldehyd und dessen Condensation. J prakt Chem [ii] 33:321–351. (b) Loew O (1888) Zur Klarstellung der Beziehungen zwischen Formose und Methylenitan. J prakt Chem [ii] 37:203–206
61. Wehmer C, Tollens B (1888) Ueber das Verhalten des Methylenitans (der sog, Formose von Loew) beim Erhitzen mit Säuren. Justus Liebigs Ann Chem 243:334–342
62. For biographical material on Breslow, see: (a) Brown JM (2019) Ronald Charles David Breslow 14 March 1931–25 October 2017. Biogr Mems Fell Roy Soc 66:53–77; (b) Joseph V, Levine M (2021) Ronald C. D. Breslow (1931–2017): A career in review. Bioorg Chem 115:104868. For an image of Breslow, see: https://commons.wikimedia.org/wiki/File:Ronald_Breslow_HD2006_Othmer_Gold_Medal_portrait.JPG
63. (a) Breslow R (1959) On the mechanism of the formose reaction. Tetrahedron Lett 1(21): 22–26. (b) Appayee C, Breslow R (2014) Deuterium studies reveal a new mechanism for the formose reaction involving hydride shifts. J Am Chem Soc 136:3720–3723
64. Fischer E (1875) Ueber aromatischer Hydrazinovderbindungen. Ber dtsch Chem Ges 8:589–594
65. Fischer E (1884) Phenylhydrazin als Reagens auf Aldehyde und Ketone. Ber dtsch Chem Ges 17:572–578
66. Fischer E (1889) Ueber einige Reactionen des Phenylhydrazins und Hydrozylamine. Ber dtsch Chem Ges 22:1930–1936
67. (a) Fischer E (1884) Verbindungen des Phenylhydrazins mit den Zuckerarten. Ber dtsch Chem Ges 17:579–584. (b) Fischer E (1887) Verbindungen des Phenylhydrazins mit den Zuckerarten. V. Ber dtsch Chem Ges 20(1):87–97
68. Images of osazones taken from: Shah T, Modi N (2016) Utility of osazone test to identify sugars. J Med Sci Clin Res 4(12):14361–14365
69. (a) Fischer E (1888) Ueber die Verbindungen des Phenylhydrazins mit den Zuckerarten. IV. Ber dtsch chem Ges 21:2631–2634. (b) Fischer E (1889) Ueber die Verbindungen des Phenylhydrazins mit den Zuckerarten. V. Ber dtsch chem Ges 22:87–97
70. Ramberg PJ (2003) Chemical structure, spatial arrangement. the early history of stereochemistry, 1874–1914. Ashgate Publishing Limited, Aldershot, ch 8:243–276
71. Hermann F (1877) Die Lagerung der Atome im Raume von Dr. J. H. van't Hoff. Nach des Verfassers Broschüre „La chimie dans l'espace." Friedrich Vieweg und Sohn, Braunschweig, pp 10–11
72. van't Hoff JH (1875) La Chimie dans l'Espace. P. M. Bazendijk, Rotterdam, p 11

73. van't Hoff JH (1874) Voorstel tot Uitbreiding der Tegenwoordige in de Scheikunde gebruikte Structuurformules in de Ruimte, benevens een daarmee samenhangende Opmerking omtrent het Verband tusschen Optisch Actief Vermogen en chemische Constitutie van Organische Verbindingen [Proposal for the Extension of Current Chemical Structural Formulas into Space, together with Related Observation on the Connection between Optically Active Power and the Chemical Constitution of Organic Compounds]. J. Greven, Utrecht.
74. Fischer E (1890) Ueber die optischen Isomeren des Traubenzuckers, der Gluconsäure und der Zuckersäue. Ber dtsch chem Ges 23(2):2611–2624
75. Fischer E (1890) Reduction des Fruchtzuckers. Ber dtsch chem Ges 23(2):3684–3687
76. Fischer E (1890) Synthese des Traubenzuckers. Ber dtsch chem Ges 23(1):799–805
77. Fischer E (1891) Ueber die Configuration des Traubenzuckers und seiner Isomeren. Ber dtsch chem Ges 24(1):1836–1845
78. Fischer E (1891) Ueber die Configuration ders Traubenzuckers und seiner Isomeren. II. Ber dtsch chem Ges 24(2):2683–2687
79. For a biography of Bijvoet, see: Groenewege MP, Peerdeman AF (1983) Johannes Martin Bijvoet 123 January 1892–4 March 1980. Biogr Mem Fell Roy Soc 29(29):27–41. For an image of Bijvoet, see: https://commons.wikimedia.org/wiki/File:JMBijvoet1931.jpg
80. Bijvoet JM, Peerdeman AF, van Bommel AJ (1951) Determination of the absolute configuration of optically active compounds by means of X-rays. Nature 168:271–272
81. https://www.ccdc.cam.ac.uk/solutions/software/csd/
82. Inexplicably, in 2008, the original atomic coordinates of Bijvoet's paper were not available in Cambridge Crystallographic Database, so the structure was redtermined with modern equipment: Lutz M, Schruers AMM (2008) Was Bijvoet right? Sodium rubidium (+)-tatrate tetrahydrate revisited. Acta Cryst C64:m296–m299
83. For biographical information on Kiliani, see: (a) Kiliani H (1932) My life and work. J Chem Educ 9(11):1908–1914. (b) Hückel W (1949) Heinrich Kiliani 30.10.1855–25.2.1945. Chem Ber 82:I–IX. For an image of Kiliani, see: https://commons.wikimedia.org/wiki/File:H._Kiliani_1920_Freiburg.jpg
84. (a) Kiliani H (1885) Ueber das Cyanohydrin der Lävulose. Ber dtsch chem Ges 18:3066–3072. (b) Kiliani H (1886) Ueber das Cyanohydrin der Lävulose. [Zweite Mittheilung] Ber dtsch chem Ges 19:221–227
85. (a) Kiliani H (1886) Ueber die Einwirkung von Blausäure auf Dextrose. Ber dtsch chem Ges 19:767–772. (b) Kiliani H (1886) Ueber die Constitution der Dextrosecarbonsäure. Ber dtsch chem Ges 19:1128–1130
86. (a) Kiliani H (1886) Ueber Arabinose. Ber dtsch chem Ges 19:3029–3036. (b) Kiliani H (1887) Vorläufige Mittheilung. Ber dtsch chem Ges 20:282–282. (c) Kiliani H (1887) Ueber die Zusammensetzung und Constitution der Arabinosecarbonsäure bezw. der Arabinose. Ber dtsch chem Ges 20:339–346
87. (a) Kiliani H (1888) Ueber die Einwirkung von Blausäure auf Galactose. Ber dtsch chem Ges 21:915–919. (b) Kiliani H (1889) Oxydation der Galactosecarbonsäure. Ber dtsch chem Ges 22:521–524
88. Fischer E (1889) Reduction von Säuren der Zuckergruppe. Ber dtsch chem Ges 22:2204–2205
89. (a) Williamson AW (1845) On the blue compounds of cyanogen and iron. Mem Proc Chem Soc 3:125-140. (b) Adie RH, Browning KC (1900) The interaction of sulphuric acid and potassium ferrocyanide. J Chem Soc Trans 77:150–160

Part IV

Exploring the Limits of Organic Structure

Historical Evolution, Stereochemistry, and Structural Limitations for the C=C Bond

10

Richard P. Johnson

Abstract

In his seminal publications during 1874–1875, Jacobus Henricus van't Hoff made the leap from *tetravalent* carbon (Kekulé's theory) to *tetrahedral* carbon. He recognized the stereochemical consequences of the different modes for joining tetrahedra, using this model to explain the existence of chemical structures which are non-superimposable mirror images. The existence of alkene stereoisomers, well known at that time, was easily explained by fusing tetrahedra along one edge. Inspired by structures suggested earlier by Kekulé, van't Hoff further extended his model to cumulated carbon structures, which were unknown in that period. He predicted the chirality of even cumulenes such as allenes (derivatives of 1,2-propadiene) and the existence of cis–trans isomers for odd cumulenes such as 1,2,3-butatrienes. This chapter describes the historical evolution and realization of these models, with a focus on allenes and cumulenes. The later development of Adolph von Baeyer's strain theory for cyclic carbon compounds led chemists to explore the synthesis of strained and highly reactive cyclic allenes and cyclic cumulenes as an extension of van't Hoff's theories. Some historical milestones in this field are described.

R. P. Johnson (✉)
Department of Chemistry, University of New Hampshire, Durham, NH, USA
e-mail: Richard.Johnson@UNH.edu

A. Greenberg and D. E. Lewis (eds.), *The Birth of the 3rd Dimension in Chemistry*,
Perspectives on the History of Chemistry,
https://doi.org/10.1007/978-3-031-97743-5_10

10.1 Introduction

At the time that Jacobus Henricus van't Hoff published his theories about chemistry "dans l'espace" in 1874–1875, [1, 2] Dalton's theory of atoms was long accepted and chemists had already deduced an impressive ensemble of information about the existence, structure, and essential properties of the compounds of carbon. Berzelius had introduced the term "organic chemistry" around 1807, followed by Wohler's synthesis of urea in 1828; this helped to disprove vitalism, the idea that only living organisms could make organic compounds. Laboratory methods for isolation, purification, and physical characterization of organic compounds were available. The current generation of chemists might be surprised that combustion analysis was already a standard tool in organic structure determination. Some basic types of organic chemical reactions were known. Kekulé's theory of tetravalent (not yet tetrahedral!) carbon in 1858 [3] had introduced carbon chains and multiple bonding; this gave chemists license to draw simple line-bond structures, much as we use today. Adolph von Baeyer's strain theory for cyclic carbon compounds was only a decade away [4] and this would lead to many new structures to challenge chemists [5–7]. Of course, revelations about electronic structure and chemical bonding based on quantum theory would have to wait until well into the next century [8–10].

Initial recognition for van't Hoff's new structural theory was slow to arrive, but this was not his only contribution to science. As a measure of his stature as a scientist, van't Hoff was awarded the first Nobel Prize in Chemistry in 1901 [11]. Contrary to what might be expected from the focus of this chapter, that award was for the discovery of the laws of chemical dynamics and osmotic pressure in solutions, not for stereochemistry! Van't Hoff died from tuberculosis in 1911 at the age of 58, long before the real impact of his stereochemical theory would be realized [12].

Van't Hoff and, independently, Le Bel [13] moved the conversation about chemical structure from two dimensions to three. This transition has its origins in Biot's discovery of optical activity in 1814 [14] and Pasteur's resolution of enantiomers, reported in 1848 [15]. Pasteur made the brilliant observation that crystals could exist as non-superimposable mirror images (*images spéculaire*) with equal but opposite optical rotations; this led to the conclusion that chemical substances made of atoms must also exist in mirror image forms. Van't Hoff and Le Bel moved from *tetravalent* to *tetrahedral*, recognizing the existence of asymmetric carbon as the molecular level source of mirror image structures and optical activity. Van't Hoff further used the fusion of tetrahedra to explain the well documented cis/trans isomers of alkenes. As will be described below, van't Hoff's extension to cumulated double bonds was especially prescient because such structures were only the subject of speculation at that time. Among the different publications which present van't Hoff's theory, the 1875 version published in French by Bazendiik is very clear, both in text and figures ("Planches": at the end of the document). A reproduction of this seminal document is easily found online [16].

The author makes no claim as a historian. This chapter is not meant as a comprehensive review but as an introduction to van't Hoff's theory as it relates to unsaturated organic structures. Connection is made to the author's own research in the chemistry of alkenes, allenes and cumulenes. Just as in the days of van't Hoff, these still challenge the limits of molecular structure!

10.2 A Brief Digression: From Alkenes to Cumulenes

Before returning to the story of van't Hoff, an introduction to the structure of cumulated double bonds is required since this more advanced topic is poorly covered in basic organic chemistry. Alkenes can be linearly extended, as shown by structures **1** – **4** (Scheme 10.1), to create the series of "cumulenes" in which carbons are shared between double bonds. Structure **2** represents an allene, a type of structure surprisingly well known in nature and commonly used in synthesis. More generally, the [n]cumulene series, where n is the number of double bonds, describes the homologous series, currently known up to [9] cumulenes [17, 18]. As might be expected, higher cumulenes become increasingly reactive. The limiting structure is a linear carbon allotrope known as "carbyne" which can be drawn as an extended cumulene (**5**) or with alternating single and triple bonds (**6**). Long chain polyynes are well known, and there is evidence for the synthesis of carbyne on a gold surface [19].

In allene and all cumulene structures, the non-terminal carbons are shared by two π bonds. Although the structure appears to be conjugated, adjacent π bonds in cumulenes are orthogonal, with attendant stereochemical consequences. For an odd number of double bonds (n = 1,3,5, …), the ligands are all in the same plane; for an even number of double bonds (2,4,6, …), the terminal carbon ligand pairs are always orthogonal. The stereochemical consequence of this odd/even switch in ligand orientation is that disubstituted ethylene (**7** and **8**) and all *odd* cumulenes such as butatriene (**9** and **10**) can exist as cis/trans (or Z/E) pairs of achiral diastereomers, while allene and all disubstituted *even* cumulenes, will exist as chiral pairs of enantiomers. In Scheme 10.2, these axially chiral structures are

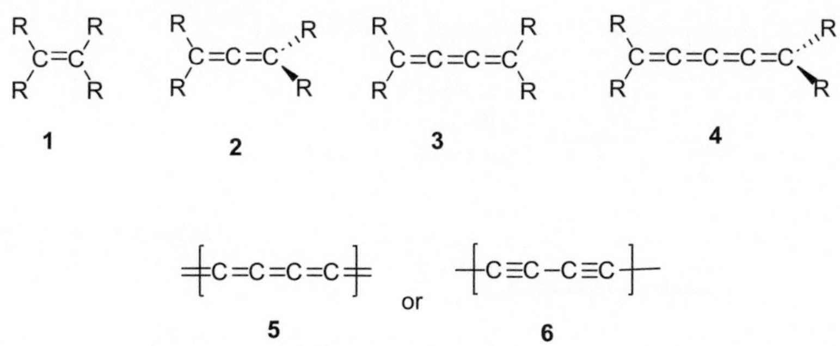

Scheme 10.1 Structures with cumulated double bonds

shown as mirror images **11** and **12**, labeled as P (plus) or M (minus) to indicate the sense of helicity.

How did a distinguished physical chemist like van't Hoff invent cumulenes? Allenes and higher cumulenes were unknown during the era of van't Hoff and his inspiration for considering such structures must have come from Kekulé. As one example, cited in "La chimie dans l'espace," van't Hoff notes that: "I have in mind pyromucic acid…" Kekulé had speculated that pyromucic acid might have a 1,2,3-butatriene structure (Fig. 10.1). This is plausible given the known molecular formula, and before alternative cyclic isomers might seem more likely. Pyromucic acid was later shown to be a furan derivative, 2-furoic acid, but the seed was planted.

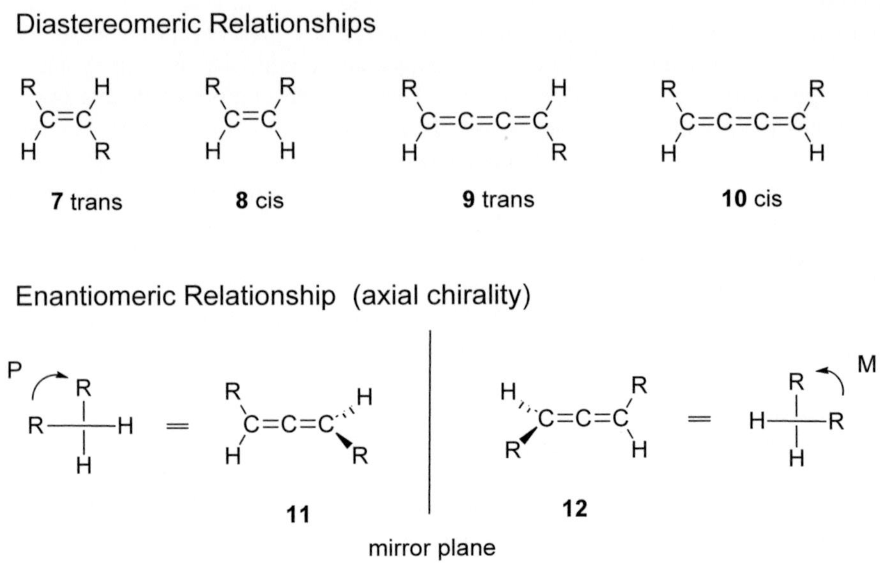

Scheme 10.2 Stereochemistry of odd and even cumulenes

Jusqu'ici nous avons traité du cas spécial de la formule:

$C(R_1\ R_2) = C_{(2\ n)} = C(R_3\ R_4)$

pour n = o; prenons en second lieu n = 1, donc:

$C(R_1\ R_2) = C = C = C(R_3\ R_4)$

ici l'hypothèse conduit également à deux isomères lorsque R_1 et R_2, R_3 et R_4 sont différents.

J'ai en vue les *acides pyromuciques.*

Déjà M. Kekulé adopta dans son livre la formule:

CH (OH) = C = C = CH. CO_2 H

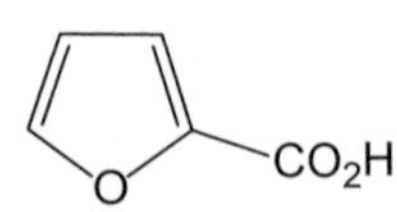

2-furoic acid
"acides pyromuciques"

Fig. 10.1 Text from La chimie dans l'espace (page 35) showing a cumulene structure attributed to Kekulé. The correct structure of pyromucic acid is shown

The cumulene field evolved slowly. The first verifiable synthesis of an allene is attributed to Burton and Pechmann in 1887, [20] but the authors misassigned the structure as an isomeric alkyne. This error was not rectified until 1954! [21] In 1924, Staudinger and Ruzicka reported isolation and structural characterization for the first natural allene, pyrethrolone [22]. The first cumulene to be reported was 1,1,4,4-tetraphenylbutatriene, described by Brand in 1921 [23]. Although Kekulé had pointed towards the existence of cumulenes, these later developments do not diminish the boldness and prescience of van't Hoff's description of the structure and stereochemistry of allenes and cumulenes!

10.3 Building with Tetrahedra

Van't Hoff moved from *tetravalent* to *tetrahedral* and this led to a simple model for the structure and stereochemistry of alkanes and alkenes. Our focus here is on structures with multiple bonds. Figures 10.2 and 10.3 reproduce the relevant text and figure for van't Hoff's treatment of alkenes.

"Beginning with the hypothesis of the equality of the two bonds…" van't Hoff represented the double bond as two tetrahedra which touch on an edge. With different R groups, "…two solids can exist represented by Figures XXIII and XXIV …" and "…the difference between these figures…announces a case of isomerism

Passons aux combinaisons à atomes de carbone doublement liés, représentés par la théorie actuelle comme suit:

$$C = C.$$

En partant de l'hypothèse d'égalité des deux liaisons l'image consiste en deux tétaèdres qui se touchent par une arrête (fig. 23): A et B représentent les liaisons des deux atomes de carbone, R_1 et R_2, R_3 et R_4 les groupes univalents par lesquels sont saturées leurs affinités libres. Si R_1 et R_2, R_3 et R_4 sont identiques, on ne peut construire par la pensée qu'une seule figure; il en est de même lorsqu'il y a seulement identité entre R_1 et R_2, ou entre R_3 et R_4; ***mais, si à la fois R_1 diffère de R_2, et R_3 de R_4 — ce qui d'ailleurs n'empêche pas que R_1 et R_3, R_2 et R_4 puissent être égaux, — deux solides peuvent exister, représentés par les fig. 23 et 24, et dont la différence tient à la situation relative des groupes R_1 et R_2 par rapport à R_3 et R_4: la dissemblance de ces figures, dont le nombre se réduit à deux, annonce un cas d'isomérie, qui n'est pas impliqué dans le mode ordinaire de représentation*** **(1).**

Fig. 10.2 Text from La chimie dans l'espace (page 13) describing alkene stereochemistry

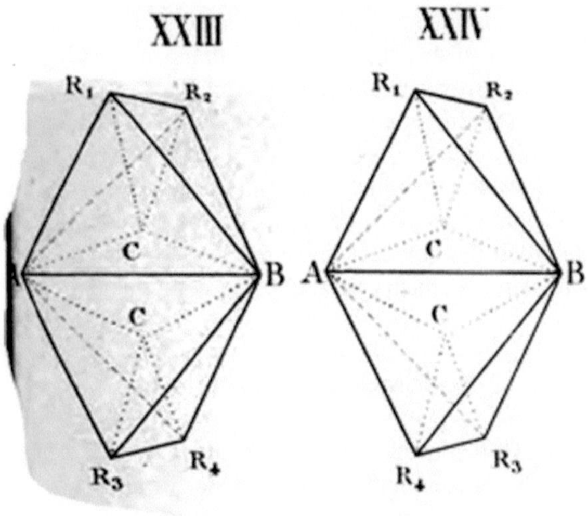

Fig. 10.3 Figure from La chimie dans l'espace (PLANCHE II) showing alkene stereoisomers

which is not implicated in the ordinary mode of representation." This short paragraph rationalized the existence of cis/trans isomeric pairs, already well known at that time.

The extension of this fused tetrahedra concept to allene and cumulenes is brilliant, especially given that such structures were unknown by experiment at this time. Figure 10.4 reproduces the brief but marvelously clear presentation. Expanding the consequences of merged tetrahedra, van't Hoff first noted it is evident that *odd* cumulenes all reduce to the structure of alkenes for which "…there always exists two stereoisomers…these isomers are not mirror images (*images spéculaire*) of each other." For their part, even cumulenes reduce to the structure of allene which has three fused tetrahedra as represented by image XXIX (Fig. 10.5). With different R groups, "…there always exist two isomers…always being mirror images of each other."

The case for triple bonds needs to be included here. The appropriate text and structure are reproduced in Figs. 10.6 and 10.7. Again, beginning with the idea of three equal multiple bonds, the faces of two tetrahedra are combined to give a triple bond, with one group free to bond. The lack of stereochemistry and linear structure are implied in this structure. "The new hypothesis is not in contradiction with the generally adopted views and we will not dwell on this point."

In a few succinct paragraphs of easily understood French, and with only a simple geometric model for atomic structure and bonding, van't Hoff had provided an explanation for the existence of cis/trans alkene isomers, predicted the existence, structure, stereochemistry, and chirality (a word introduced by Lord Kelvin in 1884) of allenes and cumulenes, and supported the linear structure of alkynes. Of course, the operative word is "model" since the electronic structure of atoms and molecules, including such important features as π bonding, would have to

Il est évident, que le cas:

$$(R_1\ R_2)\ C = C = C = C\ (R_3\ R_4)$$

ou en général:

$$(R_1\ R_2)\ C = C_{(2n)} = C\ (R_3\ R_4)$$

revient à:

$$(R_1\ R_2)\ C = C\ (R_3\ R_4)$$

donc: *il existe toujours deux isomères, lorsqu'il y a inégalité entre R_1 et R_2, comme entre R_3 et R_4; les isomères ne sont pas l'image spéculaire l'un de l'autre.*

D'autre part, le cas:

$$(R_1\ R_2)\ C = C = C = C = C\ (R_3\ R_4)$$

ou en général:

$$(R_1\ R_2)\ C = C_{(2n+1)} = C\ (R_3\ R_4)$$

revient à:

$$(R_1\ R_2)\ C = C = C\ (R_3\ R_4)$$

donc: *il existe toujours deux isomères en cas de différence entre R_1 et R_2, R_3 et R_4; isomères, étant toujours les images spéculaires l'un de l'autre.*

Fig. 10.4 Text from La chimie dans l'espace (page 14) describing cumulene stereochemistry

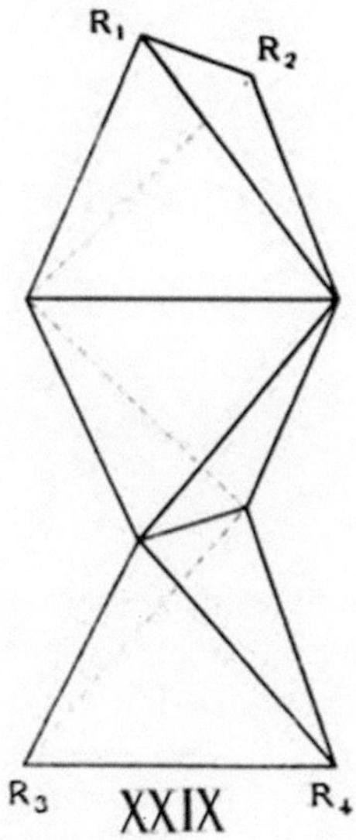

Fig. 10.5 Figure from La chimie dans l'espace (PLANCHE II) describing allene stereochemistry which results from fusing three tetrahedra

Passons enfin aux combinaisons à atomes de carbone triplement liés, représentés dans la théorie actuelle par:

$$C \equiv C.$$

L'image consiste alors (en partant toujours de l'hypothèse d'égalité des trois liaisons) en deux tétaèdres coïncidant par trois de leurs sommets, c'est-à-dire, par une de leurs faces (fig. 32): A C B est la liaison triple, R_1 et R_2 sont les groupes univalents par lesquels sont saturées les affinités restées libres dans les deux atomes de carbone.

La nouvelle hypothèse n'est pas ici en contradiction avec les vues généralement adoptées; nous ne nous étendrons donc pas sur ce point.

Fig. 10.6 Text from La chimie dans l'espace (page 14) describing alkyne bonding

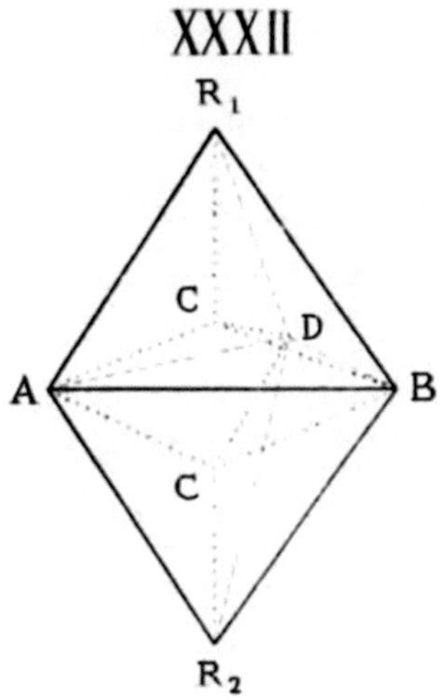

Fig. 10.7 Drawing from La chimie dans l'espace (PLANCHE II) showing alkyne bonding

wait until the development of quantum mechanics in the next century. π Bonding in allene and cumulenes remained poorly understood until the mid-twentieth century [24].

10.4 Experimental Verification of Allene and Butatriene Stereochemistry

Experimental demonstration of the stereochemistry of cumulenes would have to wait for over half a century! The chirality of allenes was first proven in 1935 when Maitland and Mills reported [25] that dehydration of racemic allylic alcohol **13**

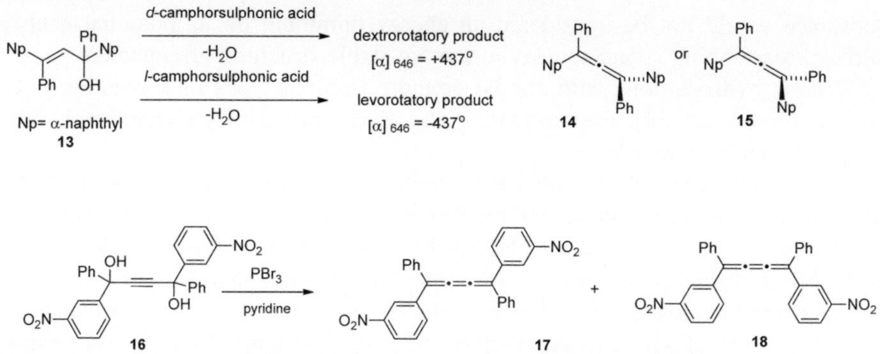

Scheme 10.3 First examples demonstrating allene and butatriene stereochemistry

with *d* or *l*-camphorsulfonic acid (Scheme 10.3) produced samples of nonracemic tetrasubstituted allene, represented as **14** or **15**, which displayed equal but opposite optical rotations. The enantiomeric excess is unknown and these experiments await repetition with modern methods. A quarter century later, Kuhn and Blum reduced diol **16** to a pair of Z/E-1,2,3-butatriene stereoisomers **17** and **18**, separated these isomers by fractional crystallization and chromatography, and demonstrated E/Z interconversion by photochemistry [26]. Cumulene isomerization remains at the forefront of science. In one recent example, Venkataraman and co-workers demonstrated that an external electric field can cause the cis-to-trans isomerization of [3] cumulene derivatives in solution, as observed in a scanning tunneling microscope [27]. The electric field lowers the barrier to rotation and is believed to selectively stabilize the trans-diastereoisomer.

10.5 Beyond Acyclic Alkenes and Cumulenes

Only a decade after van't Hoff, von Bayer used the tetrahedral carbon model to predict structure and strain in different sized hydrocarbon rings [4]. Without insight from modern structural theory, von Baeyer predicted in 1885 that small ring cycloalkanes should have increasing strain through deviation from a tetrahedral geometry. As every beginning student in organic chemistry learns, von Bayer's failure to include conformation in this analysis led to some erroneous conclusions, but this work still laid the foundation for our understanding of cyclic organic structures. Of course, the connection between ring size and strain is easily extended beyond cycloalkanes. After von Baeyer introduced the concept of strain and its relationship to ring size, generations of chemists have embraced the challenges of making increasingly strained cycloalkanes, cycloalkenes, cycloalkynes, cycloallenes, and cyclocumulenes [6]. With unsaturated structures, the series may progress from "shelf-stable" molecules in larger rings, to reactive intermediates which are not isolable under ordinary conditions, to nonexistent. A "nonexistent" substance would not be considered an energy minimum on its potential energy surface, rearranging spontaneously to a more stable structural arrangement.

Diverse hydrocarbons with simple or more tortured types of π bonding have been explored [28–30]. Representative structure types with six-membered rings are shown in Scheme 10.4.

Cis-Cyclohexene (**19**) is shelf stable, while its trans stereoisomer (**20**) is a fleeting intermediate usually generated by photochemistry. Trans-Cyclohexenes have a low barrier for π bond rotation [31] but have recently been shown by Weaver to be highly useful in synthesis [32]. We used computational modeling to show there is a concerted cycloaddition pathway (Scheme 10.5) for the S-trans-1,3-butadiene conformer to react with ethylene to make trans-cyclohexene (**20**) [33]. This stands in contrast to textbook dogma. Of course, the product strain in **20** prevents this pathway from being observed.

In the cycloalkene series, trans-cyclooctene is the smallest isolable homolog [28]. The remaining structures in this group **21–25** are all highly strained because the π bond will be bent and/or twisted. Among cycloalkynes, cyclooctyne is isolable, while cycloheptyne through cyclopentyne continue the series as increasingly reactive intermediates [34]. Computational models show that structures representing cyclobutyne and cyclopropyne should rearrange spontaneously because they do not represent minima on potential energy surfaces [35, 36]. The allene functionality in **22** is bent from linearity to ca. 120° and the ligand arrangement will be twisted from 90° toward planarity [37]. Computational models show that even [4] cumulene **24** should exist as a reactive intermediate, but efforts by the author's group to prove its intermediacy were unsuccessful [38]. Bicyclic structure **25** violates the famous Bredt's rule, [39] an early experimental observation that

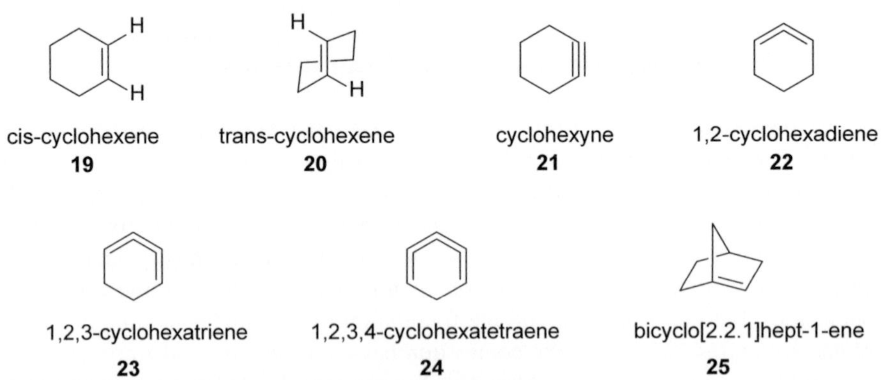

Scheme 10.4 Structures with π bonds in six-membered rings

Scheme 10.5 Conformation specificity in Diels–Alder reactions

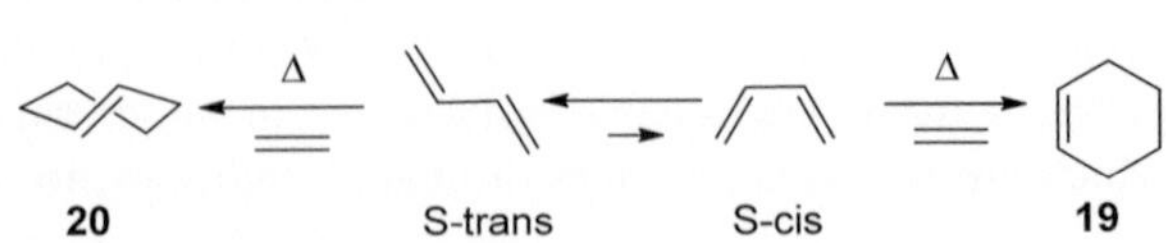

bicyclic structures with a double bond to a bridgehead carbon will not form. This is due to the twisted π bond. There is a rich history of efforts to make and understand anti-Bredt olefins ("ABOs") such as **25** [30]. One very recent development in this field reported by Garg is a method for generation of **25** and other ABOs at ambient temperature in the solution phase [40].

10.6 Cyclic Allenes and Cumulenes

Historical developments in the chemistry of cyclic allenes and cumulenes have been reviewed previously [41–43] and only a few highlights will be described here. Early researchers apparently did not fully recognize that bending the allene functional group by incorporation in a ring would lead to increasing strain and high reactivity. The first attempts to prepare cyclic allenes were reported in the mid 1930's by Favorskii in Leningrad [44]. Treatment of 1-bromo-2-chlorocycloheptene with sodium in ether gave a distillable liquid and Favorskii concluded: "The results obtained show definitively that the hydrocarbon of formula C_7H_{10} contains an allenic group in the ring and corresponds to formula II." This erroneous conclusion (Fig. 10.8) remained unchallenged in the literature for 25 years.

Among many early experiments, the first clear evidence for synthesis of a small ring allene, 1,2-cyclohexadiene, (Scheme 10.6a) is attributed to Wittig and Fritz in 1966 [45]. Treatment of 1-bromocyclohexene with KOtBu led to **27** in 7% yield. This presumably arises from stepwise dimerization of **22**. The strained allene was also trapped with styrene and diphenylisobenzofuran. A more efficient route to **22** is treatment of **26** with methyllithium, a reaction first reported by Moore and Moser [46]. The Doering-Moore-Skattebøl reaction which proceeds through carbenoid (not carbene!) intermediates, is now a standard route to many cyclic allenes [47].

What is the structure of 1,2-cyclohexadiene (**22**)? The author became interested in this question *circa* 1980 while studying allene photochemistry [48]. At that time, the model for the structure of **22** favored achiral zwitterion C (Scheme 10.6b) [6] because the chiral structure seemed too strained and this zwitterion is the electronic solution provided by closed-shell Hartree–Fock calculations [49]. This assignment as an achiral zwitterion was inconsistent with some very elegant experiments reported by Balci and Jones (Scheme 10.6c) which had supported a chiral

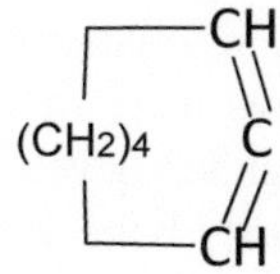

II

Fig. 10.8 1,2-cycloheptadiene (structure II) as proposed by Favorskii

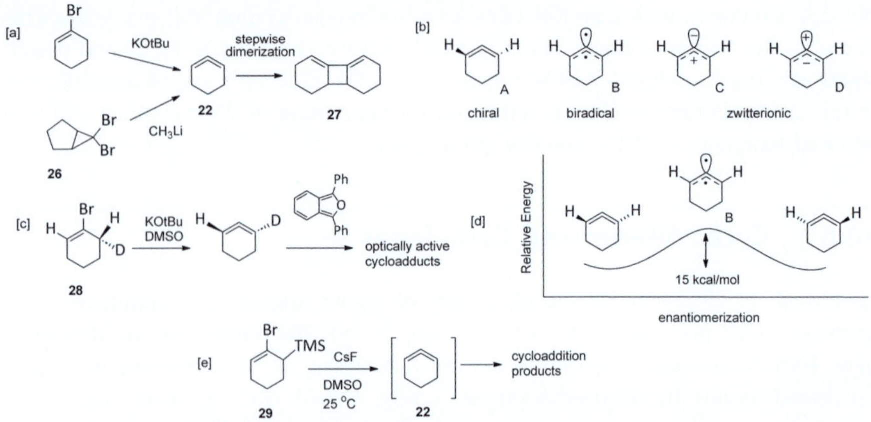

Scheme 10.6 Chemistry of 1,2-cyclohexadiene

cyclic allene intermediate [50]. In these experiments, the chirality of **28** transfers to the allene through a kinetic isotope effect. At about the same time, our photochemical studies and computations had shown that the zwitterion C represents an excited state of planar allene! [51] We soon used MCSCF (multi-configuration self-consistent field) calculations, a sophisticated method for that era, to show that **22** is indeed chiral, with a barrier of ca. 15 kcal/mol for enantiomerization through a biradical transition state represented as B [52, 53]. The accepted model for small ring cyclic allenes is now chiral, although this may be complicated by connected heteroatoms [54]. Tuononen has recently questioned whether small ring allenes might better be described as carbenes [55]. With the exception of the four membered ring (vide infra), the author favors a bent allene structure.

A small digression into the field of computational chemistry seems appropriate at this point. Van't Hoff built conceptual and physical (cardboard) models for tetrahedral carbon, which helped him to visualize the structure and stereochemistry of organic compounds. Of course, chemists still build simple ball and stick models; indeed, in his office, this author keeps a large box of well-used plastic and metal models that van't Hoff would greatly admire. What van't Hoff could not have imagined, however, is the age of computers and the powerful qualitative and quantitative models, both static and dynamic, now available to chemists. Computational chemistry is a highly advanced technology that has evolved over the past half-century with quantum leap (pun intended) advancements in computers, theory, and complex computer programs. The intimate details of mechanisms are revealed by molecular dynamics simulations, which model the time course of reactions. Artificial intelligence is now infusing this field with immense predictive power, while quantum computers, under development in labs around the world, promise future processing speeds vastly beyond current supercomputers.

With the right software, today's multicore laptop already is a powerful computational tool, useful in both research and teaching. Figure 10.9 reproduces a computer screen from the author's laptop which displays a model for the racemization of 1,2-cyclohexadiene (**22**). This is an electronically complex process which passes through a biradical transition state (biradical B in Scheme 10.6d), modeled here by density functional theory, a method valued for accurate predictions. The image shows a user-friendly graphical user interface (GUI) from the program Spartan 24 [56]. The Spartan GUI permits the rapid construction of 3D structures and connects to over a million lines of computer code which carry out the intense mathematics of quantum chemistry. This is one of many sophisticated programs now available to chemists. Computational models like this move chemistry from qualitative 3D structures and arrow-pushing mechanisms into detailed quantitative descriptions of molecular structure and chemical reactions. A few examples of the power and utility of computational models are included in this chapter. In the current era, any young chemist who leaves school without an introduction to computational modeling has been ill-served by their educational institution.

Computational strain estimates and observed reactivity of the series of cyclic allenes are summarized in Scheme 10.7 [37]. The numbers shown are derived from isodesmic computational models which are designed to compare strained and unstrained structural elements. Links between geometry, strain, and chemical reactivity are clearly demonstrated by this homologous series. With a nearly linear (169.3° predicted) allene functional group, and thus minimal strain, 1,2-cyclononadiene (**34**) is a colorless distillable liquid [57, 58]. The eight membered ring homolog (**33**) with a predicted bending 161.5° can be handled briefly in solution to measure its NMR spectrum, but dimerizes quickly [59]. Smaller cyclic allene homologs (**22**, **30**–**32**) are reactive intermediates. Only one example of a 1,2-cyclopentadiene has been reported by Balci [42].

To improve synthetic routes, we demonstrated in 1990 that 1,2-cyclohexadiene (**22**) is easily made in solution phase by treatment of **29** with CsF, as shown in

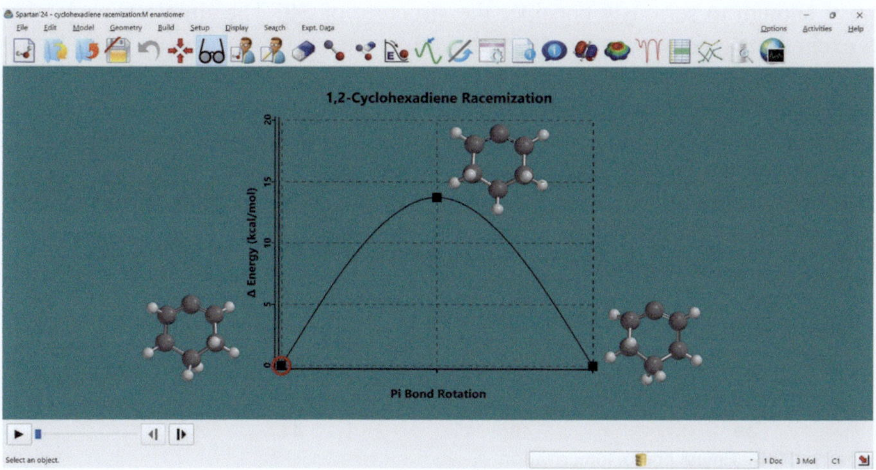

Fig. 10.9 Computational modeling (ωB97xd/6-31G*) for racemization of 1,2-cyclohexadiene

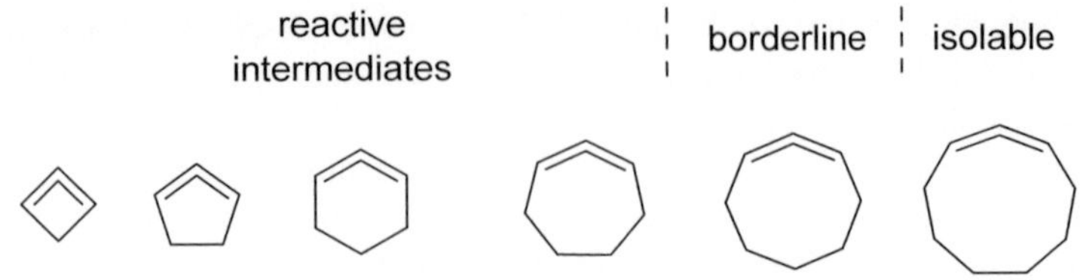

	30	31	22	32	33	34
Allene Strain	65	51	32	14	5	2
Total Strain	90	55	32	20.5	11.9	11.0

Scheme 10.7 Strain estimates for the cyclic allene series

Scheme 10.6e [60]. It is gratifying that this exotic chemistry has recently been rediscovered by the synthetic community. The strain, reactivity, and chirality of 1,2-cyclohexadienes are now being used strategically by Garg in natural product synthesis [61–63].

The most tortured substance in this series is 1,2-cyclobutadiene (**30**), with an expected bending angle of ca. 90°. This seems like an improbable structure, but we wondered if 1,2-cyclobutadiene might be made by photocyclization of conjugated enyne **35** (Scheme 10.8a), analogous to the well-known photoisomerization of 1,3-butadiene to cyclobutene. In support of this idea, Meier and Konig had reported the photoisomerization of cyclooctene-3-yne **36** to **38** shown in Scheme 10.8b, suggesting the intermediacy of **37** [64].

This led us to a beautifully simple experiment in which we demonstrated that isomers **39** and **41**, and related enyne structures, cleanly interconvert by irradiation in solution with 254 nm light and reach an equilibrium state (Scheme 10.8c). This extraordinary reaction could have been discovered decades earlier by anyone brave enough to seriously consider the structure of **30**! The simplest mechanism is excited state conversion to **40**, followed by thermal ring opening [65]. In this process, the excited state energy (ca. 110 kcal/mol) is translated into a highly strained (ca. 90 kcal/mol) ground state product, which then undergoes ring opening to release the stored strain energy, thereby transposing the methylene carbon (C1).

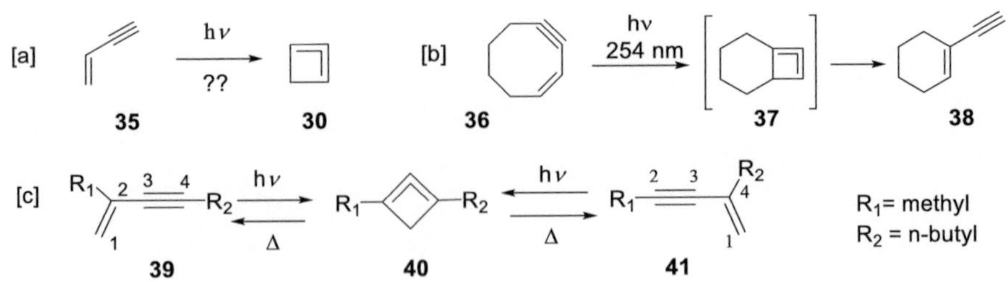

Scheme 10.8 Enyne photoisomerization

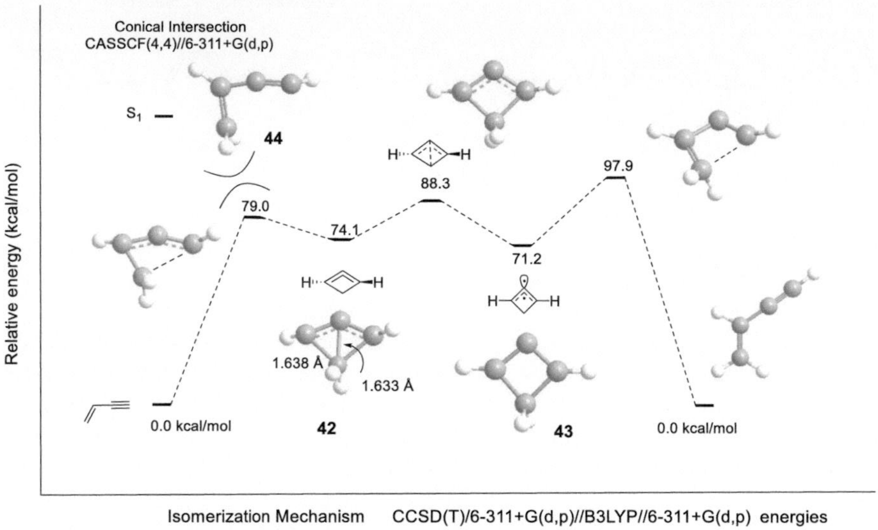

Fig. 10.10 1,2-cyclobutadiene potential energy surface

Once again, theory allows us to move beyond a simple "arrow-pushing" mechanism to help understand this complex process. Figure 10.10 summarizes the potential energy surface for the enyne photorearrangement. We were surprised to locate two structures that represent 1,2-cyclobutadiene. Biradical **43** is the logical extreme in the cyclic allene series, where structures become increasingly bent and planar in smaller rings. This has a surprisingly large barrier to ring opening. More promising is the chiral structure **42** which challenges the usual bonding scheme for carbon. This has a low barrier to ring opening and is accessible through nearby conical intersection **44**, a point which connects the ground and excited states. This study demonstrates the power of theory in understanding the chemistry of highly novel hydrocarbons [66].

Our study of enyne photochemistry led us to recognize a conceptual route to strained cumulenes and other reactive intermediates by pericyclic reactions. As one example, the ordinary Diels–Alder cycloaddition to make **19** (Scheme 10.9) can be systematically modified by removal of hydrogen atom pairs, using vinylacetylene or 1,3-butadiyne reacting with ethylene or acetylene. Some examples of "dehydro" Diels–Alder reactions were already known, but the general "dehydropericyclic" concept which we advanced presents a powerful way to design new reactions [66, 67]. We demonstrated intramolecular versions of all of the dehydro Diels–Alder variations [68, 69]. Most notable in this series is the diyne + yne cycloaddition to make benzyne (Fig. 19), first demonstrated by our group [70] and Ueda [71] in 1997. Fifteen years later, this remarkable cycloaddition reaction was rediscovered by Hoye [72] and has now become widely used in organic synthesis [73].

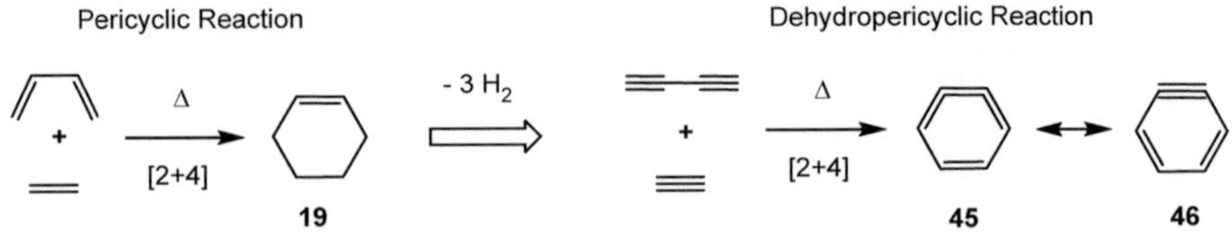

Scheme 10.9 Dehydropericyclic concept and route to benzyne

10.7 Cyclic Butatrienes

Van't Hoff anticipated the existence and stereochemistry of allenes and cumulenes, but it would be over 90 years before chemists began to explore the relationship between structure and reactivity in simple cyclic butatrienes. Scheme 10.10 summarizes strain estimates for the homologous series of cyclic butatrienes, while Scheme 10.11 presents the synthesis of homologs in this series, ordered by ring size.

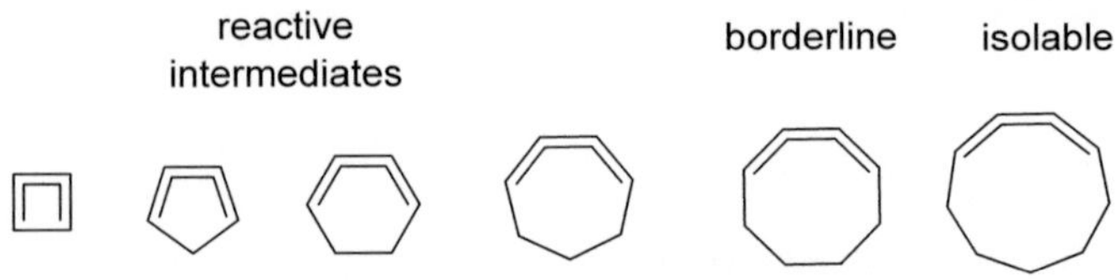

	47	48	23	49	50	51
Butatriene Strain	115	80	50	26	18	4
Total Strain	188	85	52	32	29	15

Scheme 10.10 Strain estimates for cyclic butatrienes

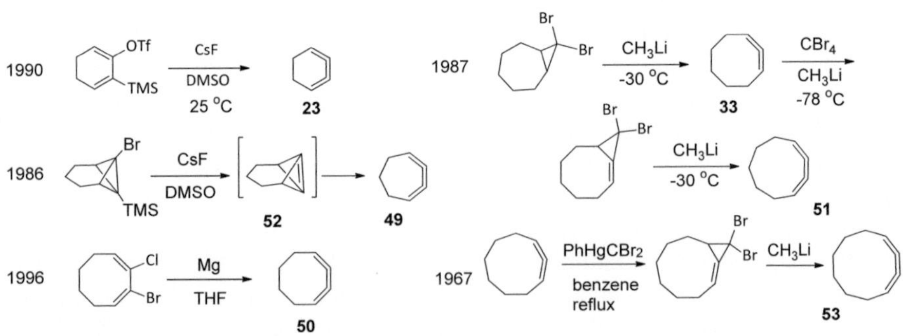

Scheme 10.11 Synthesis of cyclic 1,2,3-butatrienes

Moore and Ozretich first made the ten membered ring homolog **53** in 1967 (Scheme 10.11) using carbenoid chemistry to add one carbon to 1,2-cyclononadiene [74]. We used two carbenoid steps to make the next smaller homolog **51**, generating the unstable allene **33** in situ at low temperature [75]. Cyclocumulene **51** is isolable but easily polymerizes or reacts with oxygen. The eight membered homolog **50** was made by a vicinal halide elimination and proved too reactive to be isolated [76]. Smaller rings in this series were more challenging. Szeimies reported a novel route to 1,2,3-cycloheptatriene (**49**) by generation and strain-driven isomerization of pyramidalized alkene **52** [77]. The six membered ring cumulene **23** remained elusive until we showed in 1990 that it can be made in solution by vicinal elimination, [60] using chemistry pioneered by Kobayashi to make ortho-benzynes [78]. Cumulenes such as **23** are readily trapped in cycloaddition chemistry and are now being used strategically in natural product synthesis [79]. Efforts to make a derivative of 1,2,3-cyclopentatriene (**48**) by the same method used for **23** failed at the last step, presumably because the double helping of strain [80].

As expected, strain estimates for this series (Scheme 10.10) are significantly higher than the respective cyclic allenes because two sp-hybridized carbons are now bent from linearity. It is noteworthy that 1,2,3-cyclohexatriene, with a strain estimated to be 50 kcal/mol, is so easily generated in solution at ambient temperature. This is very similar to strain in ortho-benzyne, which can be made by the same method. The optimized structure for 1,2,3-cyclobutatriene (**47**) is badly twisted out of plane, presumably to avoid the antiaromaticity of a planar ring.

10.8 Conclusions

With Kekulé as his muse and in a few succinct paragraphs of easily understood French, van't Hoff provided an easily visualized explanation for the existence of cis/trans alkene stereoisomers, predicted the existence, structure, stereochemistry, and chirality of different types of cumulenes, and predicted the linear structure of alkynes. This was based on a simple model for tetrahedral carbon and different geometric modes of fusing tetrahedra. It is easy to imagine the *Eureka!* moment when van't Hoff (and independently, Le Bel) realized that his tetrahedral model held the key to understanding the stereochemistry of organic substances. Van't Hoff recognized the power of theories and models both to predict and to be subject to experimental verification, just as we do today. Of course, the operative word is "model" since the electronic structure of atoms and molecules, including such important features as π bonding, would have to wait until the development of quantum mechanics in the next century.

When von Baeyer introduced the concept of ring strain just a decade later, chemists soon turned to incorporating alkenes and, somewhat later, cumulated alkenes in rings. In spite of an early claim that 1,2-cycloheptadiene is a "distillable liquid," it soon became clear that cyclic allenes with a ring size of eight or smaller are increasingly reactive. All of the ring sizes have eventually yielded to synthesis. Cyclic cumulenes would come later, as the members of this homologous series and their chemical behavior would slowly be filled in. As described in examples above, computational models provide an essential tool to predict/understand structure, strain, and chemical reactivity. Tortured structures and their chemical reactions, which imitially seemed impossible, are now entering the mainstream as tools for chemical synthesis [61–63, 79].

Acknowledgements The author, whose chemical genealogy traces back to Kekulé but not van 't Hoff, is grateful to the National Science Foundation for support and to many students who contributed their talents and enthusiasm for odd molecular structures. Thanks are due to my friend and colleague Professor Arthur Greenberg for the kind invitation to have a small adventure in the field of chemical history.

References

1. van't Hoff JH (1874) Surs les formules de structure dans l'espace Archives neerlandaises des sciences exactes et naturelles 9: 445
2. Hoff JHvt (1875) La Chimie dans l'espace. P.M. Bazendijk, Rotterdam
3. Kekulé A (1858) Ueber die Constitution und die Metamorphosen der chemischen Verbindungen und über die chemische Natur des Kohlenstoffs. Justus Liebigs Ann Chem 106:129. https://doi.org/10.1002/jlac.18581060202
4. Baeyer A (1885) Ueber Polyacetylenverbindungen Berichte der deutschen chemischen Gesellschaft 18:2269. https://doi.org/10.1002/cber.18850180296
5. Wiberg KB (1986) The concept of strain in organic chemistry. Angew Chem Int Ed Engl 25:312. https://doi.org/10.1002/anie.198603121
6. Greenberg A, Liebmann J (1978) Strained organic compounds. Academic Press, New York
7. Ramsay OB (1975) Molecular models in the early development of stereochemistry. I. Van't Hoff Model. II. Kekule models and Baeyer strain theory ACS Symp. Ser 12:74
8. Lewis GN (1916) The atom and the molecule. J Am Chem Soc 38:762. https://doi.org/10.1021/ja02261a002
9. Pauling L (1960) The nature of the chemical bond, 3rd edn. Cornell University Press Ithaca, NY
10. Mulliken RS (1970) Path to molecular orbital theory. Pure Appl Chem 24:203. https://doi.org/10.1351/pac197024010203
11. https://www.nobelprize.org/prizes/chemistry/1901/summary/. Accessed 22 Nov 2024
12. https://www.nobelprize.org/prizes/chemistry/1901/hoff/biographical/. Accessed 22 Nov 2024
13. Le Bel J-A (1874) Sur les relations qui existent entre les formules atomiques des corps organiques et le pouvoir rotatoire de leurs dissolutions (On the relations that exist between the atomic formulas of organic substances and the rotatory power of their solutions. Bulletin de la Société Chimique de Paris 22:337
14. Biot JB (1815) Bulletin des Sciences, par la Société Philomatique de Paris, pp 190–192
15. Pasteur L (1848) Memoires sur la relation qui peut exister entre la forme crystalline et al composition chimique, et sur la cause de la polarization rotatoire. Compt Rend 26:535
16. https://catalog.hathitrust.org/Record/008547306. Accessed 22 Nov 2024
17. Januszewski JA, Tykwinski RR (2014) Synthesis and properties of long [n]cumulenes ($n \geq 5$). Chem Soc Rev 43:3184. https://doi.org/10.1039/C4CS00022F
18. Wendinger D, Tykwinski RR (2017) Odd [n]Cumulenes ($n = 3, 5, 7, 9$): synthesis, characterization, and reactivity. Acc Chem Res 50:1468. https://doi.org/10.1021/acs.accounts.7b00164

19. Pan B, Xiao J, Li J, Liu P, Wang C, Yang G (2015) Carbyne with finite length: the one-dimensional sp carbon. Sci Adv 1:e1500857. https://doi.org/10.1126/sciadv.1500857
20. Burton BS, von Pechmann H (1887) Ueber die Einwirkung von Chlorphosphor auf Acetondicarbonsäureäther Berichte der deutschen chemischen Gesellschaft 20:145. https://doi.org/10.1002/cber.18870200136
21. Jones ERH, Mansfield GH, Whiting MC (1954) Researches on acetylenic compounds. Part XLVII. The prototropic rearrangements of some acetylenic dicarboxylic acids. J Chemical Soc (Resumed):3208. https://doi.org/10.1039/JR9540003208
22. Staudinger H, Ruzicka L (1924) Insektentötende Stoffe I. Über Isolierung und Konstitution des wirksamen Teiles des dalmatinischen Insektenpulvers Helvetica Chimica Acta 7:177. https://doi.org/10.1002/hlca.19240070124
23. Brand K (1921) Über Untersuchungen in der Tetraarylbutan-Reihe und über das 1.1 4.4-Tetraphenyl-butatrien. (4. Mitteilung über die Reduktion organischer Halogen-verbindungen.) Berichte der deutschen chemischen Gesellschaft (A and B Series) 54:1987. https://doi.org/10.1002/cber.19210540828
24. Singh L (1957) A new electronic structure of allene Naturwissenschaften 44:536. https://doi.org/10.1007/BF00631411
25. Maitland P, Mills WH (1935) Experimental demonstration of the Allene asymmetry. Nature 135:994. https://doi.org/10.1038/135994a0
26. Kuhn R, Blum D (1959) Über Kumulene, X.cis-trans-Isomerie bei Dinitro-tetraphenylkumulenen Chemische Berichte 92:1483. https://doi.org/10.1002/cber.19590920702
27. Zang Y, Zou Q, Fu T, Ng F, Fowler B, Yang J, Li H, Steigerwald ML, Nuckolls C, Venkataraman L (2019) Directing isomerization reactions of cumulenes with electric fields. Nat Commun 10:4482. https://doi.org/10.1038/s41467-019-12487-w
28. Zanda M, Bucci R, Sloan NL, Topping L (2020) Highly strained unsaturated carbocycles. Eur J Org Chem 2020:5278. https://doi.org/10.1002/ejoc.202000512
29. Szeimies G (1981) Pyramidalized bridgehead olefins as reactive intermediates. Chimia 35:243
30. Warner PM (1989) Strained bridgehead double bonds. Chem Rev 89:1067. https://doi.org/10.1021/cr00095a007
31. Johnson RP, DiRico KJ (1995) Ab initio conformational analysis of trans-cyclohexene. J Org Chem 60:1074. https://doi.org/10.1021/jo00109a047
32. Das P, DeSpain M, Ethridge A, Weaver JD III (2023) Exploiting visible light triggered formation of trans-cyclohexene for the contra-thermodynamic protection of alcohols. Org Lett 25:7316. https://doi.org/10.1021/acs.orglett.3c02666
33. Bradley AZ, Kociolek MG, Johnson RP (2000) Conformational selectivity in the diels-alder cycloaddition: predictions for reactions of s-trans-1,3-Butadiene. J Org Chem 65:7134. https://doi.org/10.1021/jo000916o
34. Maurer DP, Fan R, Thamattoor DM (2017) Photochemical Generation of Strained Cycloalkynes from Methylenecyclopropanes. Angew Chem Int Ed 56:4499. https://doi.org/10.1002/anie.201701275
35. Johnson RP, Daoust KJ (1995) Interconversions of cyclobutyne, cyclopentyne, cyclohexyne, and their corresponding cycloalkylidenecarbenes. J Am Chem Soc 117:362. https://doi.org/10.1021/ja00106a040
36. Wiberg KB, Marquez M (1998) Effect of fluorine substitution on the energies of small ring compounds. J Am Chem Soc 120:2932
37. Daoust KJ, Hernandez SM, Konrad KM, Mackie ID, Winstanley J Jr, Johnson RP (2006) Strain estimates for small-ring cyclic Allenes and Butatrienes. J Org Chem 71:5708. https://doi.org/10.1021/jo060698k
38. Daoust KJ, Johnson RP unpublished results
39. Bredt J (1924) Sterical obstruction in pontine rings (Bredt rule) and the meso-trans position in condensed ring systems of hexamethylene. Justus Liebigs Annalen Der Chemie 437:1
40. McDermott L, Walters ZG, French SA, Clark AM, Ding J, Kelleghan AV, Houk KN, Garg NK (2024) A solution to the anti-Bredt olefin synthesis problem. Science 386:eadq3519. https://doi.org/10.1126/science.adq3519

41. Johnson RP (1989) Strained cyclic cumulenes. Chem Rev 89:1111. https://doi.org/10.1021/cr00095a009
42. Balci M, Taskesenligil Y (2000) Recent developments in strained cyclic allenes. Adv Strained Interesting Org Mol 8:43
43. Christl M (2004) Cyclic Allenes up to seven-membered rings. In: Krause N, Hashmi ASK (eds) Modern Allene chemistry. https://doi.org/10.1002/9783527619573.ch6
44. Favorskii AE (1936) Triple bond in carbon cycles and the probable structure of the simplest cyclic hydrocarbons of the composition CnH2n-4 Zh. Obshch Khim 6:720
45. Wittig G, Fritze P (1966) Intermediate occurrence of 1,2 cyclohexadiene. Angewandte Chemie 78:905
46. Moore WR, Moser WR (1970) Reaction of 6,6-dibromobicyclo[3.1.0]hexane with methyllithium. Efficient trapping of 1,2-cyclohexadiene by styrene. J Org Chem 35:908
47. Sydnes LK (2003) Allenes from cyclopropanes and their use in organic synthesis-recent developments. Chem Rev (Washington, DC, U. S.) 103:1133. https://doi.org/10.1021/cr010025w
48. Klett MW, Johnson RP (1983) Cumulene photochemistry: photoaddition of neutral methanol to phenylallenes. Tetrahedron Lett 24:1107. https://doi.org/10.1016/s0040-4039(00)86377-7
49. Dillon PW, Underwood GR (1974) Cyclic allenes. I. Electronic structure and probable deformation of the allene linkage when included in a ring. INDO-MO [intermediate neglect of differential overlap-molecular orbital] study. J Amer Chem Soc 96:779. https://doi.org/10.1021/ja00810a023
50. Balci M, Jones WM (1980) Chirality as a probe for the structure of 1,2-cycloheptadiene and 1,2-cyclohexadiene. J Am Chem Soc 102:7607. https://doi.org/10.1021/ja00545a057
51. Johnson RP (1985) Cumulene photochemistry. Org Photochem 7:75
52. Schmidt MW, Angus RO, Johnson RP (1982) Small ring cyclic allenes: an ab initio study of the structure of 1,2-cyclohexadiene. J Am Chem Soc 104:6838. https://doi.org/10.1021/ja00388a087
53. Angus RO Jr, Schmidt MW, Johnson RP (1985) Small-ring cyclic cumulenes: theoretical studies of the structure and barrier to inversion in cyclic allenes. J Am Chem Soc 107:532. https://doi.org/10.1021/ja00289a002
54. Christl M, Engels B (2009) Stable five-membered-ring allenes with second-row elements only: not allenes, but zwitterions. Angew Chem Int Ed 48:1538. https://doi.org/10.1002/anie.200803476
55. Haenninen MM, Peuronen A, Tuononen HM (2009) Do extremely bent Allenes exist? Chem Eur J 15:7287. https://doi.org/10.1002/chem.200900928
56. Spartan'24 (2024) Version 1.1.0, Mar 11 2024. Copyright 1991–2024 by Wavefunction Inc., Irvine, CA
57. Skattebol L (1961) Allenes from gem-dihalocyclopropane derivatives and alkyllithium. Tetrahedron Lett 167. https://doi.org/10.1016/s0040-4039(01)84059-4
58. Ball WJ, Landor SR (1961) The synthesis of cyclic allenes. Proc Chem Soc London 143
59. Marquis ET, Gardner PD (1966) 1,2-Cyclooctadiene, a product from the reaction of 8,8-dibromobicyclo[5.1.0]octane with methyllithium. Tetrahedron Lett 2793
60. Shakespeare WC, Johnson RP (1990) 1,2,3-cyclohexatriene and cyclohexen-3-yne: two new highly strained C6H6 isomers. J Am Chem Soc 112:8578. https://doi.org/10.1021/ja00179a050
61. Anthony SM, Wonilowicz LG, McVeigh MS, Garg NK (2021) Leveraging fleeting strained intermediates to access complex scaffolds. JACS Au 1:897. https://doi.org/10.1021/jacsau.1c00214
62. Ippoliti FM, Wonilowicz LG, Adamson NJ, Darzi ER, Donaldson JS, Nasrallah DJ, Mehta MM, Kelleghan AV, Houk KN, Garg NK (2024) Total synthesis of lissodendoric acid a angew. chem. Int Ed 63:e202406676. https://doi.org/10.1002/anie.202406676
63. Wonilowicz LG, Mehta MM, Kamecke LL, French SA, Garg NK (2024) Diels-alder cycloadditions of oxacyclic allenes and α-pyrones org. Lett 26:6465. https://doi.org/10.1021/acs.orglett.4c02294
64. Meier H, Koenig P (1986) Photochemistry of strained cycloalkynes Nouv. J Chim 10:437

65. Zheng M, DiRico KJ, Kirchhoff MM, Phillips KM, Cuff LM, Johnson RP (1993) Photorearrangements of acyclic conjugated enynes: a photochemical analog of the Bergman rearrangement. J Am Chem Soc 115:12167. https://doi.org/10.1021/ja00078a066
66. Skraba-Joiner SL, Johnson RP, Agarwal J (2015) Dehydropericyclic reactions: symmetry-controlled routes to strained reactive intermediates. J Org Chem 80:11779. https://doi.org/10.1021/acs.joc.5b01488
67. Johnson RP (2010) Dehydropericyclic routes to reactive intermediates. J Phys Org Chem 23:283. https://doi.org/10.1002/poc.1656
68. Burrell RC, Daoust KJ, Bradley AZ, DiRico KJ, Johnson RP (1996) Strained cyclic cumulene intermediates in diels-alder cycloadditions of enynes and diynes. J Am Chem Soc 118:4218. https://doi.org/10.1021/ja9601144
69. Ajaz A, Bradley AZ, Burrell RC, Li WHH, Daoust KJ, Bovee LB, DiRico KJ, Johnson RP (2011) Concerted vs stepwise mechanisms in dehydro-diels-alder reactions. J Org Chem 76:9320. https://doi.org/10.1021/jo201567d
70. Bradley AZ, Johnson RP (1997) Thermolysis of 1,3,8-nonatriyne: evidence for intramolecular [2 + 4] cycloaromatization to a benzyne intermediate. J Am Chem Soc 119:9917. https://doi.org/10.1021/ja972141f
71. Miyawaki K, Suzuki R, Kawano T, Ueda I (1997) Cycloaromatization of a non-conjugated polyenyne system: synthesis of 5H-benzo[d]fluoreno[3,2-b]pyrans via diradicals generated from 1-[2-{4-(2-alkoxymethylphenyl)butan-1,3-diynyl}]phenylpentan-2,4-diyn-1-ols and trapping evidence for the 1,2-didehydrobenzene diradical. Tetrahedron Lett 38:3943. https://doi.org/10.1016/s0040-4039(97)00785-5
72. Hoye TR, Baire B, Niu D, Willoughby PH, Woods BP (2012) The hexadehydro-Diels–Alder reaction. Nature 490:208. https://doi.org/10.1038/nature11518
73. Fluegel LL, Hoye TR (2021) Hexadehydro-diels–alder reaction: benzyne generation via cycloisomerization of tethered triynes. Chem Rev 121:2413. https://doi.org/10.1021/acs.chemrev.0c00825
74. Moore WR, Ozretich TM (1967) Cyclic cumulenes. The synthesis and some reactions of 1,2,3-cyclodecatriene. Tetrahedron Lett 8:3205. https://doi.org/10.1016/S0040-4039(01)89900-7
75. Angus RO Jr, Janakiraman MN, Jacobson RA, Johnson RP (1987) Small-ring cyclic cumulenes: synthesis and x-ray crystal structure of bis(triphenylphosphine)chloro(η2–1,2,3-cyclononatriene)rhodium. Organometallics 6:1909. https://doi.org/10.1021/om00152a013
76. Hernandez S, Kirchhoff MM, Swartz SG Jr, Johnson RP (1996) 1,2,3-Cyclooctatriene. Tetrahedron Lett 37:4907. https://doi.org/10.1016/0040-4039(96)01007-6
77. Zoch H-G, Szeimies G, Römer R, Schmitt R (1981) 1,2,3-Cycloheptatriene by Isomerization of Tricyclo[4.1.0.02,7]-hept-1(7)-ene. Angewandte Chemie Int Edition English 20:877. https://doi.org/10.1002/anie.198108771
78. Shi J, Li L, Li Y (2021) o-Silylaryl Triflates: A Journey of Kobayashi Aryne Precursors. Chem Rev 121:3892. https://doi.org/10.1021/acs.chemrev.0c01011
79. Kelleghan AV, Bulger AS, Witkowski DC, Garg NK (2023) Strain-promoted reactions of 1,2,3-cyclohexatriene and its derivatives. Nature (London, U. K.) 618:748. https://doi.org/10.1038/s41586-023-06075-8
80. Hernandez SM, Johnson RP unpublished results

Exploring Variation in Bonding and Structure for Tetravalent Carbon: [1.1.1]Propellane and Inverted Carbon

11

Kathleen M. Morgan and Kenneth B. Wiberg

Abstract

In 1874, van 't Hoff theorized that tetravalent carbon should have bonds oriented to the corners of a tetrahedron. While this is generally true, [1.1.1]propellane provides a notable exception. The synthesis and study of this remarkable compound, with comparisons to other unusual molecules including bicyclo[1.1.0]butane, will be discussed from an historical perspective.

11.1 Introduction

In 1874 at age 22, van 't Hoff proposed the theory of tetrahedral carbon, in which a tetravalent carbon should form bonds that are oriented to the corners of a tetrahedron, rather than a square planar array [1]. Van 't Hoff applied this theory to explain the number of isomers that exist for tetravalent carbon bonded to two, three and four different groups, as well as optical activity for tetravalent carbons with four unique groups. The concept of tetrahedral carbon led to major advances in our understanding of structure and bonding, especially for organic compounds. Since that time, chemists have been interested in testing the limits of carbon structure, and the influence on reactivity, through study of unusual molecules.

This chapter describes several fascinating molecules where tetravalent carbon is not tetrahedral. The examples focus on molecules studied over the years in the Wiberg group, including compounds such as bicyclobutane and [1.1.1]propellane, which

K. M. Morgan (✉)
Department of Chemistry, Xavier University of Louisiana, New Orleans, LA, USA
e-mail: kmmorgan@xula.edu

K. B. Wiberg
Department of Chemistry, Yale University, New Haven, CT, USA
e-mail: kenneth.wiberg@yale.edu

A. Greenberg and D. E. Lewis (eds.), *The Birth of the 3rd Dimension in Chemistry*, Perspectives on the History of Chemistry,
https://doi.org/10.1007/978-3-031-97743-5_11

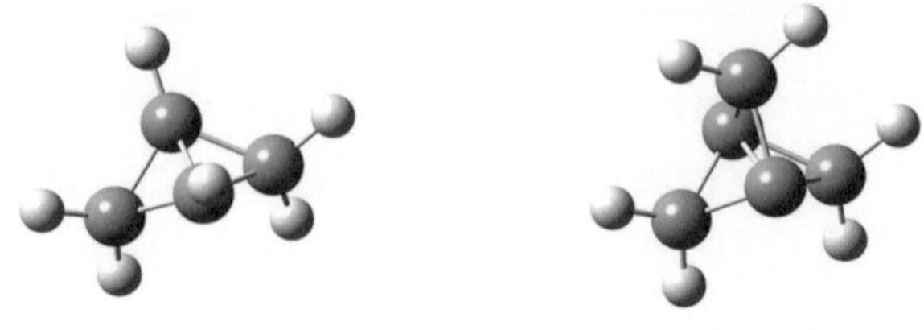

Fig. 11.1 Bicyclobutane and [1.1.1]propellane

have "inverted" bridgehead carbons (Fig. 11.1). The inverted carbons are tetravalent, but all four bonds are attached to the same hemisphere of the carbon [2].

11.2 Bicyclobutanes

The synthesis of a substituted bicyclobutane was first reported in 1913 by Thorpe, a prominent English chemist, and student Beesley, as an intermediate toward the synthesis of tetrahedrane (Fig. 11.2) [3].

Years later, Wiberg read this report as a graduate student and was intrigued by this compound. One of his first experiments as a new faculty member at the University of Washington was to try to reproduce the synthesis and study the bicyclobutane. Unfortunately, the synthesis was not successful. R. B. Woodward was also interested in synthesizing this bicyclobutane, and was also unable to make it using Thorpe's procedures. In 1950, Woodward successfully contacted Beesley, who was then working at an oil company in Mexico; by that time, Thorpe had died. Beesley was completely unaware of this publication or this research! He acknowledged working with Thorpe around that time, but had been encouraged to leave the group and join the military. Woodward's correspondence with Beesley is included in the PhD thesis of Woodward's student Larson [4].

Despite the problems with the reported synthesis, there was an enduring idea brought forward in the early Thorpe papers: a discussion of "bent bonds". In most compounds, the electrons follow the bond line. Bent bonds occur when the maximum electron density between atoms does not follow the shortest line between the nuclei, and now known to be typical for cyclopropane [5]. Shortly after Thorpe's 1913 paper, Ingold leant support to the idea and extended the concept to tetrahedrane [6].

In 1959, Wiberg and Ciula reported the first reproducible synthesis of a substituted bicyclobutane (Fig. 11.3) [7].

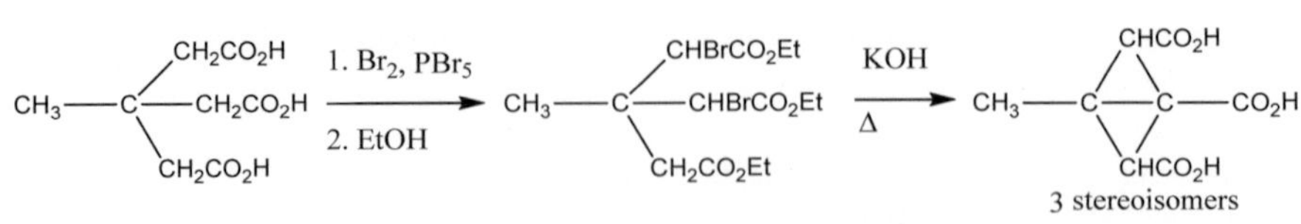

Fig. 11.2 Thorpe's published route to a substituted bicyclobutane [3]

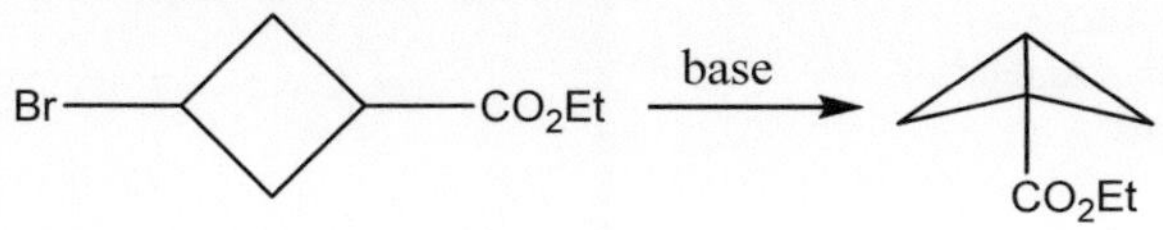

Fig. 11.3 First reproducible bicyclobutane synthesis [7]

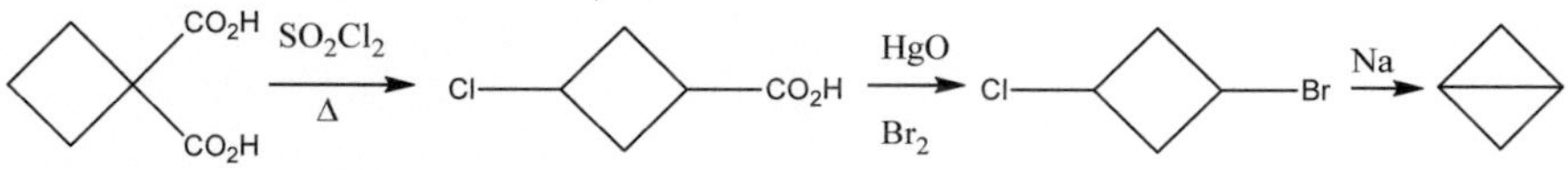

Fig. 11.4 Wiberg and Lampman's synthesis of bicyclobutane [10]

The key step is the cross-ring carbon–carbon bond formation where the anionic carbon displaces the bromine. The compound produced could be reduced via catalytic hydrogenation, which is more like an alkene reaction, yet proton NMR analysis confirmed that the molecule had no vinyl protons. The only structure consistent with the NMR spectrum is the bicyclobutane.

This bicyclobutane ester is not very stable, polymerizing over time to a compound similar to Lucite, via a radical reaction across the central bond. As an interesting aside, the US Air Force worked with DuPont to try to prepare a fabric for parachutes using a related polymer. The chemistry worked, but was too expensive to pursue [8].

In 1963, two syntheses of the parent bicyclobutane were reported. In the first, Lemal described a photochemically generated carbene that was used to create the ring system [9]. Shortly thereafter, Wiberg and Lampman reported a method more amenable to larger-scale synthesis, with a cross-ring bond formation similar to that used above (Fig. 11.4) [10]. The second procedure can easily generate 5–6 g of bicyclobutane and is included in *Organic Syntheses* [11].

The parent bicyclobutane is significantly more stable than the bicyclobutane ester, above. Fenoglio was therefore able to measure the heat of formation of bicyclobutane by flame calorimetry, allowing the strain energy to be determined: 66.7 kcal/mol, more than double that for cyclopropane [12].

In 1997, Yale held a symposium honoring Wiberg, and a poster for the event is included as Fig. 11.5 [14]. The drawing is by Harry Wasserman, another organic chemist at Yale, and shows Wiberg looking at his bicyclobutane butterflies.

Bicyclobutane has received much subsequent study, as summarized in a 2022 review [14]. Bicyclobutane has a butterfly-like geometry, with three different types of C–H bonds having three different lengths, and two types of C–C bonds, both about 1.50 Å. The central bond is clearly unusual. It has been described as a bent bond of overlapping p-type orbitals, or the overlap of the back lobes of hybrid orbitals, among others. The pKa of the bridgehead C–H is 37.9, closer to an alkyne than an alkane, and more acidic than cyclopropane (pKa 46). Bicyclobutane reacts

Fig. 11.5 Poster from Wiberg symposium (Author's personal collection)

with both nucleophiles and electrophiles, and it has become a building block for many useful compounds.

11.3 Propellanes

Propellanes are tricyclic compounds where the three rings share a central bond. Nomenclature of propellanes is like that for bicyclic compounds, where the lengths of the bridges that connect the bridgehead carbons are listed in brackets. The structure and stability of propellanes depend greatly on the sizes of the rings.

Knowing that the strain in bicyclobutane is greater than twice the strain of two cyclopropanes, it was reasonable to expect that the strain in small ring propellanes would be more than the sum of the strain of the isolated rings. The extreme example, [1.1.1]propellane, with three fused cyclopropane rings would then be expected to have a strain energy greater than 3×28 kcal/mol, approaching the strength of a carbon–carbon single bond. Could such highly strained compounds exist? Eaton's successful synthesis of cubane in 1964, [15] with a strain energy over 180 kcal/mol, suggested that synthesis and study of small ring propellanes might be possible.

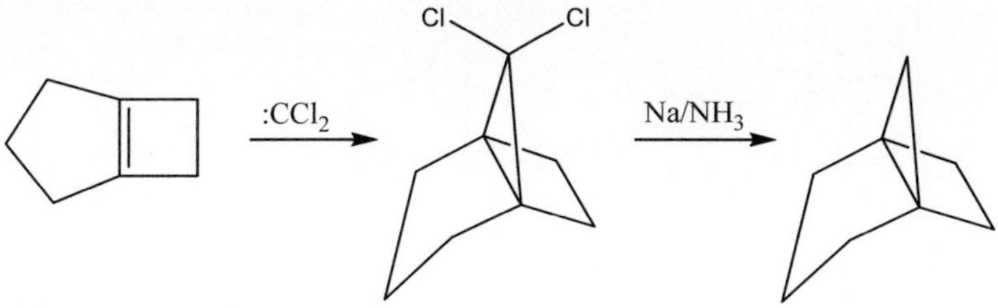

Fig. 11.6 Synthesis of [3.2.1]propellane [16]

The first of several propellanes studied in the Wiberg group was [3.2.1]propellane. This was synthesized in 1968 by Burgmaier via addition of dichlorocarbene to a bicycloalkene. Subsequent dehalogenation gave the parent compound (Fig. 11.6) [16].

The hydrocarbon is remarkably stable when heated; the half-life is 20 h at 195 °C. This is in part due to the moderate strain in this compound, approximately 67 kcal/mol [17]. More important, there is no good way to break apart the ring system. Retro [2 + 2] cycloaddition of the cyclobutane ring is forbidden. If the diradical were generated, there is no reasonable degradation pathway. However, reaction with acetic acid across the central bond occurs rapidly.

In 1972, the crystal structure of the dichloro [3.2.1]propellane was obtained [18] and reveals that the bridgehead carbons are indeed unusual. The bridgehead carbons are very close to planar (0.093 Å out of plane). The C–C central bond is very long, 1.572 Å and the four interatomic vectors (traditional bonds) all lie in the same hemisphere of the carbon. By this analysis, the bridgehead carbons are considered inverted. CNDO/2 calculations were carried out on the experimentally determined X-ray crystal structure, and were used to calculate hybridization of the various carbons. Analysis of the electron density for the bonds around the bridgehead carbons suggests that they are even more bent than in cyclopropane, and the electron density around the bridgehead carbons is not inverted.

Decreasing the size of the rings will increase the distortion at the bridgehead carbons. The limit is [1.1.1]propellane, with three fused cyclopropanes, and predicted strain energy of 101.5 kcal/mol. Early HF/6-31G* calculations suggested that [1.1.1]propellane would be kinetically stable, as no low energy pathway to degradation could be identified. This prediction was not uniformly accepted [19].

The first propellanes with only small rings that were synthesized in the Wiberg group were [2.2.1]- and [2.1.1]propellane. To this end, in the early 1980s, Walker went to U. of Utah to collaborate with Michl and carried out a newly-developed technique where a diiodo substrate was treated with alkali metal vapor (Fig. 11.7). The [2.2.1]propellane was synthesized using this method. The products were trapped in an argon matrix, allowing the IR spectrum to be obtained [20].

The [2.2.1]propellane was reasonably stable in the matrix, but oligomerizes/polymerizes on warming. It also reacts across the central bond with Br_2 added into the matrix. The strain release with central bond breaking is about 85 kcal/

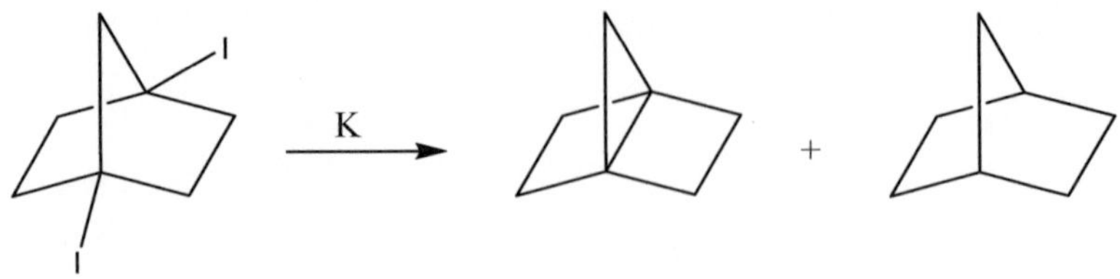

Fig. 11.7 Synthesis of [2.2.1]propellane [20]

mol, and the bond dissociation energy of the central bond was estimated to be 20 kcal/mol.

However, when the [2.1.1]propellane was synthesized in the same manner as the [2.2.1]propellane, it was found to be less stable, degrading even in the argon matrix [21]. The weak point in the structure is the four-membered ring. In this system, strain relief is an important driving force for degradation.

Back at Yale, Walker ran one final experiment in hopes of synthesizing [1.1.1]propellane. The synthesis started with a dibromobicyclopentane that was already available in the lab and used the halogen–metal exchange reaction that was successful previously (Fig. 11.8) [22].

The reaction mixture was analyzed by gas chromatography, and one product was obtained. The proton NMR spectrum shows just one singlet at 2.06 ppm, and two signals appear in the ^{13}C NMR spectrum. The [1.1.1]propellane was found to be the most stable of all the propellanes that have just 3- and 4-membered rings. In agreement with early calculations, thermal degradation is remarkably slow; rearrangement occurs in the gas phase at 114 °C, with half-life of 5 min. Subsequent experiments give the C–C central bond length as 1.594 Å (electron diffraction) or 1.60 Å (X-ray diffraction).

A final example in the propellane series studied in the Wiberg group includes a platinum bridge and was synthesized from bicyclo[2.2.0]hexene by Jason in 1974 (Fig. 11.9) [23].

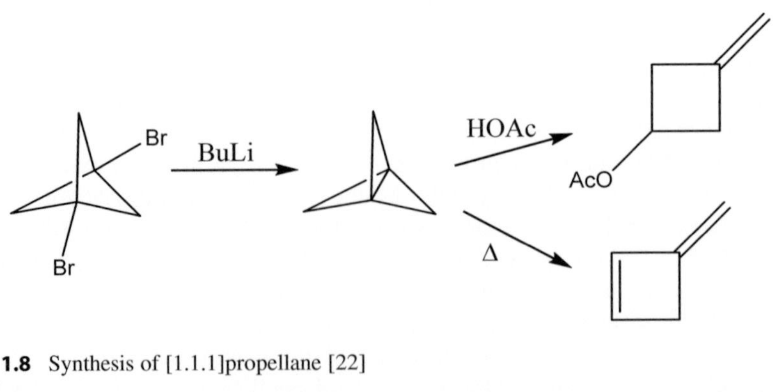

Fig. 11.8 Synthesis of [1.1.1]propellane [22]

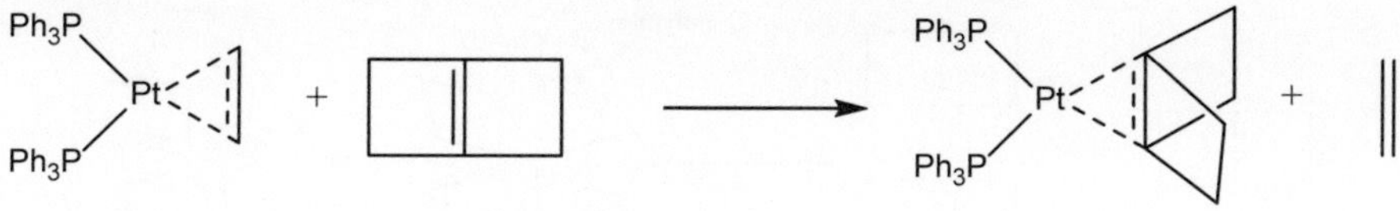

Fig. 11.9 Synthesis of a platinum [2.2.1]propellane

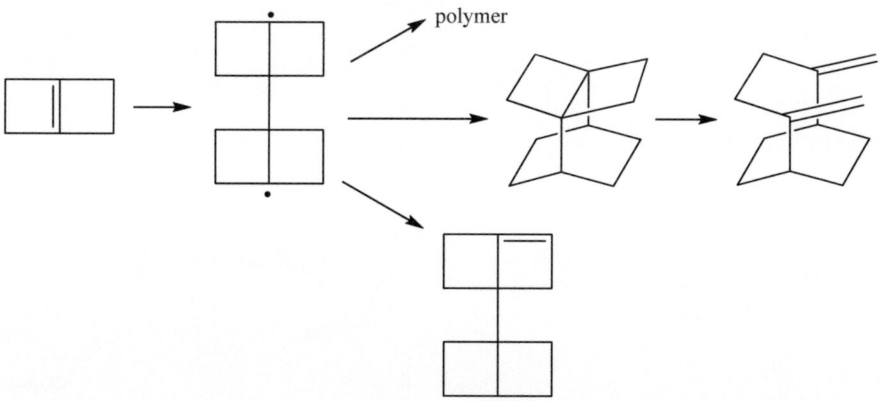

Fig. 11.10 Reactions of bicyclo[2.2.0]hexene

This hetero [2.2.1]propellane is easily prepared and significantly more stable than the hydrocarbon analog. Reaction with ethanol occurs across the central C–C bond of the propellane, rather than at the platinum, which is unexpected for a platinum compound. Formation of the heteropropellane is an effective method to separate bicyclo[2.2.0]hexene from isomeric 1,2-dimethylenecyclobutane, and to store the bicycloalkene via reversing the complexation.

The bicyclo[2.2.0]hexene itself exhibits some unusual chemistry (Fig. 11.10) [24]. Notably, there is evidence for the formation of a pentapropellane. This is thought to occur via formation of the retro 2 + 2 cycloaddition product, with a half-life of 28 min at room temperature. This reaction is consistent with that of [2.2.2]propellane studied by Eaton [25].

11.4 Fenestranes

Another highly distorted series of compounds are the fenestranes, where rings are fused together like panes in a window. Some examples are shown in Fig. 11.11; these and other strained compounds, many discussed above, were included in a 2020 computational study by Wiberg and Rablen [26].

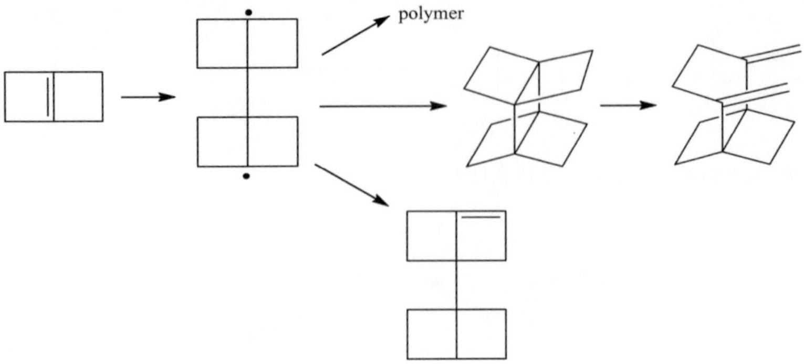

Fig. 11.11 Examples of fenestranes

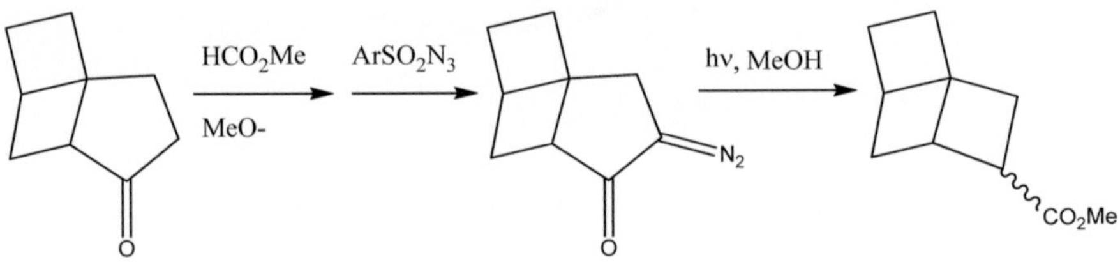

Fig. 11.12 Ring contraction toward synthesis of the third ring in windowpane [28]

[4.4.4.4]fenestrane on the left, also called windowpane, has not yet been synthesized. The carbon at the center of this compound would be very unusual and approaching planarity. The calculated strain energy is 166.3 kcal/mol, significantly higher than the [4.4.4.5] homologue, with strain energy of 105.7 kcal/mol. To synthesize windowpane, you could imagine starting with the [4.4.4.5] ketone, and carrying out a ring contraction. Whether this is possible or not remains to be seen, especially given the significant increase in strain.

In 1980, Wiberg and Olli reported the synthesis of a molecule having three of the four rings in windowpane, with the third ring created by a ring contraction of a diazoketone (Fig. 11.12) [27]. Subsequent reactions were carried out to remove the ester, affording the parent hydrocarbon.

An X-ray crystal structure was obtained on the *p*-bromoanilide derivative related to the ring-contracted ester. The central ring is planar, and the outer rings are slightly puckered, much less than cyclobutane itself. The C–C–C bond angles around the central carbon are significantly distorted from tetrahedral, with one set ranging from 132.5 to 124.4° and the other set clustered around 90°.

11.5 Conclusion

There are many interesting molecules with tetravalent carbons that are not tetrahedral. Research over the years in the Wiberg group has focused on many examples of such molecules. A complex interplay of factors influences the behavior of these compounds. There are still challenges out there, especially with regard to modeling the bonding in these unusual compounds.

Acknowledgements KBW thanks the National Science Foundation for research support over the years. KMM thanks the Keller Family Foundation for supporting participation in the symposium.

References

1. van 't Hoff JH (1874) Archives neerlandaises des sciences exactes et naturelles 9:445–454
2. Wiberg KB (1984) Inverted geometries at carbon. Acc Chem Res 17:397–386
3. Beesley RM, Thorpe JF (1913) A new series of ring compounds. Proc Chem Soc 29:346–347; Beesley RM, Thorpe JF (1920) LXIV The formation and stability of associated alicyclic systems. Part I. A system of nomenclature, and some derivatives of Methane-II-cyclopropane and of Methane-III-cyclopropane. J Chem Soc Trans 117:591–620
4. Larson HO, Woodward RB (1959) A reinvestigation of Thorpe's synthesis of bicyclobutane derivatives. Chem Ind 193; Larson HO (1950) I. A reinvestigation of Thorpe's synthesis of bicyclobutane derivatives. II. The synthesis of 3',8-Dimethyl-1,2-Cyclopentenophenanthrene, Ph.D. Thesis, Harvard, 1950. The letter from Beesley to Woodward is reproduced in the Larson thesis
5. Wiberg KB (1996) Bent bonds in organic compounds. Acc Chem Res 29:229–234
6. Ingold CK (1920) Note included in Beesley RM, Thorpe JF LXIV The formation and stability of associated alicyclic systems. Part I. A system of nomenclature, and some derivatives of Methane-II-cyclopropane and of Methane-III-cyclopropane. J Chem Soc Trans 117:603–610
7. Wiberg KB, Ciula RP (1959) Ethyl Bicyclo[1.1.0]butane-1-carboxylate. J Am Chem Soc 81:5261–5262
8. Wiberg KB (2021) My life
9. Lemal DM, Menger F, Clark GW (1963) Bicyclobutane. J Am Chem Soc 85:2529
10. Wiberg KB, Lampman GM (1963) Bicyclo[1.1.0]butane. Tetrahedron Lett 2173–2175
11. Lampman GM, Aumiller JC (1971) Bicyclo[1.1.0]butane. Org Synth 51:55
12. Wiberg KB, Fenoglio RA (1968) Heats of formation of C_4H_6 hydrocarbons. J Am Chem Soc 90:3395–3397
13. Personal collection of the author
14. Kelly CB, Milligan JA, Tilley LJ, Sodano TM (2022) Bicyclobutanes: from curiosities to versatile reagents and covalent warheads. Chem Soc 13:11721
15. Eaton PE, Cole TW (1964) Cubane. J Am Chem Soc 86:3157–3158
16. Wiberg KB, Burgmaier GJ (1969) Tricyclo[3.2.1.$0^{1,5}$]octane. Tetrahedron Lett 317–319
17. Wiberg KB (1986) The concept of strain in organic chemistry. Angew Chem Int Ed Engl 25:312–322
18. Wiberg KB, Burgmaier GJ, Shen K, LaPlaca SJ, Hamilton WC, Newton MD (1972) "Inverted" tetrahedral geometry at a bridgehead carbon. The X-Ray crystal, molecular, and electronic structure of 8,8-Dichlorotricyclo[3.2.1.$0^{1,5}$]octane ($C_8H_{10}Cl_2$) at −40 °C. J Am Chem Soc 94:7402
19. Newton MD, Schulman JM (1972) Theoretical studies of tricyclo[1.1.1.01,3]pentane and bicyclo[1.1.1]pentane. J Am Chem Soc 94:773–778; Liebman JF, Greenberg A (1976) A survey of strained organic molecules. Chem Rev 76:311–363

20. Walker FH, Wiberg KB, Michl J (1982) [2.2.1]Propellane. J Am Chem Soc 104:2057
21. Wiberg KB, Walker FH, Pratt WE, Michl J (1983) [2.1.1]Propellane. Reaction of 1,4-Diiodobicyclo[2.1.1]hexane with tert-butyllithium and with potassium atoms. J Am Chem Soc 105:3638–3641
22. Wiberg KB, Walker FH (1982) [1.1.1]Propellane. J Am Chem Soc 104:5239
23. Jason ME, McGinnety JA, Wiberg KB (1974) Preparation, structure, and reactions of an organometallic [2.2.1] Propellane, the Bis(triphenylphosphine)platinum complex of $D^{1,4}$-Bicyclo[2.2.0]hexene. J Am Chem Soc 96:6531–6532
24. Wiberg KB, Jason ME (1976) The thermal dimer and ketene cycloadduct of $D^{1,4}$-bicyclo[2.2.0]hexene. J Am Chem Soc 98:3393
25. Eaton PE, Temmee GH III (1973) The [2.2.2]Propellane system. J Am Chem Soc 95:7508–7510
26. Wiberg KB, Rablen PR (2020) Increase in strain energy during the conversion of [4.4.4.5]Fenestrane to [4.4.4.4]Fenestrane: a method for estimating the heats of formation of hydrocarbons and their derivatives from ab initio energies. J Org Chem 85:4981–4987
27. Wiberg KB, Olli LK, Golembeski N, Adams RD (1980) Tricyclo[4.2.0.01,4]octane. J Am Chem Soc 102:7467–7475

Expansion of the Definition of "Stereochemistry" to Include Quantum Chemistry. When "Funny" is Fun but not Funny

12

Jeffrey I. Seeman

Abstract

The typical definition of "stereochemistry" does not consider the centrality of quantum chemistry. Examples are provided herein involving stereochemical factors in the molecular orbital-based mechanism of pericyclic reactions—the Woodward–Hoffmann rules. Two molecular orbital tools used by Roald Hoffmann in the spring of 1965 to explain insightfully the stereospecificity of electrocyclizations and sigmatropic reactions are explicated. An especially insightful moment in Hoffmann's determination of the mechanism of sigmatropic reactions is discussed in detail. These analyses provide additional evidence that the seemingly simple determination of the orbital symmetry mechanism of pericyclic reactions from the perspective of 2024 was extraordinarily complicated from the perspective of 1965. The expansion of the definition of stereochemistry to include quantum chemistry is recommended.

12.1 "The 3rd Dimension in Chemistry"

Chemists today and for many decades have thought in terms of the four dimensions of chemistry, that is, the stereochemistry of their systems as a function of time. Indeed, the word "stereochemistry" is as such a central topic in chemist vocabulary that it is not even defined in the 2022 edition of "The Gold Book,"

Dedicated to the memory of my friends Ernest L. Eliel (1921–2008) and Samuel H. Wilen (1933–1994), experts par excellence in stereochemistry.

J. I. Seeman (✉)
Department of Chemistry, University of Richmond, Richmond, VA 23173, USA
e-mail: jseeman@richmond.edu

A. Greenberg and D. E. Lewis (eds.), *The Birth of the 3rd Dimension in Chemistry*,
Perspectives on the History of Chemistry,
https://doi.org/10.1007/978-3-031-97743-5_12

officially named "Glossary of Terms used in Physical Organic Chemistry (IUPAC Recommendations 2021)" [1]. Nonetheless, the term "stereochemistry" appears frequently in the "Glossary," and the related terms "stereoelectronic," "stereogenic center," "stereoisomers," "stereoselectivity," and "stereospecificity" are defined therein. This may be an example of "obliteration by incorporation," a wonderful concept provided by the sociologist of science, Robert K. Merton. The term "stereochemistry" has reached a level of understanding that, according to Charles L. Perrin, the corresponding author of the "Glossary," [1] wrote to me "no definition of stereochemistry per se was needed" [2].

Many definitions of "stereochemistry" appear in the literature. I shall report on two, one from what I refer to as from "The Bible of Stereochemistry" and the other from the global authority, *Wikipedia.* All of the literature definitions read somewhat alike:

> Chemistry in three dimensions, chemistry with consideration of its three dimensional aspects, but also used in relation to chemical and physical properties of cis-trans in alkenes.—The glossary in *Stereochemistry of Organic Compounds* [3] by Ernest L. Eliel and Samuel H. Wilen (1994)

and.

> Stereochemistry, a subdiscipline of chemistry, studies the spatial arrangement of atoms that form the structure of molecules and their manipulation. The study of stereochemistry focuses on the relationships between stereoisomers, which are defined as having the same molecular formula and sequence of bonded atoms (constitution) but differing in the geometric positioning of the atoms in space. For this reason, it is also known as 3D chemistry—the prefix 'stereo' means 'three-dimensionality.'—Wikipedia (2025)

What is lacking in the definition of stereochemistry is the explicit incorporation of quantum chemistry. Yes, it is true that these definitions speak to "chemistry in three dimensions" implying all of chemistry, of which quantum chemistry is foundational. But "all" is often a catchword when "all" actually means "some." Thus, Eliel and Wilen's bible states on page 2,

> Limitations of space [in their 1267-page monograph], plus the sheer vastness in which the subject of stereochemistry has grown, forces us to make choices in what will be discussed in this book... . Dynamic stereochemistry and references to calculations will be woven into the text, but reaction mechanisms will not be described in a systemic fashion. In particular, there will be no specific reference to electrocyclic reactions and orbital symmetry. While we regret these omissions... [3]

That quantum mechanics is a key feature in stereochemistry—obviously so in molecular orbital theory where orbital symmetry is controlling for pericyclic reactions—was given such unsolicitous dismissal by Eliel and Wilen is symptomatic of a real void in our thinking. (I admit, I did not sense that anything was missing in the definitions of stereochemistry during my many years of being a stereochemist.)

Indeed, the molecular orbital theory and valence bond theory inherent in stereochemistry, and the converse, were not explicitly discussed in the lectures in the symposium upon which this monograph is based and with one possible exception, and not in the chapters in this volume other than my own (Table 12.1).

12.2 Some Background on Molecular Orbital Theory as It Relates to Pericyclic Reactions and the Woodward–Hoffmann Rules

There is much involving stereochemistry embedded in molecular orbital theory. In this essay, I will focus on the stereochemistry of molecular orbitals, knowing full well that an exposition of the stereochemical consequences of valence bond theory can be treated likewise. I take my examples from the stereochemistry involved in the Woodward–Hoffmann rules and the quantum chemical mechanistic explanations for that stereochemistry. I have also utilized herein some of the discussion and figures I've presented in my own recent publications on the history of the Woodward–Hoffmann rules and the Conservation of Orbital Symmetry [4, 5].

I shall now provide you with a hint as to where we are heading. In due course, we will get to one very special page from Hoffmann's *Laboratory Notebook 16.* But first, I need to provide some background information about molecular orbital (MO) theory and specifically about MO theory of unsaturated systems like ethylene, 1,3-butadiene, allyl radical, and such. We will shortly focus on molecules that have at least one if not many carbon–carbon double bonds.

In my view, this story really starts with the reaction of two hydrogen atoms to form dihydrogen. When this reaction is considered in terms of simple MO theory, we can envision two 1*s* orbitals interacting with each other, in a bonding and antibonding fashion (Fig. 12.1). Bonding is a stabilization interaction and antibonding is destabilizing. Thus, from two atomic orbitals, two MOs are formed. If the two atomic orbitals each have one electron, following the interaction and insertion of the electrons (the last step in this sequence), a doubly-filled bonding MO and an unfilled antibonding MO result. A simple and useful way of thinking about the construction of the diagram in Fig. 12.1 is to "add" and "subtract" the AOs (or, if more complex systems, the two MOs) of the two interacting species. Addition leads to bonding; subtraction leads to antibonding. Also shown Fig. 12.1 is the generalization that the antibonding MO is destabilized more than the bonding MO stabilized relative to the nonbonding levels. In this example, the antibonding MO has no electrons, so that the potential destabilization is not realized. For an explanation of the greater destabilization than the stabilization in such systems, I point the interested reader to Hoffmann's 1971 article in *Accounts of Chemical Research* [6]. That early account is an excellent primer for qualitative perturbation theory that follows in this chapter.

Now, ethylene is the simplest molecule containing a carbon–carbon double bond. What does its π-MO look like? I learned a wonderfully simplistic and pedagogically endearing approach—at least to me—from my teacher, Roald Hoffmann

Table 12.1 Participants in the Symposium on "The Birth of the 3rd Dimension in Chemistry" and in the monograph based on that symposium along with the titles of their contributions

Authors[a]	Symposium title and chapter title
Daniel Rabinovich	"Stamps are Planar, Carbon is Not: A Philatelic History of Stereochemistry"
Nicholay V. Tsarevsky	"History of the Use of Models to Represent and Understand the Behavior of Molecules"
Pierre Laszlo	"Le Bel: Polymath Extraordinary"
Gregory S. Girolami	"The Tetrahedral Carbon Atom Four Years Before van't Hoff and Le Bel: The Life of Emanuele Paternò (1921–1932)"
David E. Lewis	"Missed it by That Much!' How Butlerov Almost Proposed the Tetrahedral Carbon"
David E. Lewis	"How Sweet it is! Emil Fischer's Stereochemical Studies of Glucose"
Gary Patterson and (in part[b]) Carmen Schmechel	"René Just Haüy and the Contribution of Crystallography to 3-Dimensional Chemistry"
Ian Bruno	"Disseminating Three-dimensional Chemistry Data and Knowledge Across Generations"
Seth C. Rasmussen	"Alfred Werner: Coordination theory and the Stereochemistry of Metal Complexes"
Bart Kahr and Miram Rossi	"Primo Levi's Undergraduate Thesis on Walden Inversion" *and* "Primo Levi's Undergraduate Thesis, University of Turin "L'Inversione di Walden (1941), Translation from Italian with Commentary"
Richard P. Johnson	"Historical Evolution, Stereochemistry, and Structural Limitations for the C = C Bond"
Kathleen M. Morgan and Kenneth B. Wiberg	"Exploring Variation in Bonding and Structure for Tetravalent Carbon: [1.1.1]Propellane and Inverted Carbon"
Jeffrey I. Seeman	"Role of Stereochemistry in the Quantum Chemical Revolution" *and* "Expansion of the Definition of 'Stereochemistry' to Include Quantum Chemistry. When 'Funny' is Fun but not Funny"
Arthur Greenberg	"Studies of Chiral Centers Other Than Carbon: William Jackson Pope and Kurt Mislow, Pioneers of Stereochemistry"
Peter Ramberg	"History of Stereochemistry: Progress and Prospects"
Marty Pomerantz, Isaiah M. Borkowski, Kay Edwards, Zeke Herman, Shayan Khan, and Joel F. Liebman[c]	"Although sp^3 Hybridization is Not Found for the Carbon in Methane nor in Isoelectronically Related Species, the Concept of Hybridization is Still Useful and Powerful"[c]

[a] Order as in the symposium. [b] Schmechel was a co-author of the symposium lecture but not of the monograph chapter. [c] Pomerantz et al. did not participate in the symposium but provided a chapter for the monograph

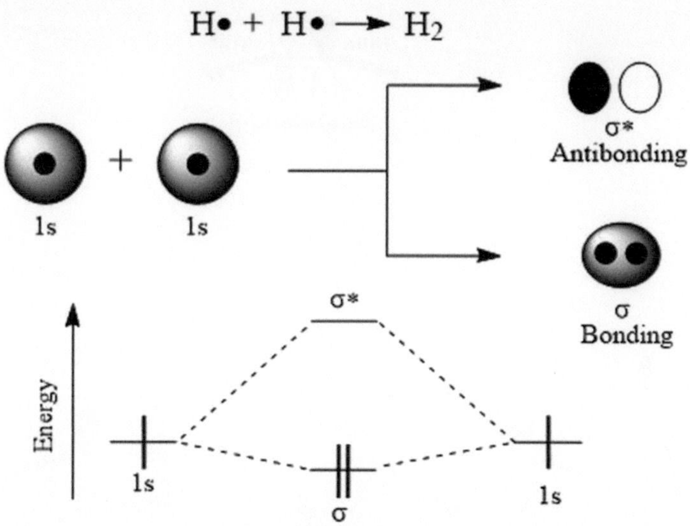

Fig. 12.1 Reaction of two hydrogen atoms to form dihydrogen as modelled by the interaction of two atomic orbitals forming the two MOs of dihydrogen. The black dots represent the nuclei, and the larger entities represent the electron clouds (i.e., contour surfaces corresponding to a particular probability of containing electrons; the difference between shaded and unshaded corresponds to a change in sign of the wavefunction). As shown in the top illustration, the two spherical structures represent the hydrogen atoms's 1*s* atomic orbitals which combine to form a bonding σ orbital and an antibonding σ* orbital, formed as described in the text. The arrows represent the spins of the electrons that have been added into the interaction diagram after its construction. Only two electrons can be placed in a MO, and they must have opposite spins. Adapted from Seeman and Tantillo [7] with permission from Springer Nature

[6, 8–10]. It appears in some of Hoffmann's early publications and very nicely in William L. Jorgenson and Lionel Salem's 1973 monograph *The Organic Chemist's Book on Orbitals*, [11] from which the following discussion of the MOs of ethylene was adapted. Consider the interaction shown in Fig. 12.2 of the *p* atomic orbitals of two carbon atoms each having two substituents and an electron in one *s*-orbital and one *p*-orbital in the depicted geometry. For simplicity, we shall focus only on the *p* atomic orbitals that will interact and form a π-bond. From these two atomic orbitals, two MOs are formed, one a doubly-occupied bonding π-orbital and the other an unoccupied π-antibonding orbital (Fig. 12.2).

Now comes the fun part. Let's construct the π MOs of 1,3-butadiene from two ethylene fragments. As shown in Fig. 12.3, the addition and subtraction of the two ethylene fragments having S symmetry along with the addition and subtraction of the two ethylene fragments having A symmetry result in the formation of the four new molecular orbitals. (Only MOs having the same symmetry can interact.) Fancy that! You start with four MOs, you must end with four MOs. These are the four MOs of 1,3-butadiene. As shown in the bottom of Fig. 12.3, the addition and subtraction of the MOs also provide an understanding as to the new bond between

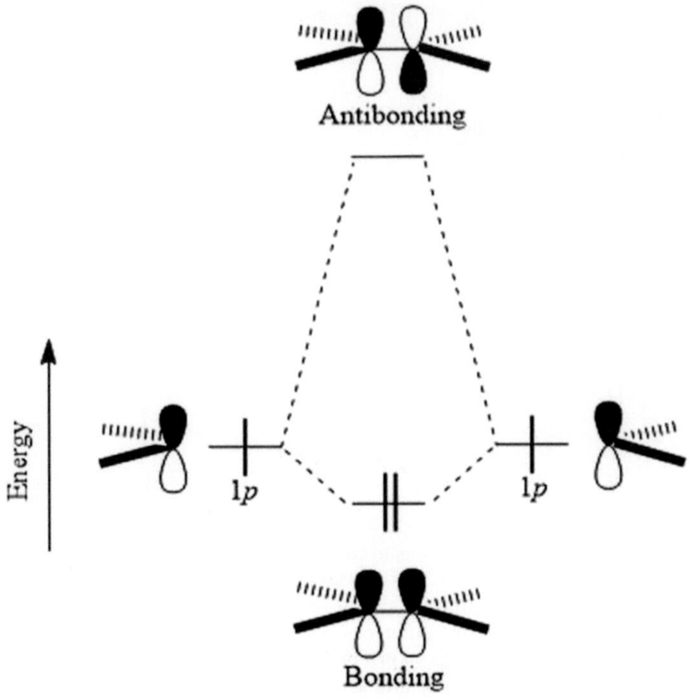

Fig. 12.2 Construction of the π-MOs of ethylene from the interaction of two carbon atoms, each having two substituents and one electron in a *p*-orbital and one in a *s*-orbital. In this interaction diagram, only the *p*-orbital interactions (mixings) are shown. In the formation of the π system, a bonding (sum) and an antibonding (difference) are taken from the individual fragments. Adapted from Jorgenson and Salem [11]

C(2) and C(3) in 1,3-butadiene and the increase in bonding in the first excited state of 1,3-butadiene, as electron density is pulled out of ψ_2 and is transferred into ψ_3.

A perusal of Figs. 12.1, 12.2 and 12.3 may suggest to some readers that there are several patterns evolving within the phases of these MOs. Figure 12.4 compares several polyenes and several polyenyl radicals. Among several patterns, one in particular is germane to the subject at hand: the relative phases of the α and ω atomic orbitals in the highest occupied molecular orbitals (HOMOs), the valence MOs, alternate within each series. Thus, in the sequence ethylene – 1,3-butadiene – 1,3,5-hexatriene, the relative phases of the α and ω atomic orbitals of the HOMOs are the same (S symmetry), opposite (A symmetry), and the same (symmetry). This pattern is emphasized by the red color in these atomic orbitals in Fig. 12.4. Had I displayed the MOs of 1,3,5,7-octatetraene, the relative phases of the α and ω atomic orbitals of the HOMO would have been the different (A symmetry). All this is due to the splitting of the interacting fragments into an in-phase (lower energy) and an out-of-phase (higher energy) pair of MOs. And this type of splitting of MOs can be traced all the way back to the simplest of MO interactions, those involving the interaction of two hydrogen atoms (Fig. 12.1)!

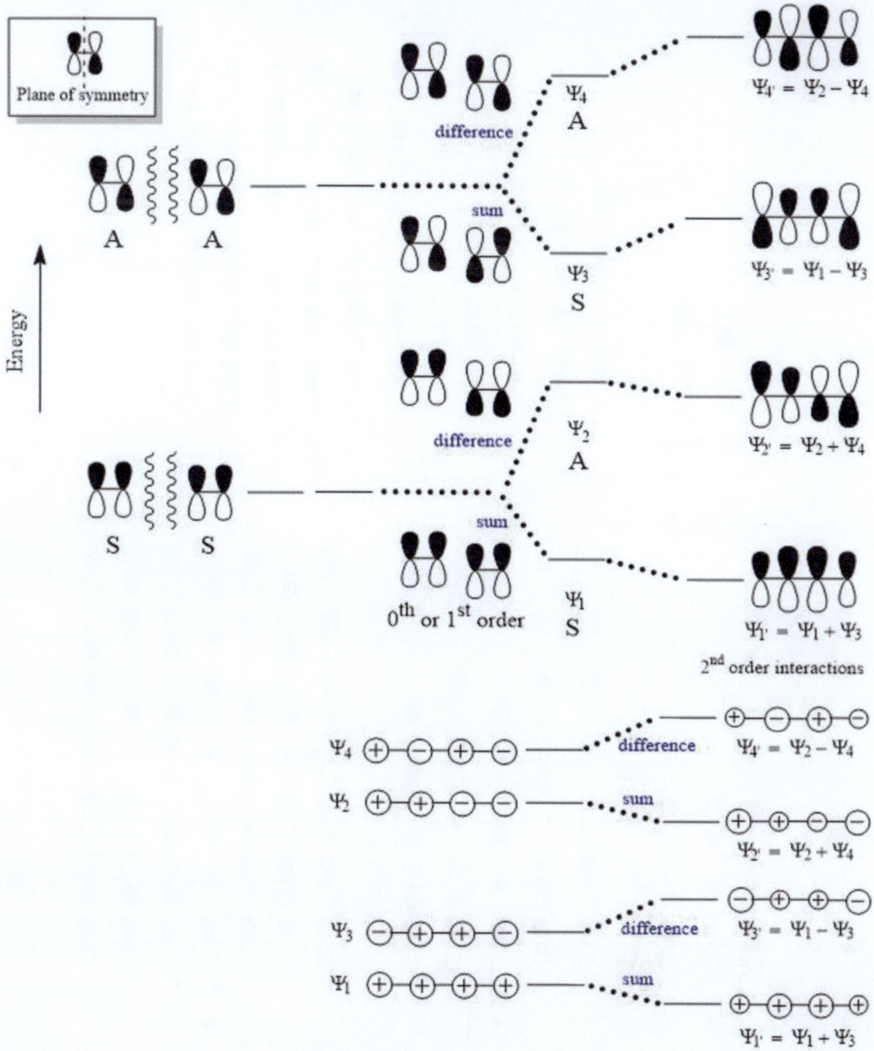

Fig. 12.3 (Top half) Construction of the π-MOs of 1,3-butadiene from the interaction of two ethylene fragments beginning at infinite distance apart and examined when they are interacting [12]. In this interaction diagram, the symmetry element is a plane of symmetry that is perpendicular to the plane containing the carbon atoms and bisects the middle carbon atoms. (Bottom half) The MOs of 1,3-butadiene are constructed by the sums and differences of the MOs having the same symmetry. Note that some of the atomic orbitals grow in size and others shrink in size, based on the second-order analyses at the bottom-right of this figure. In instances where the two coefficients of the interacting atomic orbitals have opposite signs, the growing-in or phasing-out is biased in favor of the lower-energy MO (if there are two orbitals)

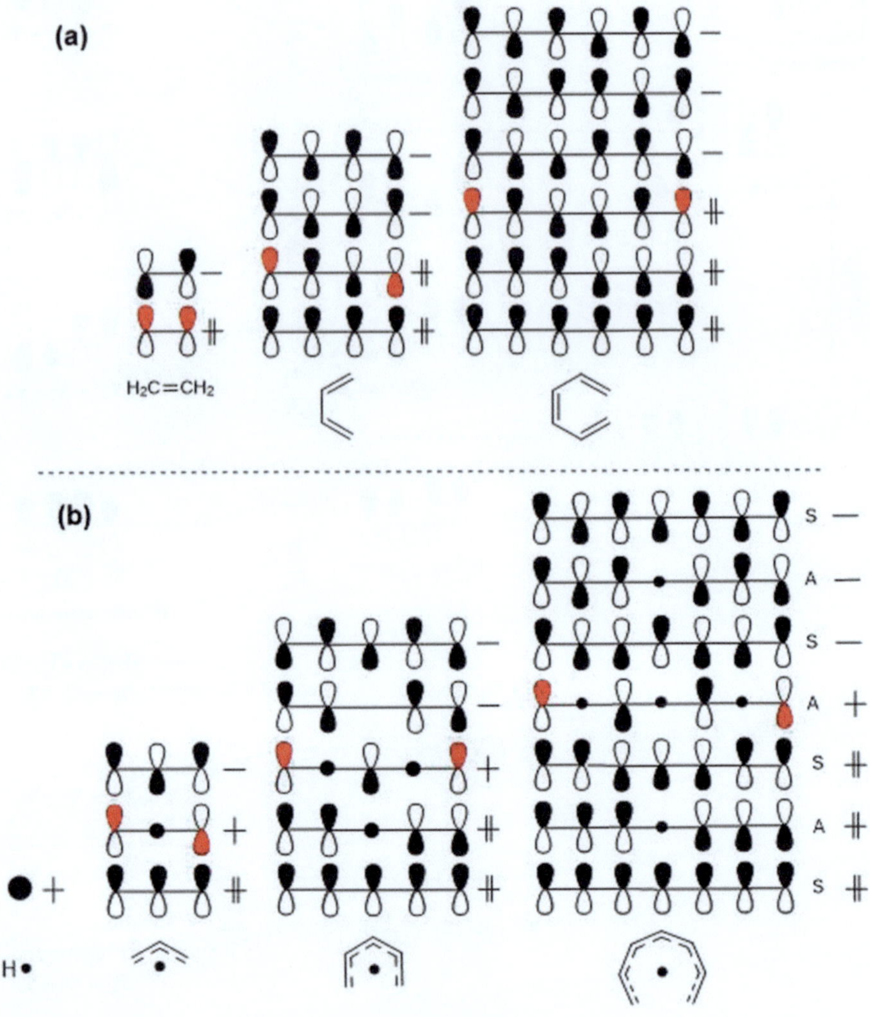

Fig. 12.4 MOs of (**a**) three olefins and (**b**) three olefinic radicals and a hydrogen atom. If one focuses on any one moiety, then the relative phases of the α and ω atomic orbitals alternate as one goes up or down in energy. Furthermore, if one focuses on the relative phases of the α and ω atomic orbitals in the highest occupied molecular orbitals in (**a**) (singly-occupied molecular orbitals (SOMOs) in the radicals in (**b**)), an alternating trend in the relative phases can be seen in each sequence: same, opposite, same, etc. in (**a**). This is emphasized by the red color of these atomic orbitals

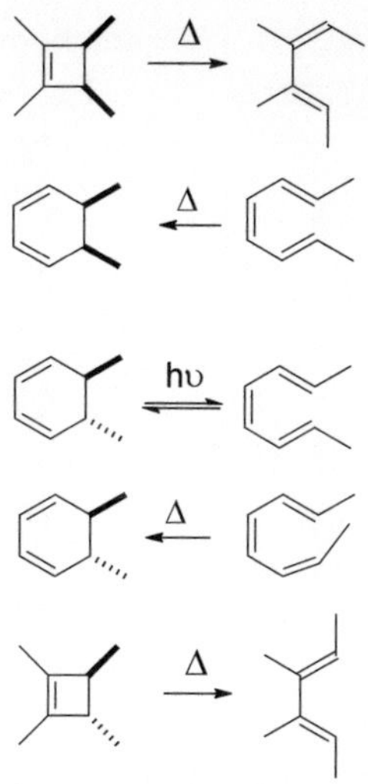

Fig. 12.5 Alternation in stereochemical properties in these stereospecific reactions. Note that the relative configurations change in going from thermal to photochemical or from four-electron to six-electron systems. These chemical reactions and related systems were reported by Gerhard J. Fonken, [27–30] Elliot Marvell, [31, 32] Rudolf Criegee [33, 34] and several studies published independently by Criegee's students, [34, 35] Emanuel Vogel, [36, 37] Rangaswamy Srinivasan, [38] and their colleagues in the early 1950s into the late 1960s. See Fig. 12.6 for the MO-based mechanism of electrocyclizations [4, 13, 16]

12.3 The Woodward–Hoffmann Rules and Electrocyclizations

As has been well known since the publication of the first three Woodward–Hoffmann (W–H) communications in *JACS* in 1965, [13–15] the alternation illustrated in Fig. 12.4 symbolizes the underlying quantum mechanical basis for the alternation found in all pericyclic reactions, i.e., electrocyclizations, cycloadditions, sigmatropic reactions, and several other minor cyclic concerted reactions and is reflected in Woodward and Hoffmann's generalized selection rules [8, 16–19]. By alternating chemistry, I mean thermal *versus* photochemical; and $4n$ *versus* $4n + 2$ electron systems, this being first generalized in terms of Hückel and Möbius systems by Howard E. Zimmerman [20–22] and also in terms of aromatic-like and antiaromatic-like transition states(TSs) by Michael J. S. Dewar [23–26]. Figure 12.5 illustrates the alternations in chemistry in electrocyclizations based on the alternations in the phases of the MOs shown in Fig. 12.4 (top) and in Fig. 12.6.

For those unfamiliar with the Woodward–Hoffmann rules, let us derive the MO explanation for electrocyclization of a four-electron system. Let us consider first the thermal cyclization embodied in Eq. 12.1c (see Fig. 12.6c) and the photochemical cyclization embodied in Eq. 12.1d (also in Fig. 12.6c). Bonding between two atoms can only occur when the phases of their incipient bonding atoms' atomic orbitals are of the same phase (see Fig. 12.2). Generally, the reactant's crucial

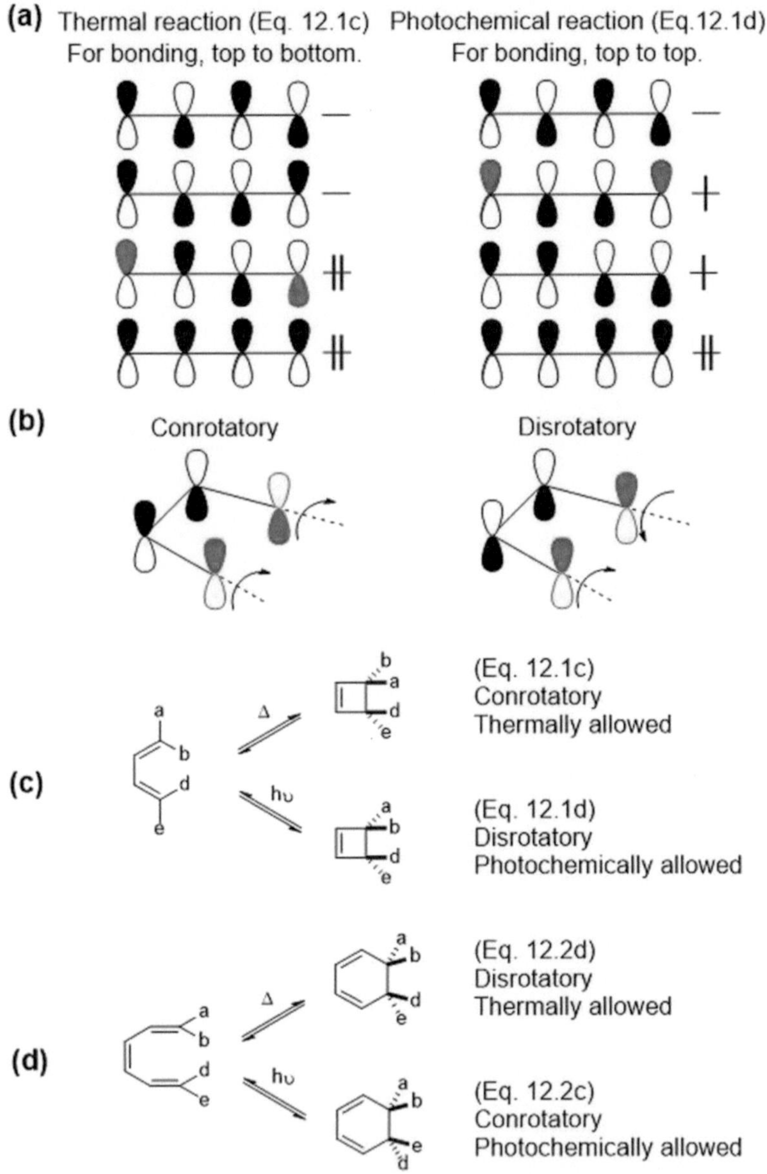

Fig. 12.6 Molecular orbital explanation for the stereospecificity observed in thermal and photochemical electrocyclizations [4, 13, 16]. See Fig. 12.5 for specific chemical examples. (**a**) The molecular orbitals of 1,3-butadiene in its ground state (left graphic) and in its first excited singlet state (right graphic). (**b**) Depiction of W–H allowed (or favored) conrotatory and disrotatory ring closures of 1,3-butadiene, ultimately to cyclobutene (not shown), via thermal and photochemical reactions, respectively. (**c**) and (**d**) Thermal and photochemical electrocyclizations involving four electrons and six electrons. (**c** and **d**) were adopted with permission from: Zhou, G. Kukier, I. Gordiy, R. Hoffmann, J. I. Seeman, K. N. Houk, "A 21st Century View of Allowed and Forbidden Electrocyclic Reactions," *J. Org. Chem.* **2024**, *89*, 1018–1034, https://doi.org/10.1021/acs.joc.3c02103, https://doi.org/https://doi.org/10.1021/acs.joc.3c02103

molecular orbital is its HOMO, that being its most reactive, valence MO. As shown in Fig. 12.6a-left, for a thermal cyclic, concerted four-electron system, the top of *p*-atomic orbital at one end of the HOMO (in green font) must bond with the bottom of the *p*-atomic orbital at the other end of the HOMO (also in green font). With this stereochemical feature, the phases are the same: green to green or white to white (or in Fig. 12.6b, green to green). In Fig. 12.6b, in order for the green portion of the α-orbital to bond with the green portion of the ω-orbital, the ends of the 1,3-butadiene molecule must rotate in the same direction. That is what Woodward and Hoffmann called a "conrotatory" motion.

Let us next consider the photochemical ring closure of 1,3-butadiene (Fig. 12.6c, Eq. 12.1d). Consideration of the MOs of the first excited singlet state of 1,3-butadiene (Fig. 12.6a—right). In order to achieve bonding between the two end carbons in Eq. 12.1d, a disrotatory motion (Fig. 6b—right) is required. Careful, stepwise, consistent analyses of Fig. 12.6a and Fig. 12.6b are necessary to correlate the relative phases of the HOMOs required for the thermal (and photochemical) reactions (Fig. 12.6c–d). Thus, using the MOs of 1,3,5-hexatriene from Fig. 12.4 with the logic just presented results in the predictions shown in Fig. 12.6d. All these predictions are in accord with the vast amount of experimental data, [39, 40] early highlights of which are recorded in Fig. 12.5.

To summarize: The relative stereochemistries of the frontier molecular orbitals in polyenes and their ring closed cycloaddition products control the direction of rotation of the reacting centers, thereby controlling the stereospecificity of the reactant–product relationships.

This analysis of electrocyclizations is based on the required overlap of incipient bonding orbitals having the same symmetry. As shown in Fig. 12.1, this is one of the oldest concepts in bonding: positive overlap of electron densities for bonding (see Fig. 12.1). Of course, before quantum chemistry, there was no concept of phases and nodes and symmetry. I shall discuss this historical observation in Sect. 10. But for now, I want to give this method a name. I have called it the "HOMO overlap interaction method" [13, 41–43].

By now, you have witnessed, if not mastered, the stereochemical factors that control the simplest reaction type within the Woodward-Hoffmann manifold of pericyclic reactions, that of electrocyclizations. Stereochemistry reigns! In doing so, you have simultaneously learned the core ideas underlying the mechanisms of all pericyclic reactions. Should you dig deeper and more broadly into the universe of pericyclic reactions, [16, 17, 40, 44, 45] you will find the stereochemistry of the underlying MOs to be the key to the stereospecificity that these reactions undergo.

12.4 Roald Hoffmann's "Funny" Laboratory Notebook Page

The third reaction type that Woodward and Hoffmann analyzed, after electrocyclizations [4, 13] and cycloadditions, [14, 46] was sigmatropic reactions [5, 15]. Sigmatropic reactions are also the most difficult type of pericyclic reaction to define, though Woodward and Hoffmann did just that [15, 16]. In their third 1965 *JACS* communication, Woodward and Hoffmann began with the following statement of scope:

> We define as a *sigmatropic change of order [i,j]* the migration of a σ-bond, flanked by one or more π-electron systems, to a new position whose termini are $i-1$ and $j-1$ atoms removed from the original bonded loci, in an uncatalyzed intramolecular process [15]. (See Eq. 12.3).

$$\overset{i}{R_2CH}\text{-}(CH{=}CH)n\text{-}CH{=}\overset{j}{C}R'_2 \xrightarrow{[i,j]} R_2CH{=}CH\text{-}(CH{=}CH)nCHR'_2 \qquad (12.3)$$

There are two types of sigmatropic reactions. Of those that are called [1, *j*], two examples are shown in Eq. 12.4a and Eq. 12.4b. In these [1, 5] cases, a hydrogen or deuterium atom rearranges to the other side of a polyolefinic system with concomitant rearrangement of the double bonds. In [*i, j*] sigmatropic rearrangements as in the Cope reaction (Eq. 12.5, a [3,3] sigmatropic reaction), a single bond migrates more than one atom distant from its original position. In my opinion, it is much easier to recognize a sigmatropic reaction than to define it.

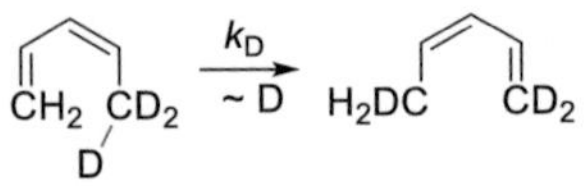

(12.4a)

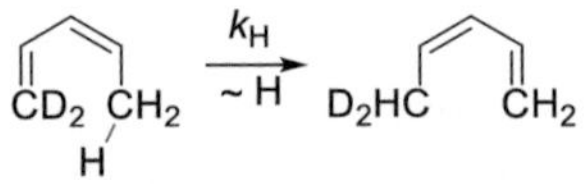

(12.4b)

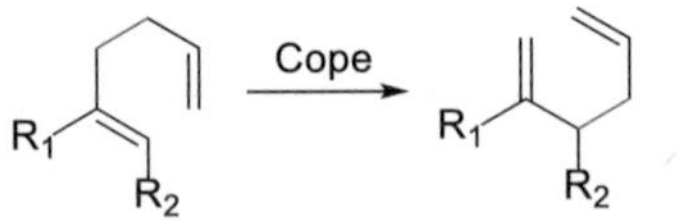

(12.5)

The mechanism of sigmatropic reactions was also difficult for Hoffmann to determine. Perhaps the most perplexing aspect of his research for Hoffmann was believing his own analyses. Indeed, Hoffmann applied five "different, but equivalent" [16, 47] molecular orbital tools before he was willing to accept the totality and unity of his proposed mechanism for sigmatropic reactions [5]. These five tools were.

a. Correlation diagrams [48–51]

b. HOMO overlap interaction method [13, 41–43]
c. Calculations using extended Hückel Theory (eHT) [13, 52–56]
d. Interaction diagram/perturbation theory method focused on the relative symmetries of the singly occupied molecular orbitals (SOMOs) in the transition state [57–60]
e. Three-center bond analyses [61–64]

The key concept in the above discussion is "different, but equivalent." Reproducibility and replicability demand that if these MO tools are valid, they *all* should give the same results for any specific application. For Hoffmann, only at the very end of a long intellectual trail did he ratify and adopt the mechanism of sigmatropic reactions that he had discovered. I will now present a single page from Hoffmann's laboratory notebook that was near the very end of Hoffmann's sigmatropic mechanism travails. This remarkable page indicates how very uncertain, perhaps even confused Hoffmann was, as he closed in on the end of his mechanistic research.

Figure 12.7 reproduces page 29 of Hoffmann's *Laboratory Notebook 17* (in shorthand, I refer to this page as NB17/29). Our focus will be on the [1, 5] suprafacial hydrogen migration shown by Eq. 12.6; structure **1** represents the symmetrical TS for this reaction. This page NB17/29 was written ca. April 6, 1965, and appears near the very end of Hoffmann's research that led to his and Woodward's *JACS* communication on sigmatropic reactions. That manuscript was received in the *JACS* editorial offices on April 30, 1965. Considering the mailing time, that's about three weeks from the day this laboratory notebook page was written to the day the manuscript was mailed. Did Hoffmann begin drafting the manuscript before he had completed his proof, sufficient for him, of the mechanism of sigmatropic reactions? Before his confusion was cleared up?

1

(12.6)

The last two columns at the top of NB17/29 (Fig. 12.7) are enlarged in Fig. 12.8a and b. Here, Hoffmann recorded his analysis of the [1, 5] suprafacial ("on the same side") hydrogen migration (Eq. 12.6) using two MO tools shown in the last and penultimate columns on the top right of NB17/29. Underneath these analyses, Hoffmann wrote, "funny." Hoffmann's amusement may have been an "ironic amusement," more indicative of his momentary confusion. Since these two MO tools are "different, but equivalent," they should provide the same result. And at first, it seems that they give opposite results.

Let's look at this reaction very carefully. I shall now analyze Eq. 12.6 applying the two MO tools used by Hoffmann. In both cases, we shall examine the properties of TS **1**, since this is the only moiety on the potential energy surface (PES) of this reaction that contains a symmetry element. That symmetry element is the plane of symmetry illustrated by graphic **2** in Fig. 12.8.

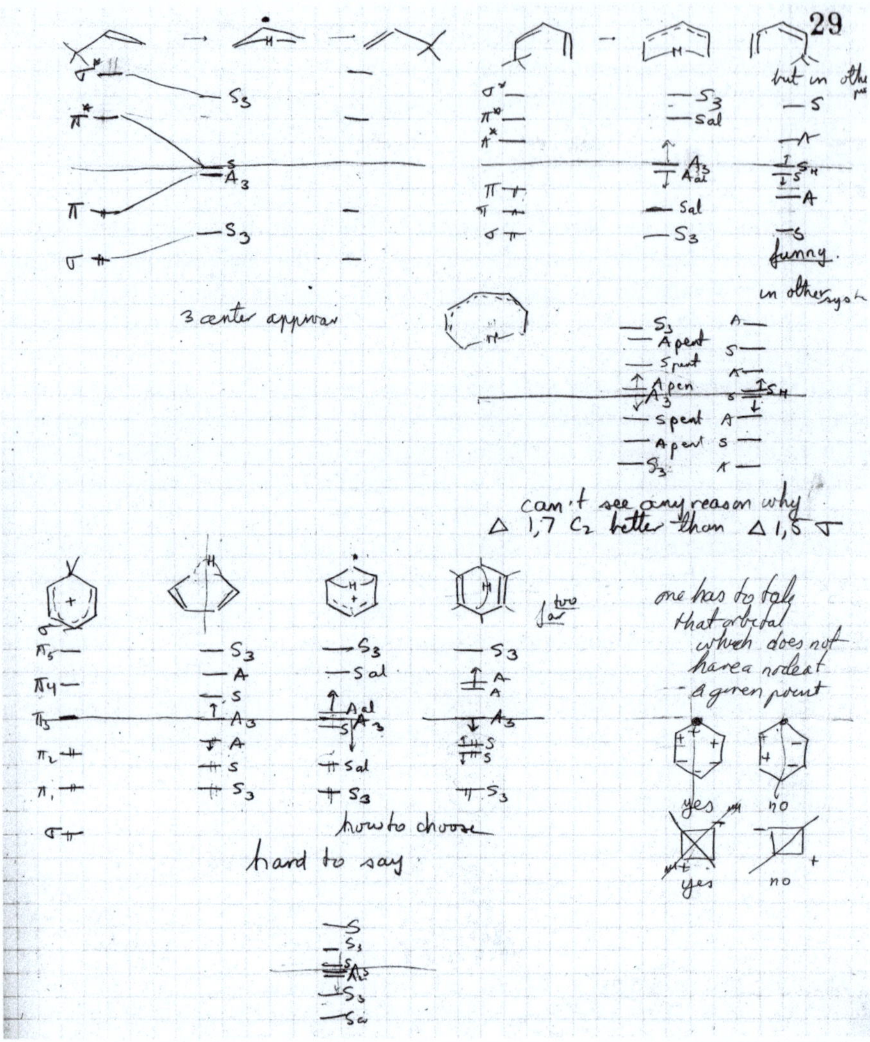

Fig. 12.7 Page 29 in Hoffmann's *Laboratory Notebook 17,* ca. April 6, 1965, about three weeks before Woodward and Hoffmann submitted their *JACS* manuscript on sigmatropic reactions. In the last three columns at the top of the page, Hoffmann examined the [1, 5] hydrogen shift in 1,3-pentadiene (Eq. 12.6). In the third column from the end at the top of the page, Hoffmann drew the MOs of 1,3-pentadiene. In the penultimate column, he drew the MOs for the three-center bond analysis of the [1, 5] suprafacial sigmatropic rearrangement. In the last column, he drew the MOs for the interaction diagram/perturbation method analysis of the [1, 5] suprafacial sigmatropic rearrangement. Just below that, he wrote, "funny." Reprinted with permission of R. Hoffmann who retains copyright

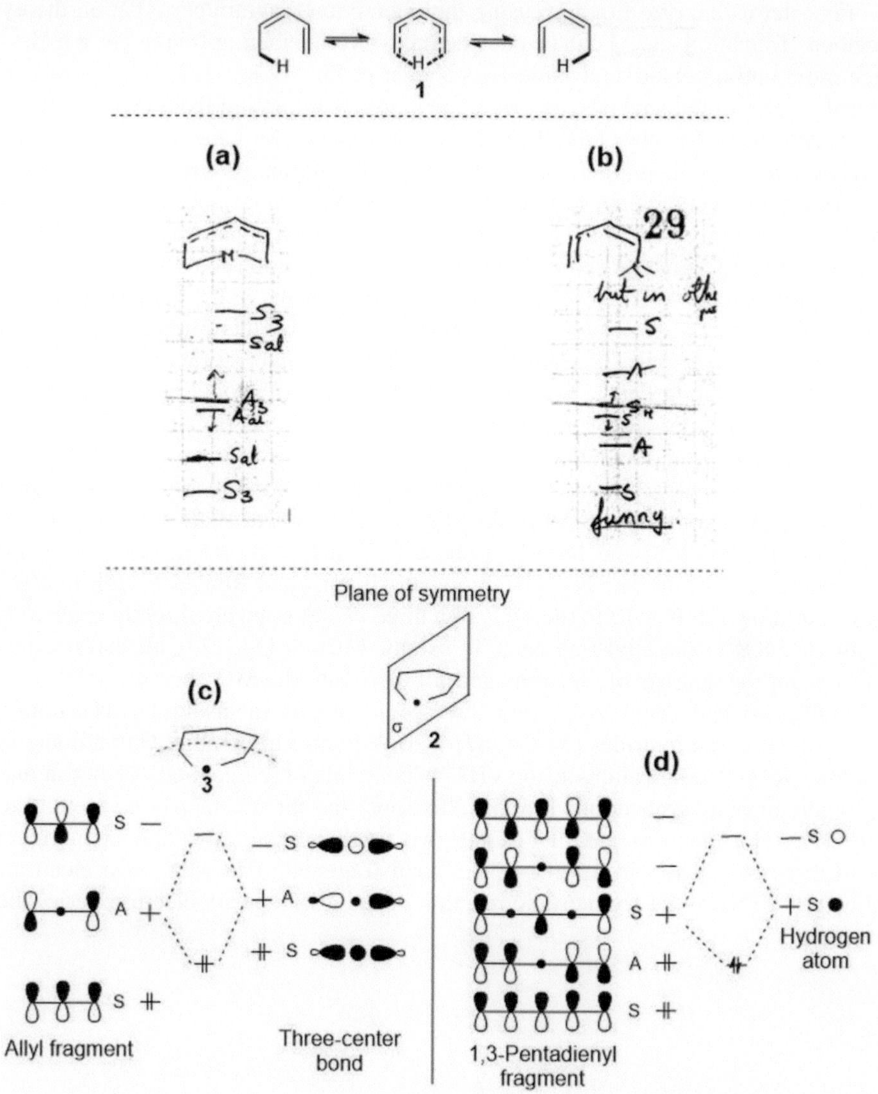

Fig. 12.8 Structure **1** is the TS of this suprafacial [1, 5] hydrogen migration. **a** and **b** are enlargements of the last two columns of the top row of NB17/29 (Fig. 12.7), for the [1, 5] suprafacial sigmatropic reaction of 1,3-pentadiene, ca. April 6, 1965. (**a**) The three-center bond analysis. **b** The interaction diagram/perturbation theory analysis. **c** A more detailed three-center bond analysis that relates to Hoffmann's analyses in (**a**). **d** A more detailed interaction diagram/perturbation theory analysis that relates to Hoffmann's analyses in (**b**). Regardless of the path of the hydrogen atom, the symmetry element for (**a**) and (**b**) is the plane perpendicular to the plane of the pentadienyl fragment that bisects the hydrogen atom and C(3). The relative energies of the MOs are not drawn to scale. Hoffmann wrote "funny." See the text for further discussions and explanations of these MO tools. Adapted with permission from a previous publication of mine [5]

First, let us analyze Eq. 12.6 using the interaction diagram/perturbation theory method. Transition state **1** can be considered as two interacting fragments, a hydrogen atom and a pentadienyl radical, as shown in Fig. 12.8d. This is a reasonable starting point in the analysis, as a hydrogen atom is migrating from one end of the C_5-backbone to its other end. The plane of symmetry **2** of the TS **1** reflects the hydrogen atom at the midpoint of its suprafacial migration. As shown in Fig. 12.8d, the HOMO of pentadienyl radical, actually a SOMO (singly-occupied molecular orbital), is symmetrical (S). The hydrogen atom, of course, also has an S symmetry. These two fragments can interact because they have the same symmetry. That interaction, called "mixing" of molecular orbitals, results in the formation of two new MOs, the doubly-occupied MO being stabilized by this interaction. (This is exactly the same practice as shown in Fig. 12.2 for the interaction of two hydrogen atoms to form dihydrogen.) This stabilization results in the qualitative assessment that Eq. 12.6 is a favorable transformation. Woodward and Hoffmann called such favorable transformation "allowed."

Now, let us analyze Eq. 12.6 using the three-center bond method. This method traces its heritage to the nonclassical carbocation of the late 1940s and early 1950s by Saul Winstein [65] and John D. Roberts [66] and even earlier, to the structure determination of diborane by H. C. Longuet-Higgins and Ronald P. Bell in 1943 [61]. As shown in Fig. 12.9, the MOs of a three-center bond are directly analogous with the MOs of the allyl fragment. To use the MOs in Fig. 12.9, all that remains is to count the number of electrons and fill them into the MO energy levels.

Hoffmann had extensively used three-center bonds in his Ph.D. dissertation research on boron hydrides [52, 64, 67] with William Lipscomb at Harvard and in his first several publications using eHT calculations on compounds of boron and nitrogen and on carbocations [55, 56]. To apply the three-center bond method to Eq. 12.6, Hoffmann divided TS **1** into two fragments, an allyl radical fragment with three electrons and a three-center bond fragment, also with three electrons (Eq. 12.7). (Note: the sigmatropic reaction in Eq. 12.7 is a six-electron pericyclic reaction.)

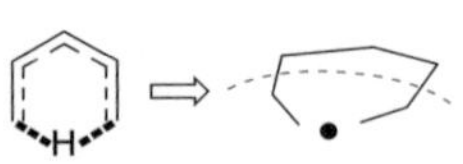

1 **3** (12.7)

The three-center bond analysis is shown in Fig. 12.8c. The HOMOs of both the allyl fragment and the three-center bond fragment are singly-occupied (SOMOs). And the symmetry of both SOMOs is A. These two fragments interact because they have the same symmetry. This stabilization results in the qualitative assessment that Eq. 12.6 is a favorable transformation, i.e., Eq. 12.6 is a Woodward–Hoffmann allowed reaction.

Indeed, both the interaction diagram/perturbation theory method and the three-center bond method result in the same result, namely that Eq. 12.6 is a W–H allowed reaction. *But wait a moment. In one method, the mixing of the two SOMOs were of MOs having S symmetry while in the other method, the mixing of the two*

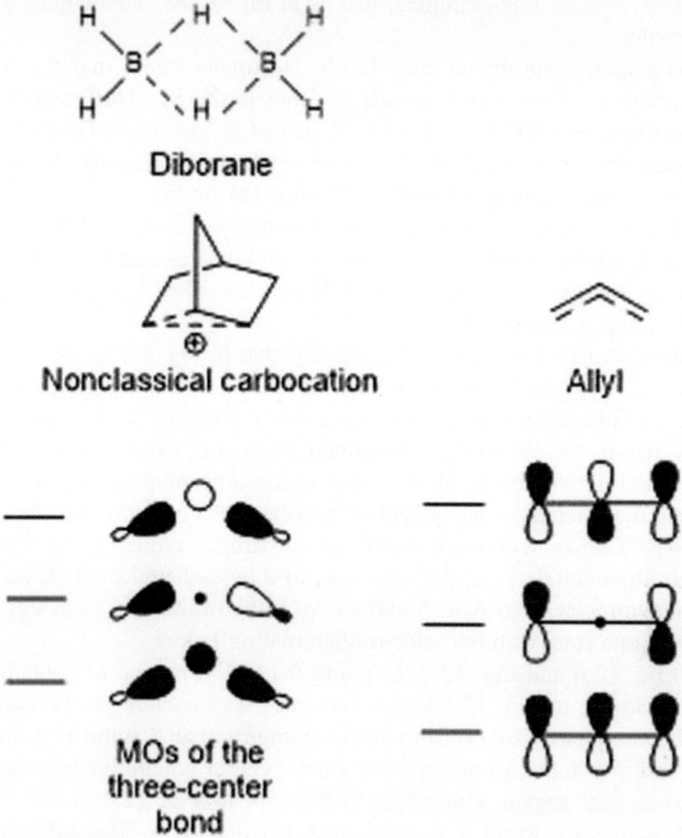

Fig. 12.9 Close relationship between the three-center bonds in diborane, in the nonclassical carbocation, and in the MOs of the allyl fragment. Electron occupation in these MOs is intentionally omitted. MO energy levels are not drawn to scale. Adapted with permission from a previous publication of mine [5]

SOMOs were of MOs having A symmetry. What is going on here? This surely is odd. Perhaps even "funny."

Ah, funny!

And now we stand exactly where Hoffmann stood in early April 1965. Let's look more carefully at what Hoffmann actually did. In Fig. 12.8a, the "A_3" and "A_{al}" refer to the three-center bond and the allyl fragment, respectively; these two "A" labels signify that the MOs are antisymmetric relative to the symmetry plane that bisects the transition state complex **1**. Hoffmann understood that A_3 and A_{al} could interact, as they have the same A symmetry. That interaction results in two new MOs, indicated by Hoffmann by the down and up arrows (as in Figs. 12.1, 12.2 and 12.3). He knew perturbation theory well by that point! The more stable

MO would become doubly-occupied, just as in Fig. 12.8c. This would be a W–H allowed reaction.

Now, let's look carefully at Fig. 12.8b. Hoffmann knew that the MO for a spherical hydrogen atom had S symmetry; that is the S_H. Hoffmann also knew how to construct the MOs for pentadienyl, just as in Fig. 12.4. Those are the five MOs, in ascending energy: S, A, S (in the nonbonding region), A, then S (the latter two in the antibonding region). Hoffmann saw, as we can see, the two MOs in the nonbonding region having the same symmetry, S and S_H. Again, Hoffmann drew a pair of opposing arrows (one up, one down) attached to S and S_H which move apart in energy. The more stable MO becomes doubly-occupied, just as in Fig. 12.8d.D, and is stabilized.

What momentarily confused Hoffmann was that in Fig. 12.8a, the two interacting MOs have A symmetry while in Fig. 12.8b, the two interacting MOs have S symmetry. This interesting phenomenon is another example of alternations in stereochemistry within the Woodward–Hoffmann story. Let us go back to Fig. 12.4b. Recall that the relative phases of the α and ω atomic orbitals in the SOMOs alternate as one either increases the system by two electrons *or* decreases the system by two electrons. That is, with each step from a hydrogen radical to an allyl radical; and then to a pentadienyl radical; and then to a heptadienyl radical, the SOMOs alternate in symmetry, S to A to S and so on. I like to think of each step as going up (or down) one rung of a two-electron/alternating ladder.

Back to Eq. 12.6 and Fig. 12.8. In going from allyl radical to pentadienyl radical (Fig. 12.8c-left to Fig. 12.8d-left), one goes *up one rung* on the two-electron ladder and consequently we go from an A symmetry to an S symmetry. And simultaneously and in parallel, in going from a three-center bond/allyl-type fragment to a hydrogen radical (again, from Fig. 12.8c-right to Fig. 12.8d-right), one goes from an A symmetry to an S symmetry. A to S, A to S. The MOs interact in both Fig. 12.8c and in Fig. 12.8d: W–H allowed. (Another interesting and salient point: In going from Fig. 12.8c-left to Fig. 12.8d-left, we move "up" the ladder by one step (i.e., two electrons). Simultaneously and in parallel, in going from Fig. 12.8c-right to Fig. 12.8d-right, we move "down" the ladder by one step (i.e., two electrons). This is another view of conservation of orbital symmetry, but in this instance, we have consistency within multiple MO tools).

I add another factor that was very important for Hoffmann in 1965. With multiple analyses giving the same results, he became more confident in his conclusions.

I momentarily leave it for the reader to discern the consequences of a [1, 7] suprafacial hydrogen migration (Eq. 12.8). Here are several hints. In Eq. 12.7 and Eq. 12.8, the symmetry element remains the plane of symmetry as in **2**. And in the interaction diagram/perturbation method, we still have a hydrogen atom (as in Fig. 12.8d) but instead of a 1,3-pentadienyl fragment, we have a 1,3,5-heptadienyl fragment. Review the bottom-far right graphic in Fig. 12.4. We have gone up the ladder by one rung for the transition from a 1,3-pentadienyl fragment to a 1,3,5-heptadienyl fragment. But we have stayed on the same rung, hydrogen atom to hydrogen atom for the second fragmentg. What happens if we go up the ladder by one rung only? Answer: a single alternation!

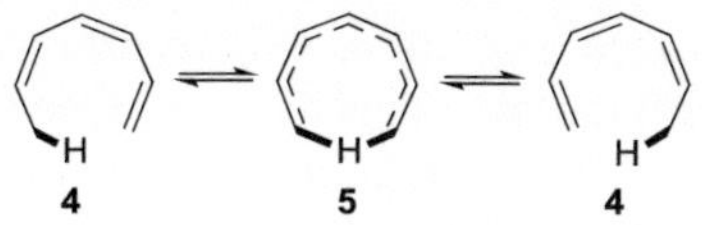

(12.8)

12.5 On [1, 7] Suprafacial Hydrogen Migrations

In the previous section, I posed the question: Is a [1, 7] suprafacial hydrogen shift as in Eq. 12.8a. A W–H allowed or forbidden reaction? Of course, an analysis can be performed just as in Fig. 12.8c or in Fig. 12.8d. As before, the symmetry of the hydrogen atom is S. The symmetry of the SOMO of heptatrienyl is A (see the graphic for the π-MOs of heptatrienyl in Fig. 12.4—bottom right). The two SOMOs do not interact as they have different symmetries (Fig. 12.10). No stabilized MO is formed; the two MOs in the TS remain in the nonbonding region as radicals; this is a W–H forbidden reaction.

The HOMO overlap interaction method gives the same result: W–H forbidden. As indicated by the blue arrow in Fig. 12.10, for a [1, 7] suprafacial hydrogen migration (where there can be no inversion of configuration in the migrating group), the top lobes of both atonic orbitals involved in the bond-breaking and bond-making processes are of different phases: W–H forbidden. But an inversion in the atomic orbital of the HOMO of the migrating moiety would make the [1,7] suprafacial migration W-H allowed, i.e., one alternation leds from forbidden to allowed.

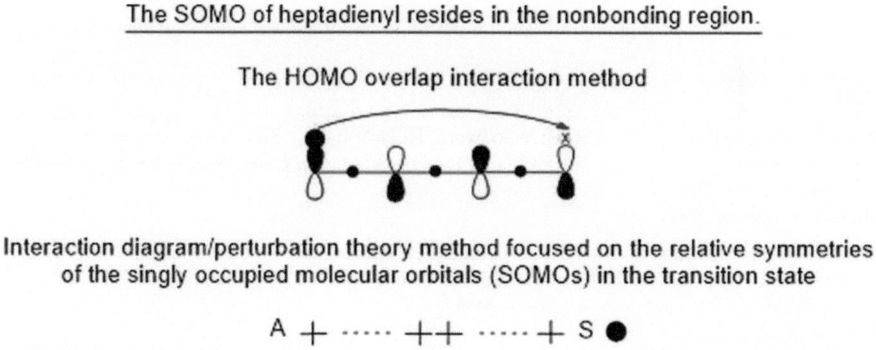

Fig. 12.10 The TS for a [1, 7] suprafacial sigmatropic hydrogen migration in 1,3,5-heptatriene (Eq. 12.8). For this reaction, the relevant symmetry element is a plane that is perpendicular to the plane containing the 1,3,5-heptatrienyl fragment and bisects C(4) and the hydrogen radical. The SOMO of heptatrienyl radical has A symmetry while the HOMO of the hydrogen radical has S symmetry. The interaction diagram of the TS reveals a non-interaction of the two SOMOs as they have different symmetries

But please wait a moment. Did we have to go through that complex analysis? Of course not. We already knew that a [1, 5] suprafacial hydrogen migration is W–H allowed. In the switch from a [1, 5] suprafacial hydrogen migration to a [1, 5] antarafacial hydrogen migration, we are exercising a single alternation. This process is like multiplying by a negative one. One switch, an alternation (as shown in Fig. 12.4). Two switches, say to a [1, 7] antarafacial hydrogen migration (or from a [1, 5] suprafacial hydrogen migration to a [3] suprafacial-suprafacial migration), negative one times negative one equals positive one, no alternation, all W–H allowed!

12.6 And Back to Hoffmann's "Funny"

What does Hoffmann's "funny" in Fig. 12.7 (and Fig. 12.8b.B) tell us about Hoffmann and his state of mind and understanding on ca. April 6, 1965? As shown in Table 12.2, by early April 1965, Woodward and Hoffmann had moved past the determination and public revelations of the orbital symmetry mechanisms for electrocyclizations and cycloadditions. And within about three weeks of April 6, 1965, they will have submitted their sigmatropic reaction manuscript. By 1964, Hoffmann was one of the world's experts in MO theory, having completed the development of the extended Hückel theory. The MOs of simple polyunsaturated molecules were trivial for him. Yet on ca. April 6, 1965, Hoffmann was still somewhat taken by surprise by aspects of the alternations in stereochemistry within simple MO theory—not within the chemistry itself but within the theory—that he uncovered in his intensive research on sigmatropic reactions.

What does Hoffmann's use of "funny" mean? I posit several considerations and possibilities.

Table 12.2 Important chronological data for the first three Woodward and Hoffmann *JACS* communications of 1965

Dates	Publication
Mailed Nov. 25, 1965 Received Nov. 30, 1964 Accepted Dec. 1, 1964 Published Jan. 20, 1965	R. B. Woodward, R. Hoffmann, "Stereochemistry of Electrocyclic Reactions," *J. Am. Chem. Soc.* **1965**, *87*, 395–397
Mailed March 19, 1965 Received March 22, 1965 Accepted March 24, 1965 Published May 1, 1965	R. Hoffmann, R. B. Woodward, "Selection Rules for Concerted Cycloaddition Reactions," *J. Am. Chem. Soc.* **1965**, *88*, 2046–2048
Mailed April 28, 1965 Received April 30, 1965 Accepted May 3, 1965 Published June 1, 1965	R. B. Woodward, R. Hoffmann, "Selection Rules for Sigmatropic Reactions," *J. Am. Chem. Soc.* **1965**, *87*, 2511–2513

- Like all researchers, Hoffmann could not always apply all that he knew at certain moments in time. When he wrote "funny," he was momentarily, yes, only momentarily, bewildered.
- In 1965, the theoretical underpinnings of the mechanisms of pericyclic reactions were *not* as simple as they seem to us today.
- The reliance on quantum chemistry in the early 1960s was the beginning of a new world, a new way to characterize complex organic reaction mechanisms in chemistry. The Quantum Chemistry Revolution was brewing [68–70].
- In 1965, what Hoffmann was doing was beyond just state-of-the-art. And it was tough going, even for him.
- At some real level for each of us, quantum chemistry is "funny." The wave-particle duality and Bohr's complementarity principle, Heisenberg's uncertainty principle, the laddering of symmetries in Fig. 12.4—all these and more, they go so far beyond our everyday ways of experiencing the tangible world around us. The seventh edition of *Roget's Thesaurus* lists the following nine words related to "funny." Surely Hoffmann was feeling odd or bewildered that day in in early April 1965 when he wrote "funny."

- humorous
- witty
- inexplicable
- ambiguous
- odd
- eccentric
- peculiar
- unbelievable
- bewildering

12.7 Revolutions in Chemistry

Let's pause a moment and reflect on the chemistry that has been explained herein. In all cases, the mechanisms were determined by MO tools used by Woodward and Hoffmann as reported in their 1965 analyses [13–15, 15, 71] of pericyclic reactions [16, 47]. These were actually called "no-mechanism reactions," [73–78] not because they were considered to have no mechanism but because their mechanisms were not known, and bewildering so. Of course, all reactions have mechanisms. Rather, it was a tongue-in-cheek terminology that illustrated the frustration that organic chemists had at that time regarding this mystery.

There was a very special reason that the mechanisms of no-mechanism reactions were unknown prior to Woodward and Hoffmann. None of the classical mechanistic explanations[79–85]—arrow pushing, steric effect, electronic effects, and yes, not stereoelectronic effects—provided the solution. Only a mechanism based on quantum chemistry was possible. This was the first time that organic chemists were faced with this absolute requirement; previously, intuitive explanations sufficed. That there were some reactions for which intuitive, heuristic explanations were totally incapable of even a hint of a mechanism was a phenomenon that *precipitated* the Quantum Chemical Revolution of the twentieth century. For a modern definition of "revolutions in science" [69] and the evaluation of the Woodward–Hoffmann rules as the precipitator of a revolution in chemistry, [70] the interested reader is referred to the primary literature referenced herein.

12.8 Conclusion

The most important conclusion from this essay, as it pertains to the theme of Greenberg and Lewis's symposium and this monograph, is as follows: The various definitions of "stereochemistry" in textbooks and in the chemistry literature should be updated to explicitly include the stereochemistry of quantum chemistry.

Sure, we all know and appreciate that the chemistry we intuit is governed by quantum chemical concepts. Many decades ago, I heard Richard Feynman begin a lecture by writing a two-foot high "ψ" on the blackboard, saying that the universal, grand wavefunction could explain and predict everything. So, we also know that all the stereochemistry we observe experimentally or calculate by any theory, all of that is undergirded by quantum mechanics.

Let me say this in another way. There is no doubt that the mechanism of the no-mechanism reactions, i.e., the mechanism of all pericyclic reactions, can *only* be obtained through quantum chemistry [68]. That and related observations imply that the fundamental mechanisms of all chemical reactions are based in quantum chemistry. To the extent that we chemists study stereochemistry and its consequences on structure and reactivity, stereochemistry naturally relies upon quantum chemistry. Should not the definition of stereochemistry explicitly include the three-dimensional imperatives of quantum chemistry? I believe so.

Acknowledgements Until developing this essay, I had not considered the definition of "stereochemistry" nor its limited scope. During the past many years, I have been tutored by Roald Hoffmann in every possible nook and cranny of the Woodward–Hoffmann story, including the use of qualitative perturbation theory and other MO tools. Hoffmann urges more researchers to follow the stereochemical trajectories of molecular orbitals using "movies" during the course of reactions. What a great opportunity for stereochemistry! While his own beginnings as a chemical physicist were steeped in the development and use of computational chemistry, Hoffmann is also omputational chemistry's first and foremost critic. He believes more in understanding and less in actual numbers [86–91]. Stereochemistry always involves the visual and mental arts. This essay could not have been written without Hoffmann's personal tutorials and our hundreds of emails, all of which involved stereochemistry and molecular orbitals. Thank you, Roald. Of course, the entire project, from the symposium itself all the way to the peer review of this monograph's chapters, was initiated by the invitation from Art Greenberg and David Lewis, and shepherded by them. I thank Art and David for all of this and Celia Arnaud and Melinda W. Davis for technical assistance. I acknowledge the use and adaptation of figures, equations, and substantial discussion from several of my recent publications, all [4, 5, 46, 92] save one [7] are open access. I also thank the staff of the Boatwright Memorial Library for their continued technical support.

References

1. Perrin CL, Agranat I, Bagno A, Braslavsky SE, Fernandes PA, Gal J-F, Lloyd-Jones GC, Mayr H, Murdoch JR, Nudelman NS, Radom L, Rappoport Z, Ruasse M-F, Siehl H-U, Takeuchi Y, Tidwell TT, Uggerud E, Williams IH (2024) Glossary of terms used in physical organic chemistry (IUPAC Recommendations 2021). Pure Appl Chem 2022(94):353–534. https://doi.org/10.1515/pac-2018-1010;accessedDecember7
2. Perrin C, email to Seeman JI, La Jolla CA (November 2, 2024)
3. Eliel EL, Wilen SH (1994) Stereochemistry of organic compounds. John Wiley & Sons, New York
4. Seeman JI (2015) Woodward–Hoffmann's stereochemistry of electrocyclic reactions: from day 1 to the *JACS* Receipt Date (May 5, 1964 to November 30, 1964). J Org Chem 80:11632–11671. https://doi.org/10.1021/acs.joc.5b01792; https://pubs.acs.org/doi/abs/. Accessed 7 Dec 2024
5. Seeman JI (2024) Going beyond Woodward and Hoffmann's electrocyclizations and cycloadditions: Sigmatropic rearrangements. Chem Rec 24:e202400050. https://doi.org/10.1002/tcr.202400050
6. Hoffmann R (1971) Interaction of orbitals through space and through bonds. Acc Chem Res 4:1–9
7. Tantillo DJ, Seeman JI (2023) On a unified theory of acids and bases: Hasok Chang, Eric R. Scerri, modern theoretical chemistry, and the philosophy of chemistry. Found Chem 25:1–22. https://doi.org/10.1007/s10698-022-09456-5
8. Hoffmann R, Woodward RB (1968) The conservation of orbital symmetry. Acc Chem Res 1:17–22. https://doi.org/10.1021/ar50001a003
9. Hoffmann R (1982) Building bridges between inorganic and organic chemistry (Nobel Lecture). Angew Chem Int Ed 21:711–724. https://doi.org/10.1002/anie.198207113; https://onlinelibrary.wiley.com/doi/abs/
10. Hoffmann R (1981) Buidling bridges between inorganic and organic chemistry. In: Nobel Prize presentation. Stockholm, Sweden, December 8, 1981. https://www.nobelprize.org/uploads/2018/06/hoffman-lecture.pdf. Accessed 5 Aug 2024
11. Jorgenson WL, Salem L (1973) The organic chemist's book of orbitals. Academic Press, New York/London
12. Salem L (1966) The molecular orbital theory of conjugated systems. W. A. Benjamin, New York
13. Woodward RB, Hoffmann R (1965) Stereochemistry of electrocyclic reactions. J Am Chem Soc 87:395–397. https://doi.org/10.1021/ja01080a054
14. Hoffmann R, Woodward RB (1965) Selection rules for concerted cycloaddition reactions. J Am Chem Soc 87:2046–2048. https://doi.org/10.1021/ja01087a034
15. Woodward RB, Hoffmann R (1965) Selection rules for sigmatropic reactions. J Am Chem Soc 87:2511–2513. https://doi.org/10.1021/ja01089a050
16. Woodward RB, Hoffmann R (1969) The conservation of orbital symmetry. Angew Chem Int Ed 8:781–853. https://doi.org/10.1002/anie.196907811
17. Hoffmann R, Woodward RB (1970) Orbital symmetry control of chemical reactions. Science 167:825–831. https://doi.org/10.1126/science.167.3919.82
18. Hoffmann R, Woodward RB (1972) Das konzept von der erhaltung der orbitalsymmetrie. Chem unserer Zeit 6:167–174
19. Hoffmann R, Woodward RB (1973) Den reaktionskontrollerande orbitalsymmetrin. Kemisk Tidskrift *85*:28–30:33–34, 37
20. Zimmerman HE (1966) On molecular orbital correlation diagrams, the occurrence of Mobius systems in cyclization reactions, and factors controlling ground- and excited-state reactions. I. J Am Chem Soc 88:1564–1565

21. Zimmerman HE (1966) Molecular orbital correlation diagrams, Möbius systems, and factors controlling ground- and excited-state reactions. II. J Am Chem Soc 88:1566–1567
22. Zimmerman HE (1971) Möbius-Hückel concept in organic chemistry. application of organic molecules and reactions. Acc Chem Res 4:272–280. https://doi.org/10.1021/ar50044a002
23. Dewar MJS (1966) A molecular orbital theory of organic chemistry -VIII. Aromaticity and electrocyclic reactions. Tetrahedron 8(I):75–92
24. Dewar MJS (1967) Some recent developments in the theory of conjugated systems. In: Aromaticity: an international symposium held at Sheffield on 6th-8th July 1966, Special Publication No 21. The Chemical Society (London) and the University of Sheffield, London, pp 177–215
25. Dewar MJS (1971) Aromaticity and pericyclic reactions. Angew Chem Int Ed 10:761–776
26. Dewar MJS (1974) Comments on two early papers on the theory of pericyclic reactions. In: Simmons HE, Bunnett JF (eds) Orbital symmetry papers: comprising articles from Accounts Chemical Research. J Am Chem Soc Journals. American Chemical Society, Washington, D.C, pp 16–19
27. Fonken GJ (1961) Photo-isomerisation of muconic anhydride. Chem Ind 39:1575
28. Fonken GJ (1962) Photochemical cyzlization of acyclic trienes. Tetrahedron Lett 3:549–551
29. Fonken GJ, Mehrotra K (1964) Photolysis of cyclic dienes. An example of a dual light-induced valence tautomerism. Chem Ind (London, U. K.) 1025
30. Shumate KM, Neuman PN, Fonken GJ (1965) Thermal cyclization of 1,3-Dienes to cyclobutenes. J Am Chem Soc 87:3996. https://doi.org/10.1021/ja01095a047
31. Marvell EN, Caple G, Schatz B (1965) Thermal valence isomerizations: Stereochemistry of the 2,4,5-octatriene to 5,6-dimethyl-1,3-cyclohexadiene ring closure. Tetrahedron Lett 6:385–389
32. Marvell EN (1973) The *cis*-hexatriene electrocyclization. Attempted calculation of the transition state geometry. Tetrahedron 29:3791–3796
33. Criegee R, Noll K (1959) Umsetzungen in der Reihe des 1,2,3,4-Tetramethyl-cyclobutans. Liebigs Ann Chem 627:1–14
34. Criegee R, Seebach D, Winter RE, Börretzen B, Brune H-A (1965) Valenzisomerisierungen Von Cyclobutenen. Chem Ber 98:2339–2351
35. Winter REK (1965) The preparation and isomerization of cis- and trans-3,4-dimethylcyclobutene. Tetrahedron Lett 6:1207–1212
36. Vogel E (1954) Uber Die Stabilltat Des Ungesattigten Kohlenstoff-Vierringes. Angew Chem 66:640–642
37. Vogel E (1955) Kleine kohlenstoffringe. Fortschr chem Forsch 3:430–502
38. Srinivasan R (1969) Photochemical valence isomerization. Pure Appl Chem 20:263–270
39. Marvell EN (1980) Thermal electrocyclic reactions. Academic Press, New York
40. Magid R (2024) Orbital symmetry control of pericyclic reactions. Books and review articles. http://web.utk.edu/~rmagid/650sp02rev.pdf; https://web.utk.edu/%7Ermagid/650sp02rev.pdf. Accessed 13 Dec 2024, Accessed 16 May 2023
41. Hoffmann R (1965) Letter to W. G. Dauben. Ithaca, NY, December 14, 1965
42. Hoffmann R (1966) Letter to R. B. Woodward. Ithaca, NY, July 27, 1966
43. Woodward RB (1966) Letter to M. Gates. Cambridge, MA, August 30, 1966
44. Houk KN, Liu F, Yang Z, Seeman JI (2021) Evolution of the diels-alder reaction mechanism since the 1930s: Woodward, Houk with Woodward, and the influence of computational chemistry on understanding cycloadditions. Angew Chem Int Ed 60:12660–12681. https://doi.org/10.1002/anie.202001654
45. Zhou Q, Kukier G, Gordiy I, Hoffmann R, Seeman JI, Houk KN (2024) A 21st century view of allowed and forbidden electrocyclic reactions. J Org Chem 89:1018–1034. https://doi.org/10.1021/acs.joc.3c02103
46. Seeman JI (2024) On cycloadditions and the origination of ideas: Woodward-Hoffmann's second publication. The transformation of roald hoffmann from a "calculator" to an "explainer". Chem Rec 24. https://doi.org/10.1002/tcr.202300181. https://onlinelibrary.wiley.com/doi/
47. Woodward RB, Hoffmann R (1970) The conservation of orbital symmetry. Verlag Chemie, Weinheim

48. Hund F (1927) Zur deutung der molekelspektren. I. Z Phys 40:742–764. https://doi.org/10.1007/BF01400234
49. Mulliken RS (1932) The interpretation of band spectra. III. Electron quantum numbers and states of molecules and their atoms. Rev Mod Phys 4:1–86
50. Pople JA (1953) Electron interaction in unsaturated hydrocarbons. Trans Faraday Soc 49:1375–1385
51. Walsh AD (1953) The electronic orbitals, shapes, and spectra of polyatomic molecules. Part I. Ah_2 molecules. J Chem Soc 2260–2266
52. Hoffmann R, Lipscomb WN (1962) Boron hydrides: LCAO-MO and resonance studies. J Chem Phys 37:2872–2883
53. Hoffmann R (1963) An extended Hückel theory. I. hydrocarbons. J Chem Phys 39:1397–1412. https://doi.org/10.1063/1.1734456
54. Hoffmann R (1964) Extended Hückel theory. II. Orbitals in the Azines. J Chem Phys 40:2745
55. Hoffmann R (1964) Extended Hückel theory. III. Compounds of Boron and Nitrogen. J Chem Phys 40:2474–2480
56. Hoffmann R (1964) Extended Hückel theory. IV. Carbonium ions. J Chem Phys 40:2480–2488. https://doi.org/10.1063/1.1725551
57. Hoffmann R, Imamura A, Zeiss GD (1967) The spirarenes. J Am Chem Soc 89:5215–5220
58. Gleiter R, Hoffmann R (1968) The stabilization of centers with six valence electrons through electron-rich three-center bonding. Tetrahedron 24:5899–5911
59. Gleiter R, Hoffmann R (1968) On stabilizing a singlet methylene. J Am Chem Soc 90:5457–5460
60. Hoffmann R (1970) Geometry changes in excited states. Pure Appl Chem 24:567–584
61. Longuet-Higgins HC, Bell RP (1943) The structure of the boron hydrides. J Chem Soc 250–255. https://doi.org/10.1039/jr9430000250
62. Lipscomb WN (1957) Valence in the boron hydrides. J Phys Chem 61:23–27
63. Fukui K, Morokuma K, Yonezawa T, Nagata C (1960) Charge transfer mechanism of reaction of conjugated molecules. J Chem Phys 32:1743–1747
64. Hoffmann R, Lipscomb WN (1962) Theory of polyhedral molecules. III. Population analyses and reactivities for the carboranes. J Chem Phys 36:3489–3493. https://doi.org/10.1063/1.1732484
65. Winstein S, Trifan DS (1949) The structure of the bicyclo[2,2,1]2-heptyl (norbornyl) carbonium ion. J Am Chem Soc 71:2953–2953. https://doi.org/10.1021/ja01176a536
66. Roberts JD, Lee CC (1951) The nature of the intermediate in the solvolysis of norbornyl derivates. J Am Chem Soc 73:5009–5010
67. Hoffmann R, Lipscomb WN (1963) Intramolecular isomerization and transformations in carboranes and substituted polyhedral molecules. Inor Chem 2:231–232
68. Seeman JI, Tantillo DJ (2022) Understanding chemistry: from 'heuristic (soft) explanations and reasoning by analogy' to 'quantum chemistry.' Chem Sci 13:11461–11486. https://doi.org/10.1039/D2SC02535C
69. Seeman JI (2023) Revolutions in science. revolutions in chemistry. Found Chem 25:321–335. https://doi.org/10.1007/s10698-023-09467-w
70. Seeman JI (2023) Revolutions in chemistry: assessment of six 20th century candidates (the Instrumental Revolution; Hückel molecular orbital theory; Hückel's 4n+2 Rule; the Woodward-Hoffmann rules; quantum chemistry; and retrosynthetic analysis). J Am Chem Soc Au 3:2378–2401. https://pubs.acs.org/doi/; https://doi.org/10.1021/jacsau.3c00278
71. Hoffmann R, Woodward RB (1965) Orbital symmetries and *endo-exo* relationships in concerted cycloaddition reactions. J Am Chem Soc 87:4388–4389. https://doi.org/10.1021/ja00947a033
72. Hoffmann R, Woodward RB (1965) Orbital symmetries and orientational effects in a sigmatropic reaction. J Am Chem Soc 87:4389–4390. https://doi.org/10.1021/ja00947a034
73. von E. Doering W, Roth WR (1962) The overlap of two allyl radicals or a four-centered transition state in the Cope rearrangement. Tetrahedron 18:67–74. https://doi.org/10.1016/0040-4020(62)80025-8

74. Huisgen R (1963) Kinetics and mechanism of 1,3-dipolar cycloadditions. Angew Chem Int Ed 2:633–645. https://doi.org/10.1002/anie.196305651. https://onlinelibrary.wiley.com/doi/abs/
75. Rhoads SJ (1963) Rearrangements proceeding through 'no mechanism' pathways: The Claisen, Cope, and related rearrangements. In: de Mayo P (ed) Molecular Rearrangements, vol 1. Interscience, New York, pp 655–706
76. Hill RK, Edwards AG (1964) Stereochemistry of 'no-mechanism' reactions. II. Asymmetric induction in the Claisen rearrangement. Tetrahedron Lett 5:3239–3243
77. Marvell EN, Stephenson JL, Ong J (1965) Stereochemistry of the Claisen rearrangement. J Am Chem Soc 87:1267–1274
78. Seeman JI (2022) History of the Woodward-Hoffmann rules. The no-mechanism puzzle. Chem Rec 22. https://doi.org/10.1002/tcr.202100212
79. Gould ES (1962) Mechanism and structure in organic chemistry. Holt, Rinehart and Winston, New York
80. Wheland GW (1955) Resonance in organic chemistry. John Wiley and Sons, New York
81. Wheland GW (1960) Advanced organic chemistry, 3rd. John Wiley & Sons
82. Fieser LF, Fieser M (1956) Organic chemistry, 3rd. Reinhold Publishing Corporation, New York
83. Fieser LF, Fieser M (1961) Advanced organic chemistry. Reinhold Publishing Corporation, New York
84. Ingold CK (1953) Structure and mechanism in organic chemistry. Cornell University Press, New York City, NY
85. Branch GEK, Calvin M (1941) The theory of organic chemistry. An advanced course. Prentice-Hall, New York, NY
86. Hoffmann R (2007) What might philosophy of science look like if chemists built it? Synthese 155:321–336
87. Hoffmann R (1998) Qualitative thinking in the age of modern computational chemistry-or what Lionel Salem knows. J Mol Struct 424:1–6
88. Kovac J, Weisberg M (eds) (2012) Roald Hoffmann on the philosophy, art, and science of chemistry. Oxford University Press, Oxford, England
89. Hoffmann R, Malrieu J-P (2020) Simulation vs. understanding: a tension, in quantum chemistry and beyond. Part A. Stage setting. Angew Chem Int Ed 59:12590–12610
90. Hoffmann R, Malrieu J-P (2020) Simulation vs. Understanding: a tension, in quantum chemistry and beyond. Part B. The march of simulation, for better or worse. Angew Chem Int Ed 59:13156–13178. https://doi.org/10.1002/anie.201910283; https://onlinelibrary.wiley.com/doi/abs/
91. Hoffmann R, Malrieu J-P (2020) Simulation vs. understanding: a tension, in quantum chemistry and beyond. Part C. Toward consilience. Angew Chem Int Ed 59:13798–13814
92. Seeman JI (2025) Woodward and Hoffmann on secondary orbital interactions. How to make a fine two-course meal from left-overs. Chem Res, https://doi.org/10.1002/tcr.202400051

Although sp^3 Hybridization is Not Found for the Carbon in Methane Nor in Isoelectronically Related Species, the Concept of Hybridization is Still Useful and Powerful

13

Martin Pomerantz, Isaiah M. Borkowski, Kathleen F. Edwards, Zelek S. Herman, Shayan Khan, and Joel F. Liebman

Abstract

A brief history of the development and utility of hybridization theory and valence bond theory is presented. Qualitative reasons and a simple mathematical model are offered as to why the assertion of sp^3 hybridization in methane and isoelectronic species is not correct. Nonetheless, we acknowledge that simple hybridization is a convenient way when looking at the bonding in molecules.

M. Pomerantz
Department of Chemistry and Biochemistry (Emeritus), University of Texas at Arlington, Arlington, TX, USA
e-mail: pomerantz@uta.edu

I. M. Borkowski · S. Khan · J. F. Liebman (✉)
Department of Chemistry and Biochemistry, University of Maryland, Baltimore County, Baltimore, MD, USA
e-mail: jliebman@umbc.edu

I. M. Borkowski
e-mail: iisaiahb1@umbc.edu

S. Khan
e-mail: skhan21@umbc.edu

K. F. Edwards
Doctor of Business Administration Program, School of Business, University of Maryland Global Campus, Largo, MD, USA
e-mail: kathleen.edwards@faculty.umgc.edu

Z. S. Herman
Herman Scientific Consulting, Palo Alto, CA, USA
e-mail: zelek.herman@att.net

A. Greenberg and D. E. Lewis (eds.), *The Birth of the 3rd Dimension in Chemistry*, Perspectives on the History of Chemistry,
https://doi.org/10.1007/978-3-031-97743-5_13

13.1 Introduction

In 1931, Linus Pauling published a seminal paper in the *Journal of the American Chemical Society* entitled "The Nature of the Chemical Bond. Applications of Results Obtained from the Quantum Mechanics and from a Theory of Paramagnetic Susceptibility to the Structure of Molecules" [1]. In this ground-breaking treatise, Pauling employed the recently developed quantum mechanical theory [2] to explain the nature of the chemical bond in complex molecules by making the assumption of radial equivalence in the character of atomic eigenfunctions to focus on the angular dependence of atomic orbitals in bond formation. The results came to be known as the hybridization theory. Among other things, the hybridization theory rationalized the tetrahedral arrangement of the C-H bonds in methane and enabled organic chemists to picture the bonding geometries in simple molecules. Hybridization has since generally become an essential part of the most often qualitative language of descriptive organic chemistry for both researchers and students alike.

A thorough presentation of the role of symmetry, hybridization, and bonding in molecules has been published by Maksić [3]. In this paper, Maksić discusses how the hybridization model describes the directional features of covalent bonding as well as a number of other properties not directly related. He emphasizes that the hybridization model's reflection of molecular properties can be traced to the fact that hybrid bond orbitals conform to the symmetry of an atom in a molecular environment.

13.2 Accuracy of the Hybridization Theory for Carbon in Methane and Isoelectronically Related Species

For some 90 years, chemists have employed hybridization theory and valence bond theory to conceptualize the bonding in molecules. Almost 50 years ago, two of the present authors (M.P., J.F.L.) published a brief note entitled "Is 100% *s* character necessary?" in which the molecular and electronic structure of organic compounds was discussed [4]. In this study, it was asserted that summing over all four bonds attached to a "central" tetracoordinate carbon, as found in a generic CWXYZ species, need not result in 100% *s* character, so the carbon of methane need not be sp^3 hybridized. In this earlier paper, a brief theoretical explanation for this observation was given, as well as the documentation of this by use of atomic hybridizations determined from NMR coupling constants. In the present paper, additional reasons for this assertion are presented. These new insights are not being presented because of subsequent conflicting thoughts about the first paper, but rather, owing to the general observation that use of multiple, distinct documentation, verification, proofs, and modes of understanding often give additional insights not found from the understanding derived from but one observation. Note that, not uncommonly, tetracoordinate carbon is labeled sp^3, and tricoordinated carbon sp^2, regardless of the local geometry of the species. The authors do not wish to encourage such simplification of the language, the use of such mnemonics, and verbiage that is too simple, misleading and incorrect.

Now consider the even simpler eight valence electron AH_4 species and naturally start with methane CH_4, with 4 C-H bonds of equal length, 1.091 ± 0.002 Å, and 6 equal H-C-H angles of ca. 109.5°, i.e., precisely arccos $(-1/3)$ [5]. For this situation, a regular tetrahedral geometry prevails, with a hydrogen atom at every other corner of a cube and situated equidistant from the central carbon atom. While the bond length cannot be mathematically determined from stoichiometry and symmetry alone, the bond angle is precisely determined. Now it remains to determine the atomic hybridization of the central carbon atom, or more precisely, the *s*/*p* ratio of the atomic orbitals for the electrons involved in the bonding of molecular methane. Related to the conceptual analysis and quantum chemical calculations of ref. [6], consider a hypothetical CH_4 molecule wherein again there are four equivalent C-H bonds and six equal H-C-H angles, but where the C-H bonds are significantly longer. In other words, the molecule is stretched such that its strictly tetrahedral geometry is maintained. Simultaneously, the molecule is allowed to electronically relax at this new geometry. The molecule is *en route* to dissociating and is getting ever closer to its description as unbound C + 4H, i.e., the separate atomic dissociation products. The local atomic orbital description of such a free unbound carbon is $2s^2\ 2p^2$. Following conventional organic chemists' implicit practice, but not that of all theoretical chemists [6], ignore the inner $1s^2$ electrons and their negligible contribution to the bonding of CH_4, resulting in an atomic electron configuration $2s^2\ 2p^2$. The *s*/*p* ratio is 2:2 or more now simply written as 1:1, which in turn, can be written as sp. Admittedly, this is not the same as the construction of sp hybridized orbitals, but instead speaks to the *s*/*p* ratio, from which the more conventional and convenient sp hybrids may ultimately be constructed.

Ref. [6] gives the total *s* and *p* orbital occupancies for carbon, as found in CH_4, at a wide variety of carbon-hydrogen distances. Now subtract 2 from the former value classically corresponding to the two electrons in the carbon 1 *s* orbital, and equate the derived ratio of the *p* orbital occupancy to the corrected *s* orbital occupancy with the hybridization of the carbon. At a C-H distance of 5.3 Å (10 a_0) [about five times the experimentally observed value], the *s*/*p* ratio was found to be 2.03:2.00, very nearly the free atom ratio of 1:1 found for $1s^2\ 2s^2\ 2p^2$ (^3P) atomic carbon [7], and the earlier enunciated sp hybridization. Now consider a C-H distance of ca. 1.6 Å (3 a_0), 50% too large for a C-H bond and comparable to that of conventional C–C bonds. The calculated ratio is 1.57:2.41 corresponding roughly to $sp^{1.5}$. At the C-H bond distance taken of ca. 1.1 Å (2.25 a_0), the ratio is 1.57:2.55 corresponding to roughly $sp^{1.6}$.

Alternatively, consider a scrunched CH_4 molecule in which there is a decreased C-H bond length, but the tetrahedral H-C-H angles are still maintained. Neglect the ever-increasing C-H and H–H internuclear repulsions, which are always present in all molecules and at all internuclear distances. Observe that the total repulsion always exceeds the chemical bond energies of interest. The new species so generated is an atom with a nuclear charge of 10, thereby approaching an isotope of neon (^{16}Ne). The electron configuration of neon is $1s^2\ 2s^2\ 2p^6\ {}^1S$ [7]. Once again, ignore the $1s^2$ electrons, resulting in an electron configuration of $2s^2\ 2p^6$ with an *s*/*p* ratio of 2:6 or 1:3. This can be written as sp^3.

Now ask, "What is the *s*/*p* ratio of CH_4 at the 'real' C-H bond distance of ca. 1 Å?" From the calculational studies [6], we find this quantity to equal 1.61: 2.57 or 2: 1.6. It might be felt 1that since 1 Å is far closer to the latter case of 0 Å than the former with infinite C-H separation, the hybridization for the carbon is closer to the latter assignment of sp^3 rather than the former of *sp*. However, is 1 far closer to 0 than to ∞? Most chemists, including the authors, prefer to describe molecules in terms of their component atoms, and thus at infinite separation, rather than in terms of their "united atoms" with their scrunched nuclei. That is, a C-H infinite bond distance corresponds in this view more than of 0 Å, and so this again suggests the carbon hybridization at a C-H 1 Å bond distance is closer to sp than sp^3.

In fact, the present mathematical analysis does not reveal anything. Instead, assume that the *s*/*p* ratio varies monotonically, always decreasing as the C-H bond length increases, so that the *s*/*p* ratio increases from 1/1 to 1/3, and the n in sp^n is always increasing, but the precise value of n is unknown if not unknowable. However, it is impossible to prove the assumption about the monotonic behavior of the *s*/*p* ratio—indeed the values derived from ref. [6] shows it is not monotonic, leading to greater ignorance than previously. Ref. [6] tells us that the ratio is 1.61:2.57 or 1.1.6, quite distant from 1:3.

Consider $[BH_4]^-$, isoelectronic to CH_4, in terms of the above analysis. Pulling $[BH_4]^-$ apart generates $B + 3H + H^-$, one of the hydrogen atoms having the negative charge as opposed to the boron naturally ascribed from assignments of formal charges. After all, the electron affinity of B is less than that of H [8]. The electron configuration of B, again ignoring the inner $1s^2$ electrons, is $2s^2\ 2p^1$ (^2P), with an *s*/*p* ratio of 2:1 or 1:0.5. Allowing for noninteger hybridization values, as is very often done, one can characterize the boron as being $sp^{0.5}$. Again ignoring isotopic concerns (i.e., the number of neutrons, as opposed to the number of protons and electrons) now, scrunching $[BH_4]^-$ gives a structure which has a $2s^2\ 2p^6$ electron configuration and again an sp^3 hybridization. Thus, the hybridization of B is unknown for $[BH_4]^-$, with its B-H bond distance presumably larger than that of CH_4.

Now examine $[NH_4]^+$, isoelectronic to CH_4, and apply the above analysis. Pulling $[NH_4]^+$ apart generates $N + 3H + H^+$, one of the hydrogen atoms having the positive charge as opposed to the nitrogen atom naturally ascribed from assignments of formal charges. The ionization potential of H is greater than that of almost all of the elements, but is nonetheless smaller than that of N [7]. The electron configuration of ground sate N, ignoring the inner $1s^2$ electrons, is $2s^2\ 2p^3$ (^4S) with an *s*/*p* ratio of 2:3 or 1:1.5. Allowing for noninteger hybridization values as before, allows one to speak of the nitrogen atom being $sp^{1.5}$. Ignoring isotopic concerns once again, scrunching $[NH_4]^+$ yields Na^+, which has a $2s^2\ 2p^6$ (^1S) electron configuration [7], and again an sp^3 hybridization. Once more, one really does not know what the hybridization of N is for $[NH_4]^+$, with its N–H bond distance presumably smaller than that of CH_4.

The analysis for SiH_4 is the same as for CH_4, where now the *s*/*p* ratio is unknown for this species, but it is known that the well-measured Si–H bond length is larger than the C-H bond length in CH_4 [9]. Likewise, the analysis for $[AlH_4]^-$ is the same as for $[BH_4]^-$ because the electron affinity of Al, while larger than that of B, is nonetheless still less than that of H [8]. What then about $[PH_4]^+$? Because the ionization potential of P is less than that of H [7], its separation product is $P^+ + 4H$. The electron configuration of P^+ is $3s^2\ 3p^2\ (^3P)$ [7], with a s:p ratio of 2:2 or 1:1 and a corresponding sp description. The scrunched product of $[PH_4]^+$ is now an isotope of Ar with a $3s^2\ 3p^6$ electron configuration, and a 2:6 = 1:3 ratio and accordingly sp^3 hybridization So, what are the hybridizations of Si, Al, and P in SiH_4, $[AlH_4]^-$, and $[PH_4]^+$? By inference, the hybridization is unknown for carbon in CH_4 with its ca. 1 Å C-H bond distance, or that found in any other organic compound.

At this point, let us construct a CH_4 molecule by placing the central C at the origin, coordinates (0, 0, 0), of a cube of length 2. The four hydrogen atoms are placed at the opposite corners of the cube, i.e., with coordinates (x, y, z) = (1, 1, 1), (1, −1, −1), (−1, 1, −1) and (−1, −1, 1). On the carbon, now put a 2 *s* atomic orbital, as well as a $2p_x$, $2p_y$ and $2p_z$ atomic orbital, and put no other on this central atom. Call these four atomic orbitals φ_s, φ_x, φ_y, and φ_z Put an s orbital on each hydrogen atom, and call these atomic orbitals φ_1, φ_2, φ_3, φ_4. For CH_4, there are four bonding occupied molecular orbitals. The lowest lying is composed of φ_s and an equal amount of the hydrogen orbitals, and so may be written as $\varphi_s + b(\varphi_1 + \varphi_2 + \varphi_3 + \varphi_4)$ where the coefficients a and b have whatever value they do. There is a set of three equal energy orbitals composed of φ_x, φ_y and φ_z, again making use of the four hydrogen orbitals. There are an infinite number of ways these orbitals can be written because these three equal energy orbitals are linearly independent, orthogonal, degenerate and implicitly normalized. One such description consists of $(\underline{\varphi_1} + \varphi_2 - \varphi_3 - \varphi_4)$, which has the same symmetry as φ_x, and so is written χ_x; $(\underline{\varphi_1} - \varphi_2 + \varphi_3 - \varphi_4)$, which has the same symmetry as φ_y and so is written as χ_y; and $(\overline{\varphi_1} - \varphi_2 - \varphi_3 + \varphi_4)$ and so is written as χ_z.

The three next higher-energy molecular orbitals likewise maintain their linear independence, orthonormality, and degeneracy. They can be written as $c\varphi_x + d\chi_x$, $c\varphi_y + d\chi_y$, and $c\varphi_z + d\chi_z$. The *s*/*p* hybridization reflects only the orbitals from the central atom and so the hybridization, the s/p ratio, equals $a^2/(c^2 + c^2 + c^2) = a^2/3c^2$. Can this equal 1/3 corresponding to sp^3? Of course it can. Does this equal 1:3? Yes, it does, if $|a| = |c|$ but for no other choice of values. Does $|a| = |c|$? It need not be the case—there are no mathematical reasons nor any from the fundamental physics of organic compounds. Even if $|a| = |c|$ for CH_4, it all but guarantees it would not be for $[BH_4]^-$, $[NH_4]^+$ or any other 10 valence electron AH_4 species, and even less likely for tetracoordinate carbon in some organic compound, with tetracoordinated carbon CWXYZ.

In recent years, there has been active dispute between those supporting and those opposing the use of using hybrid bond orbitals as a useful model to describe the bonding in molecules, e.g., refs. [10–27]. In fact, it has been argued that hybridization and valence bond theory and molecular orbital based theory are conceptually complementary, if not fundamentally indistinguishable in their limiting cases, e.g., [27–29].

We may single out the paper [11] from the above that used molecular orbital calculations with gross atomic population and overlap density criteria to derive s:p ratios for our AH_4 species, namely: CH_4, $[BH_4]^-$, $[NH_4]^+$, SiH_4, $[AlH_4]^-$, $[PH_4]^+$, and some of their substituted derivatives. These studies affirmed the s:p ratios were definitively not 1:3 and were, in fact, related to the varying electronegativities of the central atom A. However, in that these two criteria resulted in quite numerically different s:p ratios for all of these species, we hesitate to quantitatively discuss any further the results from this study.

For many years, molecular orbital theory was more amenable to computer calculations on the electronic properties of complex molecular systems than valence bond theory. Nevertheless, in recent years, with the advent of vastly more powerful computers, several software packages have become available for computing valence bond functions. The generalized valence bond theory and its identical twin, spin-couple valence bond theory allow one to calculate bonding properties free of ad hoc assumptions [27]. Carried out to their penultimate extremes, valence bond theory, with the inclusion of all possible valence bond structures, and molecular orbital theory, with full configuration interaction, are indeed, equivalent [27–29].

Does this mean that the concept of hybrid orbitals has now been obsoleted? We painfully recall reference [16] that proffers these words "Even though hybrid orbitals might have fulfilled a role as crude models in the twentieth century, we predict that they will not survive in this new century as more complicated models are required and applied." We disagree.

Tetrahedral, tetracoordinate carbon and sp^3 hybridization are inseparable concepts, while admittedly the former is a generally experimentally observed phenomenon and the latter is a faulted, but still highly useful, component of our understanding of chemistry.

13.3 Conclusion

Thus, one concludes that sp^3 hybridization for methane, and by inference other organic compounds, is not to be found. To speak of sp^3 hybridization is convenient because to specify more than one type of hybridization for one atom in one type of molecule is much more complicated, especially if quantitation is desired. In methane, there are four C-H bonds of equal length and strength. There is no reason to distinguish one bond from another. It would violate the sense of atomic and molecular equality. "Things that look different usually are" is not a corollary of "things that look the same usually are the same." We also spoke of central atoms and their valence electrons even though all electrons are the same. And now we note that it is not uncommon to read a study in the chemical literature wherein any four coordinated carbon is called *sp*^{3}. In such studies, tetrahedral, tetracoordinate and *sp*^{3} are synonymous, even if the geometry is most geometrically irregular. The word "tetrahedron" is an indelible part of the chemist's vocabulary and understanding almost throughout the chemical community. Indeed, one can recall that the name of the journal in which the original study of 50 years ago appeared is *Tetrahedron Letters.*

13.4 A Mathematical Addendum

We can be more mathematical. Let us derive the energy levels for a "completely connected tetrahedral array." That is, each of the four vertices or corners are connected, equally strongly, to the other three. Relatedly, each vertex is not connected to itself. Using Huü Molecular Orbital (HMO) theory like logic, we construct the related 4×4 matrix T where the matrix element Tij = 1 when the two vertices/corners are connected and $T_{ij} = 0$ when they are not connected.

[0 1 1 1

1 0 1 1

1 1 0 1

1 1 1 0]

There is one normalized eigenvector that has the eigenvalue 3,

[1/2
1/2

1/2

1/2]

Or more simply written, and certainly less space consuming, as [1/2, 1/2, 1/2, 1/2]. In other words, all four vertices/corners are equivalent and symmetrically disposed to each other. That is, interchanging two vertices or matrix element does not changing the eigenvector. Furthermore, So labeling the vertices of a tetrahedron results in no change upon exchanging two vertices. All of this is reminiscent of the symmetry properties of an s orbital.

There are three more eigenvectors all of which have eigenvalue of −1. These three eigenvectors may be written together as [1/2, 1/2, −1/2, −1/2], [1/2, −1/2, 1/2, −1/2] and [1/2, −1/2, −1/2, 1/2]. Alternatively, they are may be written together as [$2^{-1/2}$, $-2^{-1/2}$, 0, 0], [0, 0, $2^{-1/2}$, $-2^{-1/2}$] and [1/2, 1/2, −1/2, −1/2]. Either set of these eigenvectors has the same sign reversal property. From both the threefold degeneracy and sign-reversal behavior as p-orbitals.

We may thus equate the presence of nondegenerate eigenvector and a triply degenerate eigenvector to a set of sp^3 orbitals. However, if we wish to have the minimum energy for a total of four electrons, we would have two electrons using the [1/2, 1/2, 1/2, 1/2] eigenvector and using two electrons chosen from [1/2, 1/2, −1/2, −1/2], [1/2, −1/2, 1/2, −1/2] and [1/2, −1/2, −1/2, 1/2], or equally as correct, two chosen from [$2^{-1/2}$, $-2^{-1/2}$, 0, 0], [0, 0, $2^{-1/2}$, $-2^{-1/2}$] and [1/2, 1/2, −1/2, −1/2]. This corresponds to a set of degenerate s^2p^2, or should we say sp, orbitals. Do we dare say we have sp hybridization?

Should we wish to have two electrons apiece in either set of four orbitals, {[1/2, 1/2, 1/2, 1/2], [1/2, 1/2, −1/2, −1/2], [1/2, −1/2, 1/2, −1/2], [1/2, −1/2, −1/2, 1/2]} or {[1/2, 1/2, 1/2, 1/2], [$2^{-1/2}$, $-2^{-1/2}$, 0, 0], [0, 0, $2^{-1/2}$, $-2^{-1/2}$], [1/2, 1/2, −1/2, −1/2]}, we have s^2p^6, or should we say sp^3, hybridization? This is tempting to do so, but then all of these species CH_4, $[BH_4]^-$, $[NH_4]^+$, SiH_4, $[AlH_4]^-$, $[PH_4]^+$ have the same hybridization. In other words, we recover the traditional understanding of tetrahedral species having sp^3 hybridization. This is also the case for all tetrahedral AX_4 species with eight bonding electron such as CF_4 and $[BF_4]^-$. It is then remembered how much the orbital of the affixed X atoms mixes with the s orbital of the central atom A is different from how much the affixed atoms X mixes with the p orbital of the central atom A. This depends on both the A and the X. We may show this using a just constructed 8 × 8 matrix. Let us call this new matrix, M.

$$\begin{bmatrix} M_{11} & M_{12} & M_{13} & M_{14} & M_{15} & M_{16} & M_{17} & M_{18} \\ M_{21} & M_{22} & M_{23} & M_{24} & M_{25} & M_{26} & M_{27} & M_{28} \\ M_{31} & M_{32} & M_{33} & M_{34} & M_{35} & M_{36} & M_{37} & M_{38} \\ M_{41} & M_{42} & M_{43} & M_{44} & M_{45} & M_{46} & M_{47} & M_{48} \\ M_{51} & M_{52} & M_{53} & M_{54} & M_{55} & M_{56} & M_{57} & M_{58} \\ M_{61} & M_{62} & M_{63} & M_{64} & M_{65} & M_{66} & M_{67} & M_{68} \\ M_{71} & M_{72} & M_{73} & M_{74} & M_{75} & M_{76} & M_{77} & M_{18} \\ M_{81} & M_{82} & M_{83} & M_{84} & M_{85} & M_{86} & M_{87} & M_{98} \end{bmatrix}$$

We may now make some simplifications. The first is that $M_{ij} = M_{ji}$ for all i and j. For i = 1, 2, 3, 4, the subscript i in M_{ij} refers to the central atom while for the subscript 5. 6. 7. 8, the subscript refers one apiece to the four affixed atoms to form the new 8 × 8 matrix. We now let $M_{ii} = 0$ for i = 1, 2, 3, 4 but M_{ii} equals some different constant, $M_{55} = M_{66} = M_{77} = M_{88} = a$. (The letter "a" was chosen to remind all of us of the "α" used in Hückel Molecular Orbital theory.) Let $M_{ij} = 1$ for unequal i and j which are chosen among 1, 2, 3, 4. We assume that each affixed atom is attached but one of the vertices of the "completely connected tetrahedral array." Let us assume that 1 and 5 are so connected, as are 2 with 6, 3 with 7, 4 with 8. Accordingly, $M_{15} = M_{26} = M_{37} = M_{48}$ (also all equal to $M_{51} = M_{62} =$

$M_{73} = M_{84}$) Let us call this last quantity b. (The letter "b" was chosen to remind everyone of the "β" used in Hückel Molecular Orbital theory.)

We thus have the following 8 × 8 matrix. M for the tetrahedral AX_4, species of interest.

$$\begin{bmatrix} 0 & 1 & 1 & 1 & b & 0 & 0 & 0 \\ 1 & 0 & 1 & 1 & 0 & b & 0 & 0 \\ 1 & 1 & 0 & 1 & 0 & 0 & b & 0 \\ 1 & 1 & 1 & 0 & 0 & 0 & 0 & b \\ b & 0 & 0 & 0 & a & 0 & 0 & 0 \\ 0 & b & 0 & 0 & 0 & a & 0 & 0 \\ 0 & 0 & b & 0 & 0 & 0 & a & 0 \\ 0 & 0 & 0 & b & 0 & 0 & 0 & a \end{bmatrix}$$

Admittedly, most tetrahedral species do not have four identical groups attached to the central atom. Said differently, where we spoke of CX_4 and the more general AX_4 we should have more generally spoken of AWXYZ and most often of carbon-containing species such as CH_3Li and CH_3F to accompany CH_4 and for that matter CLi_4 and CF_4. The reader may share our hesitancy, however temporarily for the current study, of investigating of the eigenvalues and eigenvectors for the necessarily even more complicated 8 × 8 matrices wherein the four identical matrix elements denoted by "b" need be written as "b_W, b_X, b_Y, and b_Z". Relatedly, the four identical matrix elements denoted by "a" need be written as "a_W, a_X, a_Y, and a_Z". That is, we have the new even more complicated 8 × 8 matrix.

$$\begin{bmatrix} 0 & 1 & 1 & 1 & b_W & 0 & 0 & 0 \\ 1 & 0 & 1 & 1 & 0 & b_X & 0 & 0 \\ 1 & 1 & 0 & 1 & 0 & 0 & b_Y & 0 \\ 1 & 1 & 1 & 0 & 0 & 0 & 0 & b_Z \\ b_W & 0 & 0 & 0 & a_W & 0 & 0 & 0 \\ 0 & b_X & 0 & 0 & 0 & a_X & 0 & 0 \\ 0 & 0 & b_Y & 0 & 0 & 0 & a_Y & 0 \\ 0 & 0 & 0 & b_Z & 0 & 0 & 0 & a_Z \end{bmatrix}$$

This increase of complexity is found in the computational quantum chemistry literature with its ever increasing variety of methods. Depending on the species of interest, these approaches both calculational and conceptual constructs alike, range from semiempirical through simple molecular orbital and valence bond studies, to those contemporary studies that obsolete orbitals. Indeed, in an increasing number of studies, "chemical accuracy" has been achieved wherein calculated and measured enthalpies of formation are in agreement to 1 kcal mol^{-1}. And for some species, even the more precise "subchemical accuracy" has been achieved.

As we have no intent of providing details for any of the literature studies, let us hereby simply return to our matrices. The reader may share our comfort with the earlier eigenvalues and eigenvectors for 4×4 matrices, and even easier to understand, 2×2 matrices. Consider now the simple 2×2 matrix.

$$\begin{matrix} [0 & 1 \\ 1 & 0] \end{matrix}$$

Forgetting about normalization and all multiplicative constants, it is easy to show that there are two normalized eigenvectors, (0.71, 0.71) and (−0.71, 0.71) wherein the 0.71 being used as a numerical simplification for the rigorously correct $2^{-1/2}$. We can write these eigenvectors even more simply as (+, +) and (+, −), respectively, where the + and − refer to numerically positive and negative components of these eigenvectors. In turn, they are recognized as being respectively symmetric and antisymmetric. Indeed, s and p atomic orbitals are simple exemplars.

Consider now the 4×4 matrix.

$$\begin{matrix} [0 & 1 & 1 & 0 \\ 1 & 0 & 0 & 1 \\ 1 & 0 & 0 & 0 \\ 0 & 1 & 0 & 0] \end{matrix}$$

The reader may recognize as the HMO matrix for 1,3-butadiene with admittedly using a somewhat strange atom numbering, $C^3 = C^1 - C^2 = C^4$. Quite sensibly, we confirm the atom numbering does not affect the associated eigenvalues [30]. Consider now the coefficients for the components of the associated molecular orbitals, or equivalently associated eigenvectors, (k_1, k_2, k_3, k_4). Now, if k_1 and k_2 have the same sign, and likewise k_3 and k_4 have the same sign, this is recognized as symmetric. Even simpler, we say that both (+, +, +, +) and (−, +, +, −) are symmetric. Conversely, if k_1 and k_2 have opposite signs, and likewise k_3 and k_4 have opposite signs, this is recognized as antisymmetric. We say that both (+, +,−, −) and (+, +, −, −) are symmetric This is analogous to a triatomic molecule AX_2, or more properly XAX, wherein the central atom has two atomic orbitals, one symmetric and antisymmetric apiece and so, one s and p atomic orbitals apiece. Likewise, the two X atoms have one orbital apiece from which w form two new molecular orbitals, one symmetric and one antisymmetric.

We thus have four new orbitals, two symmetric and two antisymmetric, of which we choose the one of each that has the most positive eigenvalue.

We thus consider the matrix or we should matrices wherein the "c" relates to the details the eigenvalue or we may say energy levels of the central atom. In turn, this relates to the electronegativity (χ) of the central atom when using Allen's definition of this quantity [31].

$\chi_{spec} = (m\varepsilon_s + n\varepsilon_p)/(m + n)$ for an atom with a $s^m p^n$ electron configuration. Since for the current case of AX_2 (XAX) species the central atom has only two valence s electrons, $\chi_{spec} = \varepsilon_s$.

$$
\begin{bmatrix} c & 1 & 1 & 0 \\ 1 & c & 0 & 1 \\ 1 & 0 & 0 & 0 \\ 0 & 1 & 0 & 0 \end{bmatrix}
$$

Let us now vary the value of c in our evaluation of the desired eigenvalues and eigenvectors [30].

For c = 1, the 1st eigenvalue is 1.62 with the eigenvector (1.62, 1.62, 1, 1).

"Amount of s" from atoms 1 and 2 = $[(1.62)^2 + (1.62)^2)]/ [(1.62)^2 + (1.62)^2 + 1^2 + 1^2)]$.

$= (2.64 + 2.64)/(2.64 + 2.64 + 1 + 1) = 5.28/7.28 = 0.73$.

The 2nd eigenvalue is 0.62 with the eigenvector (−0.62, 0.62, −1, 1).

"Amount of p" on atoms 1 and 2 = $[(-0.62)^2 + (0.62)^2)]/[[(-0.62)^2 + (0.62)^2) + ((-1)^2) + 1^2\}]$.

$= (0.38 + 0.38)/(0.38 + 0.38 + 1 + 1) = 0.76/2.76 = 0.28$.

We have s/p = (0.73/0.28) = 2.6, and so the hybridization would be said to be $sp^{1/2.6} = sp^{0.4}$

For c = 2, the 1st eigenvalue is 2.41 with the 1eigenvector (2.41, 2.41, 1, 1).

"Amount of s" on atoms 1 and 2 = $[(2.41)^2 + (2.41)^2])/[[(2.41)^2 + (2.41)^2 + 1^2 + 1^2)$.

$= (5.8 + 5.8)/(5.8 + 5.8 + 1 + 1) = 11.6/13.6 = 0.85$.

The 2nd eigenvalue is 0.41 with the eigenvector −0.41, 0.41, −1, 1).

"Amount of p" on atoms 1 and 2 = $[(-0.41)^2 + (0.41)^2)/[(-0.41)^2 + (0.41)^2 + (-1)^2 + 1^2\}$.

$= (0.17 + 0.17)/(0.17 + 0.17 + 1 + 1) = 0.34/2.34 = 0.07$.

We have $sp^{(0.07/0.85)} = sp^{0.1}$.

In both of the above cases, the 1st eigenfunction is symmetric and accordingly from our results for 2 × 2 matrices, it has the symmetry of an s orbital. The 2nd eigenfunction is antisymmetric. Relatedly, from our results for 2 × 2 matrices, and so has the symmetry of a p orbital. Should c = 0, again using chemist's words, the s and p-like orbitals then would be degenerate. Linear combinations of these orbitals are also degenerate while identification with s and p orbitals is lost. To guarantee we have the ability to visualize the description of s and p orbitals, let us choose a small value of c.

For c = 0.1, the 1st eigenvalue is 1.05 with the eigenvector (1.05, 1.05, 1, 1).

"Amount of s" on atoms 1 and 2 = $[(1.05)^2 + (1.05)^2])/[\,[(1.05)^2 + (1.05)^2 + 1^2 + 1^2)$.

$= (1.10 + 1.10)/(1.10 + 1.10 + 1 + 1) = 0.2.20/4.20 = 0.55.$

The 2nd eigenvalue is 0.95 with the eigenvector (−0.95, 0.95, −1, 1).

"Amount of p" on atoms 1 and 2 = $[(-0.95)^2 + (0.95)^2)/[(-0.95)^2 + (0./95)^2 + (-1)^2 + 1^2\}$.

$= (0.90 + 0.90)/(0.90 + 0.90 + 1 + 1) = 1.80/3.80 = 0.47.$

We have $s^{0.55}p^{0.47} = sp^{0.85}$.

We conclude, again using chemical words, as c increases, the s-p energy gap increases, the s orbital of the central atom contribution to the bonding increase, the hybridization becomes increasingly "pure" s. Equivalently, as c approaches 0, the s-p gap shrinks, and the hybridization approaches the sp that we initially expected. We conclude that the A atom in AX_2 species are most generally not sp hybridized. Extrapolating to AX_4 species such as CH_4 and its necessary 8×8 matrix, we hereby conclude that the A is not sp^3, hybridized. From the "s" and "p" orbitals on the central A atom and the orbitals on the two attached atoms, we have constructed two new orbitals, or more properly have constructed four and used but two. That both the "s" and "p" orbitals on the central atom are used may be understood, however qualitatively, as two "sp" orbitals. Extrapolating to the tetrahedral AX_4 molecules, we expect to find in use the "s" and all three "p" orbitals on the central atom. This could be understood as sp^3 hybridization. we are convinced that sp^3, hybridization remains a useful concept, Alternatively said, that a carbon atom bonded to four other atoms is (usually only roughly) tetrahedral, not planar, is an essential part of our understanding of chemistry.

Acknowlegement Authors M. P. and J. F. L. wish to thank Prof. Arthur Greenberg for introducing them, thereby catalyzing both papers 4 and the current study. We wish to also thank Profs. Igor Alabugin and Frank Weinhold for numerous discussions.

References

1. Pauling L (1931) The nature of the chemical bond. Application of results obtained from the quantum mechanics and from a theory of paramagnetic susceptibility to the structure of molecules. J Am Chem Soc 53:1367–1400
2. Heitler W, London F (1927) Interaction of neutral atoms and homopolar binding according to the quantum mechanics. Z Physik 44:455–472
3. Maksić ZB (1986) Symmetry, hybridization and bonding in molecules. Comp Maths Appls. 12B: 697–723. (Reprinted with unchanged pagination as a chapter in Hargittai, I (ed.) "Symmetry Unifying Human Understanding", Pergamon Press, NY, 1986)
4. Pomerantz M, Liebman JF (1975) Is 100% s-character necessary? Tetrahedron Lett 16:2385–2388
5. Stevenson GP, Ibers JA (1960) The equilibrium C-H distance in methane. J Chem Phys 33:762–763

6. Clementi E, Popkie H (1971) Electronic structure of molecules, XVI. Analysis of the formation of the methane molecule in the Hartree-Fock model. J Am Chem Soc 94:4057–4068
7. Moore CE (1970) Ionization potentials and ionization limits derived from the analysis of optical spectra. Nat Stand Ref Data Ser, Nat Bur Stand (US), NSRDS-NBS 34
8. Andersen T, Haugen HK, Hotop H (1999) Binding energies in atomic negative ions. III. J Phys Chem Ref Data 28:1511–1533
9. Kattenberg HW, Oskam A (1974) Infrared and laser Raman gas spectra of silane and silane-d_4. J Mol Spectrosc 49:52–69
10. Herman ZS (1983) Recent advances in simple valence-bond theory and the theory of hybrid bond orbitals. Intl J Quantum Chem 23:921–943
11. Magnusson E (1984) Sp Hybridization reconsidered: the composition of orbitals in main-group hydrides. J Am Chem Soc 106:1177–1185
12. Grushow A (2011) Is it time to retire the hybrid atomic orbital? J Chem Educ 88:860–862
13. Tro NJ (2012) Retire the hybrid atomic orbital? Not so fast. J Chem Educ 89:567–568
14. DeKock RL, Strikwerda JR (2012) Retire hybrid atomic orbitals? J Chem Educ 89:569–560
15. Landis CR, Weinhold F (2012) Comments on "Is it time to retire the hybrid atomic orbital?" J Chem Educ 89:570–572
16. Truhlar DG (2012) Are molecular orbitals delocalized? J Chem Educ 89:573–575
17. Hiberty PC, Volatron F, Shaik S (2012) In defense of the hybrid atomic orbitals. J Chem Educ 89:575–577
18. Grushow A (2012) In response to those who wish to retain hybrid atomic orbitals in the curriculum. J Chem Educ 89:578–579
19. Matxain JM, Piris M, Mercero JM, Lopez X, Ugalde JM (2012) Sp3 Hybrid orbitals and ionization energies of methane from PNOF5. Chem Phys Lett 531:272–274
20. Alabugin IV, Bresch S, Manoharan M (2014) Hybridization trends for main group elements and expanding the Bent's rule beyond carbon: More than electronegativity. J Phys Chem A 118:3663–3677
21. Alabugin IV, Bresch S, Gomes GP (2015) Orbital hybridization: A key electronic factor in control of structure and reactivity. J Phys Org Chem 28:147–162
22. Xu LT, Thomson JVK, Dunning TH (2019) Spin-coupled generalized valence bond description of group 14 species: The carbon, silicon and germanium hydrides, XH_n ($n = 1$-4). J Phys Chem A 123:2401–2419
23. Lamoureux G, Ogilvie JF (2019) Hybrid atomic orbitals in organic chemistry. Part 1: Critique of formal aspects. Quim Nova 42:812–816
24. Lamoureux G, Ogilvie JF (2019) Hybrid atomic orbitals in organic chemistry. Part 2: Critique of practical aspects. Quim Nova 42:817–822
25. Glendening ED, Weinhold F (2021) Pauling's conceptions of hybridization and resonance in modern quantum chemistry. Molecules 26: paper 4110
26. Lamoureux G, Ogilvie JF (2022) A critical history of hybrid atomic orbitals and hybridization. J Chem Rev 4:120–146
27. Slater JC (1932) Note on molecular structure. Phys Rev 41:255–257
28. Hoffmann R, Shaik S, Hiberty PC (2003) A conversation on VB vs MO theory: a never-ending rivalry? Accts Chem Res 36:750–756
29. Galbraith JM, Shaik S, Danovich D, Braida B, Wu W, Hiberty P, Cooper DL, Karadakov B, Dunning TH Jr (2021) Valence bond and molecular orbital: two powerful theories that nicely complement one another. J Chem Educ 98:3617–3620
30. All eigenvalue and eigenvector calculations presented in the current study were obtained through the use of the Wolfram Alpha Widget: "4x4 Eigensystem with Numeric Answers", Wolfram Research, Inc., Mathematica, Version 14.1, Champaign, IL (2024). a Free Mathematics Widget (wolframalpha.com), It is now acknowledged that we were also given results with either 6 or 7 significant figures, and even more exact, closed form solutions. We opted for much less numerical "accuracy" as not to obscure the fundamental chemistry by the formal mathematics. For example, when we wrote the eigenvector (1.62, 1.62, 1, 1) for $c = 1$, and that the 1st

eigenvalue is 1.62, we had earlier been given the more accurate but still unnormalized approximate solution (1.61803, 1.61803, 1, 1) and the likewise unnormalized mathematically exact solution (½ (1 + $5^{1/2}$), ½ (1 + $5^{1/2}$). 1, 1)
31. Allen LC (1989) Electronegativity is the average one-electron energy of the valence-shell electrons in ground state free atoms. J Am Chem Soc 111:9003–9014

Index

A. Greenberg and D. E. Lewis (eds.), *The Birth of the 3rd Dimension in Chemistry*, Perspectives on the History of Chemistry, https://doi.org/10.1007/978-3-031-97743-5

MIX
Papier aus verantwortungsvollen Quellen
Paper from responsible sources
FSC® C105338

If you have any concerns about our products,
you can contact us on
ProductSafety@springernature.com

In case Publisher is established outside the EU,
the EU authorized representative is:
Springer Nature Customer Service Center GmbH
Europaplatz 3, 69115 Heidelberg, Germany

Printed by Libri Plureos GmbH
in Hamburg, Germany